Agripreneurship, Digital Inclusion and Sustainability: Rural Innovation and Growth in the Global South

Agripreneurship, Digital Inclusion and Sustainability: Rural Innovation and Growth in the Global South

Edited by

Dr P. K. Dutta
Associate Professor
Amity University Kolkata, India

Dr Sonal Trivedi
Associate Professor
Manav Rachna University, Faridabad, India

CABI

CABI is a trading name of CAB International

CABI
Nosworthy Way
Wallingford
Oxfordshire OX10 8DE
UK

Tel: +44 (0)1491 832111
E-mail: info@cabi.org
Website: www.cabi.org

CABI
200 Portland Street
Boston
MA 02114
USA

T: +1 (617)682-9015
E-mail: cabi-nao@cabi.org

A catalogue record for this book is available from the British Library, London, UK.

ISBN-13: 9781836992622 (hardback)
9781836992639 (ePDF)
9781836992646 (ePub)

DOI: 10.1079/9781836992646.0000

Commissioning Editor: David Hemming
Editorial Assistant: Theresa Regueira

Typeset by Exeter Premedia Services Pvt Ltd, Chennai, India
Printed and bound in the UK by CPI Group (UK) Ltd, Croydon, CR0 4YY

Contents

Contributors

AbdulKareem, Hauwah K.K., Department of Economics and Development Studies, Kwara State University, Malete, Nigeria

Admire, Mthombeni, Graduate School of Business and Leadership, University of Kwazulu Natal, Durban, South Africa

Aggarwal, Pooja, Assistant Professor, School of Business Management and Commerce, Jagat Guru Nanak Dev Punjab State Open University, Patiala, India

Akpan, Uduak S., Department of Development Studies, SOAS University of London, London, UK

Annamalai, Sumathi, Associate Professor, School of Business and Management, Christ University, Bangalore, India

Anugrah, Iwan Setiajie, Research Center for Social Welfare, Villages, and Connectivity – National Research and Innovation Agency (BRIN), Jakata, Indonesia

Assoua, Joe E., Department of Agricultural Economics and Agribusiness, Faculty of Agriculture and Veterinary Medicine, University of Buea, Buea, Cameroon; Trade Policy Training Centre in Africa (TRAPCA), Arusha, Tanzania

Atrey, Ishan, Faculty Lecturer, Integrated Programme in Law, Indian Institute of Management Rohtak, Rohtak, India

Basuki, Seno, Research Center for Social Welfare, Villages, and Connectivity – National Research and Innovation Agency (BRIN), Jakarta, Indonesia

Bhatia, Swati, Professor, Asian Business School, Noida, India

Binuomote, Samuel O., Professor, Department of Agricultural Economics, Ladoke Akintola University of Technology, Ogbomoso, Nigeria; Visiting Professor of Agricultural Economics, Department of Agricultural Economics and Management, University of Eswatini, Luyengo Campus, Eswatini

Bomdzele, Eric, Jr., Visiting Researcher, Alliance Bioversity and CIAT, Cali, Colombia; PhD Research Fellow, Department of Agricultural Economics and Agribusiness, Faculty of Agriculture and Veterinary Medicine, University of Buea, Buea, Cameroon; Founder, Sustainable Outreach Foundation (SOF), Buea, Cameroon

Boxall, Katherine, Programme Leader Fashion Business, Istituto Marangoni, London, UK

Bronson, Mutanda, Lecturer, Department of Applied Business Sciences, Manicaland State University of Applied Sciences, Mutare, Zimbabwe

Carolina, Carolina, Research Center for Social Welfare, Villages, and Connectivity – National Research and Innovation Agency (BRIN), Jakarta, Indonesia

Centobelli, Piera, Associate Professor, Department of Industrial Engineering, University of Naples Federico II, Naples, Italy

de Lange, Annet H., Professor, Open University, Faculty of Psychology, Heerlen, the Netherlands

Devi, Nisha, Assistant Professor, Department of Commerce, Shri Ram College of Commerce, University of Delhi, Delhi, India

Dhir, Manisha, Lecturer in Finance, Coventry Business School, Coventry University, Coventry, UK

Enet, Mukurazita, College of Education and Human Development, Department of Organisational Leadership, Policy, and Development, University of Minnesota, Minneapolis, Minnesota, USA

Enwa, Sarah, Department of Agricultural Economics, Delta State University, Abraka, Nigeria

Goyal, Anil Kumar, Head of Department, Institute of Hotel Management, Catering and Nutrition Pusa, New Delhi, India

Guleria, Sandeep, Professor, Department of Tourism and Hospitality Management, Chandigarh University, Mohali, India

Gunasekaran, Angappa, Professor of Supply Chain Management, School of Business Administration, Penn State Harrisburg, Middletown, Pennsylvania, USA

Hakim, Lukman Nul, Faculty of Education, Universitas Islam Internasional Indonesia (UIII), Depok, Indonesia; Research Center for Social Welfare, Villages, and Connectivity – National Research and Innovation Agency (BRIN), Jakarta, Indonesia

Hassan, Shuaibu Chiroma, Department of Home Economics, Isa Kaita College of Education, Dutsinm, Nigeria

Hooda, Renu, Assistant Professor, Department of Business Administration, Maharaja Surajmal Institute, New Delhi, India

Husmiati, Husmiati, Research Center for Social Welfare, Villages, and Connectivity – National Research and Innovation Agency (BRIN), Jakarta, Indonesia

Jeffasi, Muzondo Pardon, Department of Supply Chain Management, Marondera University of Agricultural Sciences and Technology, Marondera, Zimbabwe

John, Suja, Professor, Department of Tourism Management, Christ University, Bangalore, India

Julius, Tapera, Vice Chancellor Office, Lupane State University, Lupane, Zimbabwe

Khandagale, Vidyanand S., Assistant Professor, Department of Education, Shivaji University, Kolhapur, India

Kumar, Suneel, Professor, Department of Commerce, Shaheed Bhagat Singh College, University of Delhi, Delhi, India

Kumari, Madhu, Research Scholar, Department of Tourism and Hospitality Management, Chandigarh University, Mohali, India

Malik, Punita, Assistant Professor, Department of Commerce, Government College for Women, Kulana, Jhajjar, India

Margono, Tri, Research Center for Social Welfare, Villages, and Connectivity – National Research and Innovation Agency (BRIN), Jakarta, Indonesia

Mohanan, Smitha Amarail, Research Scholar, Department of Tourism Management, Christ University, Bangalore, India

Mujawar, Nigar K., Assistant Professor, Department of Pharmaceutics, Womens College of Pharmacy, Peth Vadgaon, Kolhapur, India

Nirmale, Rekha L., I/C Principal, College of Education, Peth Vadgaon, Kolhapur, India

Ode-Omenka, Loveth Chiduma, Lecturer, Department of Economics, Accounting and Finance, Bells University of Technology, Ota, Nigeria; Ogun State and Research Associate, DePECOS Institutions and Development Research Centre (DIaDeRC), Ota, Nigeria

Okunade, Solomon O., Lecturer, Department of Economics, Chrisland University, Abeokuta, Nigeria; Research Lead, DePECOS Institutions and Development Research Centre (DIaDeRC), Ota, Nigeria

Omosehin, Olanrewaju O., Lecturer, Department of Agricultural Economics, University of Ibadan, Ibadan, Nigeria

Osabuohien, Evans, Professor of Economics, Department of Economics and Development Studies, Covenant University, Ota, Nigeria; Founding Chair, DePECOS Institutions and Development Research Centre (DIaDeRC), Ota, Nigeria

Pandey, Sonam, Asian Business School, Noida, India

Purity, Hamunakwadi, Post-Doctoral Research Fellow, Department of Building and Human Settlement Development, Nelson Mandela University, Gqeberha, South Africa

Purwasantana, Dian, Research Center for Social Welfare, Villages, and Connectivity – National Research and Innovation Agency (BRIN), Jakarta, Indonesia

Rahabhi, Mashapure, Lecturer, Department of Entrepreneurship and Business Management, Chinhoyi University of Technology, Chinhoyi, Zimbabwe

Reichwein, Wilko, Hamburg Institute for Vocational Education and Training (HIBB), Hamburg, Germany

Rojas-Rivas, Edgar, Department of Tourism, School of Business and Economics, Universidad de las América's Puebla (UDLAP), Puebla, Mexico

Ruvimbo, Mamutse, Department of Agribusiness and Management, Marondera University of Agricultural Sciences and Technology, Marondera, Zimbabwe

S., Vivek, Assistant Professor, Department of Social Work, Amrita Vishwa Vidyapeetham, Amritapuri, India

S.R., Deepika, Assistant Professor, School of Business and Management, Christ University, Bangalore, India

S.S., Sreejith, Assistant Professor, Department of Management Studies, National Institute of Technology Calicut, Kozhikode, India

Samanta, Navajyoti, Associate Professor, Leicester Law School, University of Leicester, Leicester, UK

Sandhu, Komal, Associate Professor, Department of Law, Gurugram University, Gurugram, India

Sathiyaseelan, Anuradha, Professor, Department of Psychology, Christ University, Bangalore, India

Sufi, Tahir, Professor, Department of Tourism, School of Business and Economics, Universidad de las América's Puebla (UDLAP), Puebla, Mexico

Suresh, Veena, Research Scholar, Department of Management Studies, National Institute of Technology Calicut, Kozhikode, India

Valeri, Marco, Associate Professor of Organizational Behavior, Faculty of Economics, Niccolo Cusano University, Rome, Italy

Valerio, Erika, Research Fellow, School of Environmental and Rural Science, University of New England, Armidale, Australia

Vasishta, Prihana, Research Scholar, Centre of Management and Humanities (CMH), Punjab Engineering College, Chandigarh, India

Preface

The intersection of agriculture, digital innovation, and sustainability represents one of the most transformative forces shaping the future of the Global South. *Agripreneurship, Digital Inclusion, and Sustainability: Rural Innovation and Growth in the Global South* explores this confluence with a multidisciplinary lens, bringing together diverse perspectives from Africa, Asia, and Latin America. The book underscores how rural entrepreneurship – rooted in agricultural traditions yet invigorated by digital inclusion – is redefining local economies, enhancing livelihoods, and promoting sustainable growth.

This volume presents an insightful collection of scholarly contributions that examine agripreneurship not merely as a means of income generation but as a catalyst for systemic transformation. It captures the entrepreneurial spirit of small-scale farmers, women innovators, youth leaders, and community organizations that are leveraging technology, education, and social capital to overcome barriers in markets, finance, and information access. Through empirical studies, comparative analyses, and practical frameworks, the chapters unravel how digital tools and inclusive business models are reshaping value chains, promoting food security, and empowering marginalized populations.

The book's structure traverses multiple dimensions of rural transformation. The initial chapters investigate the evolving role of small agripreneurs and youth-led enterprises in countries such as Zimbabwe and Indonesia, highlighting community-based innovations that sustain both livelihoods and ecosystems. Subsequent sections delve into the empowerment of rural women, the role of technology in agribusiness, and the emergence of new financial models such as farmer producer organizations and rural fintech ecosystems. The concluding chapters explore tourism, cultural heritage, and education as complementary drivers of sustainable rural economies, offering an integrated understanding of rural resilience in the 21st century.

By combining grounded research with visionary perspectives, this volume seeks to inspire policy makers, researchers, educators, and practitioners to reimagine rural development through the lens of entrepreneurship and digital equity. It provides actionable insights for achieving the United Nations' sustainable development goals (SDGs), particularly those related to poverty reduction, gender equality, decent work, and responsible consumption.

Ultimately, *Agripreneurship, Digital Inclusion, and Sustainability* is both a scholarly resource and a call to action. It invites readers to recognize the vast potential that resides in the rural landscapes of the Global South – where tradition meets technology, and where innovation thrives at the intersection of community, culture, and sustainability.

1 Traditional Value Addition of Grain Legumes to Small-Scale Agripreneurs: Practices, Hindrances, and Effects on Profitability

Mashapure Rahabhi[1*], Mamutse Ruvimbo[2], Hamunakwadi Purity[3], Mthombeni Admire[4], Mutanda Bronson[5], Tapera Julius[6], Mukurazita Enet[7] and Muzondo Pardon Jeffasi[8]

[1]Department of Entrepreneurship and Business Management, Chinhoyi University of Technology, Chinhoyi, Zimbabwe; [2]Department of Agribusiness and Management, Marondera University of Agricultural Sciences and Technology, Marondera, Zimbabwe; [3]Department of Building and Human Settlement Development, Nelson Mandela University, Gqeberha, South Africa; [4]Graduate School of Business and Leadership, University of Kwazulu Natal, Durban, South Africa; [5]Department of Applied Business Sciences, Manicaland State University of Applied Sciences, Mutare, Zimbabwe; [6]Vice Chancellor Office, Lupane State University, Lupane, Zimbabwe; [7]College of Education and Human Development, Department of Organisational Leadership, Policy, and Development, University of Minnesota, Minneapolis, Minnesota, USA; [8]Department of Supply Chain Management, Marondera University of Agricultural Sciences and Technology, Marondera, Zimbabwe

1.1 Introduction

Small-scale entrepreneurs play a crucial role globally in economic development and they form the backbone of the emerging economy with small to medium enterprises comprising approximately 94% of businesses (Adeola *et al.*, 2021), with agripreneurs playing a vital role in countrys' food security and nutrition. However, small-scale agripreneurs face various challenges, including limited access to markets, low productivity, and income generation. Traditional value addition techniques have been used for centuries and could provide a solution to these challenges. Traditional value addition can enhance the value of legumes, increase production efficiency, and provide markets for small-scale entrepreneurs. However, the effects of traditional value addition techniques on small-scale agripreneurs in developing nations like Zimbabwe are not well understood. This chapter investigates the effects of traditional

*Corresponding author: cmashapure29@gmail.com

DOI: 10.1079/9781836992646.0001

value addition on small-scale agripreneurs in developing nations.

Legume crops have a long and diverse history associated with the development of agricultural practices and meeting human nutritional needs (Affrifah *et al.*, 2023). Common types of legumes include dried beans, broad beans, peas, chickpeas, cowpeas, lentils, lupins, peanuts and soybeans (Hughes *et al.*, 2022). These food crops have different origins of domestication. Many pulses (i.e. lentils and chickpeas) were generally derived and cultivated in the Middle East regions, whereas common or dry beans (Siddiq *et al.*, 2021) are referred to as "new world" crops, with small-seeded types (navy, black, pinto) coming from the Mesoamerican region and large-seeded types (kidney, cranberry, yellow Mayacoba) of Andean origins (Miklas *et al.*, 2021). Geographic distribution of these domesticated gene pools has contributed to distinctive global regions of predominant usage (Affrifah *et al.*, 2023). These crops have distinctive sizes, shapes, colors, and flavors that contribute to consumer appeal. Thus, products are prepared and cooked in many regionally specific cuisines (Amin and Borchgrevink, 2021). Legumes are particularly well suited to cultivation in tropical and humid climates, whereas pulses are more readily adapted to semi-arid areas. Sustainable productivity, storage stability, and their nutritional profile distinguish these crops as nutrient- and energy-dense foods (Miklas *et al.*, 2021).

The incorporation of legume-based ingredients in different food products can potentially expand the utilization of legumes beyond traditional uses and consumption patterns (Affrifah *et al.*, 2023). There is an increased potential in the food industry for using legume ingredients in various food systems (Dhull *et al.*, 2023). Legume-based ingredients are highly suitable for developing diverse food products including composite mixes and doughs, meat alternatives and extenders, gluten-free products, baked products, snack foods, dairy products, and regional or ethnic products (Affrifah *et al.*, 2023). The development, promotion, and consumption of protein-rich, legume-based meat alternatives provide consumers with a range of options to the energy-intense production and processing of animal-based foods (Affrifah *et al.*, 2023). Affrifah *et al.* (2023) highlighted that in order to successfully market new legume-based products, it is essential to improve consumer acceptance through education and communication about the benefits of legumes related to health, economics, and the environment.

Grain legumes such as beans, lentils, chickpeas, and peas are staple crops for many small-scale agripreneurs across the world, especially in developing countries. These crops are vital for food security and nutrition, providing a significant source of protein, fiber, vitamins, and minerals. However, their potential for improving livelihoods extends beyond their nutritional value through the practice of value addition. Traditional value addition encompasses a range of practices aimed at enhancing the market value and utility of grain legumes, ultimately impacting the profitability of small-scale agripreneurs. Grain legumes play a crucial role in sustainable agriculture due to their ability to fix nitrogen in the soil, thus improving soil fertility and reducing the need for synthetic fertilizers. They are also adaptable to various agro-ecological conditions and can be intercropped with other staple crops, providing a diversified source of income for small-scale farmers. Despite their potential, the marketability and profitability of these crops remain limited without proper value addition.

Mainly processed at the micro- and small-scale household level, grain legumes represent an important economic opportunity in particular for women and the marginalized (Kpossilande *et al.*, 2020). Entry barriers are low, start-up costs are low, no possession of particular assets is required, and value addition is combinable with domestic responsibilities (Materia *et al.*, 2021). With appropriate access to inputs, knowledge, and education (Kebede, 2020), value addition of legumes can allow the most marginalized in society to contribute to the food security of their families and communities, and increase their independence through the generation of income (Materia *et al.*, 2021).

Currently, value addition and preservation processes of traditional legume products have mostly not been standardized or formalized due to a lack of documented knowledge and process formalization and optimization (Affrifah *et al.*, 2023). For standardization, and subsequently formalization, product group-specific evidence is needed on the nutritional benefits and functional properties that lead to value addition to raw materials. Evidence is also needed on base interventions in support of local traditional food processing (Materia *et al.*, 2021). Even though local and traditional foods are often more or equally nutritious and as microbiologically safe as their global counterparts, these functional properties have often not been measured, limiting the inclusion of such local foods in interventions or policies to promote food security (Mendonca *et al.*, 2020).

Agripreneurial opportunities are currently hampered by the lack of effective value chains, which often prevents highly nutritious local traditional products from meeting consumers' needs and from reaching a wide range of consumers in both informal and formal markets (Materia *et al.*, 2021). Addressing the problem of food distribution, diversity, consumption, and commercialization through active engagement of the informal sector represents an effective strategy to strengthen livelihoods for the poor (Lorato *et al.*, 2023). Materia *et al.* (2021) highlighted the need to put small local processors (usually women working in informal settings) at the center stage of food system transformations when exploiting the tremendous agripreneurial potential of traditional fermented foods. This chapter argues that the industrial upscaling of traditional value-addition processes leaves current traditional processors behind, impeding long-term development of the rural poor. Generating, identifying, and exploiting business opportunities related to traditional foods in rural areas provides incentives for women in particular to build their future there rather than migrating to larger urban areas (Kwiatkowski *et al.*, 2024). This mitigates the pressure on rapidly growing cities and facilitates modern development and better infrastructure in rural areas, which are among the drivers for the transformation of food systems globally (Materia *et al.*, 2021).

In this chapter, researchers discuss, in particular, local traditional value-addition products of grain legumes. Despite their contribution to tackling hunger and malnutrition (Mendonca *et al.*, 2020), few studies have investigated the value addition practices relating to these agricultural commodities (Mazenda *et al.*, 2022). However, none of this research has gavin much focus on the effects of the value addition of grain legumes on small-scale agripreneurs specifically in the Zimbabwean situation. While acknowledging the fact that value addition generally played a role in the enhancement of small-scale agripreneurship in other countries, its contribution in the Zimbabwean context is unknown (Maphosa and Maphosa, 2020). Therefore, this chapter aims to investigate the effects of traditional value addition for grain legumes on small-scale agripreneurs in Harare Metropolitan Province, Zimbabwe.

1.2 Theory Guiding the Study

Under the passive learning model (PLM), a firm enters a market without knowing its own potential growth (Adebanwi, 2024). Only after entry does the firm start to learn about its profitability based on information from profits realized (Kirtley and O'Mahony, 2023). By continually updating such learning, the firm can decide to expand, contract, or exit (Lowe *et al.*, 2020). This learning model states that firms and managers of firms learn about their efficiency once they are established in the industry (Materia *et al.*, 2021). The theory suggests that, for the firm to understand its growth potential, it has to undergo some induction and awareness-creating activities that business development service providers (BDSPs) can provide. The PLM theory explains the contribution of BDSPs in creating awareness among firms, and that managers need to learn and establish how BDSPs can facilitate learning and growth processes. It is only when BDSPs know the market needs of small-scale agripreneurs and can deliver competent services that the impact of their services will translate into agripreneurship growth. In relation to this study, this theory describes how certain small-scale agripreneurs are able to determine the profitability of adding value to their grain legume goods. Some agripreneurs might not be able to determine whether value addition in legumes would increase profitability until they actually start the business. Instead of conducting market research, the parties involved just enter the market with the intention of making money; they do not know the outcome until they have done so. Through the responses of their target market, these small-scale agripreneurs can learn the extent to which traditional value addition has impacted their businesses.

1.2.1 Traditional value addition and small-scale agripreneurial opportunities

The global popularity of legume foods exemplifies the opportunities for diversification in traditional value addition. Well-known (Western) examples of "artisanal" fermented products that are now produced at large/industrial scale include beer made by microbreweries and French cheese like Roquefort made from raw milk (Materia *et al.*, 2021). As an economic activity, value addition of legumes has potential in providing a diversified stream of income for a rural household. Although for several well-known grain legume products such as roasted groundnuts and roundnuts, roasted nuts, roasted maize meal, cheese, beer, and wine the process has been industrialized, the traditional ones (those whose processing has not been formalized, standardized, or upscaled to an industrial scale) maintain important cultural values that consumers want to preserve and restore, often being considered part of the national heritage of a country (Rampa *et al.*, 2020). The indigenous knowledge of producing these products is an integral component of many cultures (Rampa *et al.*, 2020).

In most cases, traditional value-addition foods are processed at the household level for home consumption using locally available raw materials (Materia *et al.*, 2021). Excess product is sold at local markets by some processors, and a small fraction of processors may specifically process for sale (Materia *et al.*, 2021). Most of traditional value-addition products are acidic, some are alcoholic. Processing typically starts by preprocessing raw materials (Famuwagun *et al.*, 2024). In the case of cereals as the raw material, these may be ground and mixed with water, and then heated and boiled for some time. Raw milk may be checked for quality using simple tests and be sieved to eliminate large particles (Materia *et al.*, 2021). The pre-processed raw materials are then placed in a fermentation vessel, usually a plastic container, metal can, or calabash, and kept at ambient temperature (Materia *et al.*, 2021). Fermentation vessels may be reused between fermentation cycles. In this way, part of the microbial community underlying the fermentation is transferred to fresh raw material to kick-start a new round of fermentation (Materia *et al.*, 2021).

Traditional value-addition processing needs to be upscaled to reach a wider share of consumers (Aderibigbe *et al.*, 2022). To maximize the beneficial functional properties, a diverse microbial community should be used for fermentation (Szutowska, 2020). In most industries, a microbial starter of only one or two strains is most commonly used since this makes the processing more reproducible with respect to product properties. A single strain fermentation is easier to control, which is a key feature in most industrial fermentations, than a fermentation with a mix of microbes (Materia *et al.*, 2021). The diverse nature of the microbial community in traditional fermentation has a higher potential to maximize the full benefits, because of higher levels of resilience and more diverse functionality (Materia *et al.*, 2021). Opportunities for upscaling traditional fermentation rather than upscaling using fewer microbial strains are feasible. There are examples of industrially made fermented foods using undefined, mixed microbial communities, such as sourdough breads made with mixed communities of wild yeasts (De Vuyst *et al.*, 2023). Moonga *et al.* (2020) previously showed that the standardization of processing conditions and procedures results in uniform product properties, even when undefined microbial communities consisting of at least 15 different bacterial species were used. Since other traditional foods rely on similar microbial communities, this most likely also applies to other traditionally fermented foods (Alekseeva *et al.*, 2021). This is promising, since higher levels of production can be achieved using high levels of microbial diversity, which promote the unique functional properties of traditional fermented foods (Yilmaz *et al.*, 2022). In turn, the specific nutrition and health benefits of consuming traditional fermented foods such as mabisi, munkoyo, mahewu, and akpan can be part of promotion and sensitization campaigns (Materia *et al.*, 2021).

1.2.2 Benefits of traditional value addition in legume-based products

Legume-based products in traditional value addition have unique properties that distinguish them from other foods in the human diet. Legumes are intrinsically produced in an environmentally sustainable manner, thereby constituting an economical source of proteins as compared with animal sources (Affrifah *et al.*, 2023). Regular consumption of legumes offers a variety of well-documented health benefits (Albuquerque *et al.*, 2020). These attributes of legumes make them highly suitable for developing value-added products, which can potentially increase the per capita consumption of legumes beyond current levels, especially in the emerging economies. The emphasis on increased consumption of food legumes is reflected in government-issued dietary guidelines.

These benefits include the following.

- *Increased shelf life*. Traditional value addition can help reduce food losses (Mokoena *et al.*, 2021). For example, fermenting microbes may excrete various compounds, such as bacteriocins, that specifically inhibit pathogenic bacteria (Materia *et al.*, 2021).
- *Food security*. By enhancing the shelf life and nutritional value of legumes, traditional value addition can contribute to food security, particularly in resource-poor settings where legumes are a major source of protein and other essential nutrients (Kebede, 2020). Odeku *et al.* (2024) add that traditional processing can lead to a variety of legume-based products, encouraging dietary diversity and improving nutritional outcomes.
- *Improved taste and other organoleptic properties*. Traditionally, products with value addition have a richer taste pallet than most industrially produced fermented foods (Nanda *et al.*, 2021). Diverse microbial communities commonly generate more diverse aroma profiles.
- *Increased nutritional value and improved standard of living*. Legumes possess very rich nutritional profiles, being high in protein and fiber content and having lower carbohydrate levels compared with cereals (Siddiq *et al.*, 2021).
- *Health benefits to consumers*. Traditional processing helps to enhance diet quality in diets low in (bio)available micronutrients and rich in antinutritional factors (Agregán *et al.*, 2023).
- *Cultural preservation*. Traditional value-addition techniques often preserve cultural culinary practices, which contributes to the maintenance of local food traditions and knowledge systems (Okech and Timothy, 2023). The continuation of traditional food-processing methods can strengthen community bonds as these practices are often communal activities (Kennedy *et al.*, 2022).
- *Preservation of raw materials*. Song *et al.* (2022) highlighted that traditional value addition methods help in the preservation of raw materials. For example, roasting and sun drying can remove food-borne pathogens from raw materials, which inhibits growth of spoilage organisms and prolongs the shelf life of the legumes after processing.
- *Environmental sustainability*. Traditional methods often rely on low-tech, energy-efficient processes that have a smaller environmental footprint compared with modern industrial processing (Nwaila *et al.*, 2022). These methods typically make use of locally available resources, minimizing the need for imported inputs and reducing transportation-related emissions (Tanguy *et al.*, 2023).
- *Enhanced income generation*. Traditional value addition can create opportunities for small-scale farmers and local communities by increasing the market value of legume products (Branca *et al.*, 2021). This can lead to the development of cottage industries and the generation of income (Desire *et al.*, 2021).
- *Improved marketability*. Traditional value addition improves the marketability of products as their functions and benefits are improved, unlike selling legumes from the field (Kebede, 2020; Mwadzingeni *et al.*, 2021).

1.3 Methodology

Positivism research philosophy was used since it allows the production of explanatory linkages or causal relationships that, in turn, enable prediction and control of the phenomenon under consideration (Mbanaso *et al.*, 2023). A structured questionnaire with a Likert scale was used to determine the relationship of value addition practices and profitability. The data were collected from small-scale legume agripreneurs in the Harare Metropolitan Province. A sample size of 377 small-scale agripreneurs was used, using the Raosoft calculator of the unknown population. The study employed a stratified sampling procedure to select the respondents. With this approach, the population was first separated into subgroups (or strata) that all had a common trait (legume crop). In these subgroups, the researcher used simple random sampling to allow the estimation of statistical measures for each subpopulation. The distribution of the strata was: beans 30.84%, soya beans 8.77%, groundnuts 22.72%, bambara 16.56%, and cowpeas 21.10%.

Table 1.1. Demographic data of small-scale agripreneurs in the study.

Characteristic	Percentage
Gender:	
Male	38.7
Female	61.3
Age (years):	
Less than 30	7.2
30–39	20.1
40–49	41.9
Above 50	30.8
Marital status:	
Single	21.6
Married	40.0
Widow	38.4

1.4 Results and Discussion

A total of 377 questionnaires were administered to small-scale grain legume entrepreneurs in Warren Park district, Zimbabwe. A total of 308 respondents completed the questionnaire, giving a response rate of 81.70%. Sixty-nine questionnaires were not returned, hence a non-response rate of 18.30%. Hendra and Hill (2019) suggested that very high response rates are acceptable for scientific research and an 80% response rate has been long considered satisfactory. This is consistent with Chang *et al.* (2022) who stipulated that a response rate of less than 60% is unacceptable, whilst a 70% response rate is considered as minimally acceptable. Thus, the response rate was acceptable.

The results in Table 1.1 show that females are more engaged in small-scale traditional legume entrepreneurship at 61.3%, compared with 38.7% male participation. The results also show that the biggest group involved in traditional value addition of grain legumes are people aged from 40 to 49 years (41.9%). This is a dominant age group because in Zimbabwe they are the ones who have gained more knowledge on how to traditionally add value to legumes. The marital status results are similar to those of Mohsin *et al.* (2020) and Olateju *et al.* (2017) who found that the majority of participants in small-scale entrepreneurship were married as indicated by the figure of 40%, followed by widows with a percentage of 38.4% and singles of 21.6% in the study. This signifies that as the respondents grow older and get married, they tend to be more entrepreneurially active than when they were single and young.

According to the data in Table 1.2, the most practiced traditional value-addition technique is boiling and freezing beans, indicated by a percentage of 30%. Boiling and freezing beans have proven to be the most popular techniques in Zimbabwe since they can be more readily substituted as a major meal compared with other products (Siddiq *et al.*, 2021). It thus tends to have a greater market demand. The next most common technique is mixing boiled Bambara and cowpeas (22.05%), followed by roasting and salting groundnuts (12.01%), boiling and sun-drying groundnuts (10%), boiling cowpeas (10.06%), boiling Bambara (5%), roasting soybeans (4%), and boiling soybeans (3.9%). Very few legume entrepreneurs were traditionally adding value to soybeans because in Zimbabwe soybeans are mostly used for oil extraction and soy milk rather than consuming it in a traditionally value-added way (Canaan *et al.*, 2022). Only 3% of the respondents were using fermentation.

The results in Table 1.3 show that the coefficient of determination (R^2) is 0.723, indicating that approximately 72.3% of the variation in value-addition output is explained by the explanatory variables included in the model. This suggests that the model has a strong explanatory power, while the remaining 27.7% of the variation is due to other factors not captured by the model. The adjusted R^2 value of 0.010 indicates a slight adjustment for the number of predictors, while the standard error of the estimate reflects the average distance that the observed values fall from the regression line.

The model shows that the study is significant as the *P* value should be less than 0.05 and the *F* ratio should be greater than one in order for it to be an efficient model (Table 1.4). This also shows that the model was correctly specified.

In Table 1.5, out of ten independent variables, only three variables were statistically significant, as indicated by their *P* values of less than 0.05 at the 5% level of significance. These are age, educational level, and income level. Some of the figures for the estimated coefficients were expected, while in some cases they were contrary to expectation. The significant variables are discussed here.

A coefficient for age of 0.477 is significant at the 1% level as indicated by $P = 0.001$. The results also show that educational level is negatively related to traditional value addition, as shown by the unstandardized coefficient of −0.371, which is significant at the 5% level ($P = 0.031$). Lastly, the coefficient of income level (−0.350) is significant at the 1% level ($P = 0.003$).

The results also show that the educational level is negatively related to traditional value addition as shown by the unstandardized coefficient of −37.1%. This is significant at 5% level. This means that as the education level increases by a unit, traditional value addition of grain legumes decreases by 0.371 units, which translates to about 37.1%. This is because as people become highly educated, they do not want to engage in traditional ways of value addition but are rather interested in technical ways. This supports what Brown *et al.* (2021) said in the literature, that farmers with tertiary education derive other incomes from other sources and are therefore not motivated to add value to products.

Table 1.2. Traditional value-addition practices for grain legume products by small-scale agripreneurs.

Characteristic	Percentage
Boiling and freezing beans	30
Roasting and salting groundnuts	12.01
Boiling and sun-drying groundnuts	10
Boiling Bambara groundnuts (nyimo)	5
Boiling cowpeas	10.06
Roasting soybeans	4
Boiling soybeans	3.9
Mixing boiled Bambara and cowpeas	22.03
Fermentation (beans, soybeans)	3

Table 1.3. Diagnostic test results.

Model	R	R^2	Adjusted R^2	Standard error of estimate
1	0.721	0.723	0.010	2.763

Table 1.4. Goodness of fit of the model: analysis of variance (ANOVA).[a]

Model	Sum of squares	DF	Mean square	*F*	Significance (*P*)
Regression Residual Total	68.906	12	5.742	17.052	0.000[b]
	2252.377	295	7.635		
	2321.282	307			

[a]Dependent variable: value addition.
[b]The overall regression model is statistically significant ($P = 0.000$), indicating that the independent variables jointly have a significant effect on value addition

Table 1.5. Results of the significant variables on effects of traditional value addition coefficients.

Model	Unstandardized coefficients		Standardized coefficients		Significance (*P*)
	B	Standard error	Beta	*T*	
(Constant)	4.263	2.452		1.739	0.083
Gender	0.090	0.436	0.015	0.207	0.836
Age	0.477	0.282	−0.197	11.693	0.001*
Educational level	−0.371	0.199	0.011	0.188	0.031**
Years in business	0.395	0.357	0.114	1.107	0.269
Access to credit	0.127	0.158	0.047	0.804	0.422
Age of first entrepreneur	0.040	0.206	0.016	0.192	0.848
Income level	−0.350	0.236	−0.009	0.148	0.003*
Inflation rate	0.096	0.170	−0.039	−0.566	0.572
Proximity of market	−0.227	0.209	−0.081	−1.083	0.280
Labor availability	0.028	0.230	0.009	0.120	0.905
Other factors	−0.113	0.271	−0.034	−0.418	0.676

Dependent variable: value addition ($R^2 = 0.721$ indicates that approximately 72.1% of the variation in value addition is explained by the independent variables).
The overall regression model is statistically significant ($P = 0.000$), implying that the independent variables collectively have a significant effect on value addition.
*, ** Significance at the 1 and 5% levels, respectively.

Table 1.6. Gross margin (GM) per bucket for small-scale legume entrepreneurs ($N = 308$).

GM/50 kg	Minimum (US$)	Mean (US$)	Maximum (US$)
Non-value addition	−350.00	250.40	657.50
Value addition	−233.00	292.09	1310.75
Method	DF	*P* value	Profitability
T-test	80	0.7926	0.4319

Lastly, a coefficient of income level of −0.35 is significant at 1%. This suggests that a unit increase in the income level leads to a decrease of about 0.35, which translates to 35% in traditional value addition of legumes. This can be because as income increases the affordability rate also increases; therefore they engage more in other more profitable businesses than traditional value addition of legumes.

Overall, the results confirmed that age has the greatest impact on traditional value addition as indicated by its coefficient value of 0.477, followed by income level with a coefficient of −0.35, and lastly educational level with a coefficient of −0.371. According to statistical significance, gender, years in business, access to credit, inflation rate, proximity of the market, and availability of labor are not significant, meaning that they have no impact on the traditional value addition of fabacea legumes.

Profitability of value addition in grain legumes was determined using gross margin analysis and the results are presented in Table 1.6. The gross margin of value-added and non-value-added fabacea legumes were compared and the results indicate that there is no statistically significant difference on the gross margin between value addition and non-value addition of legumes as represented by $P = 0.7926$.

However, non-value adders made a profit of around US$250.40 per 50 kg and value adders earned a profit of $292.09 per 50 kg. The findings show that although there is profit in both value-added and non-value-added products, value addition has greater profit. The findings corroborate those of Hassan *et al.* (2021), who found that the more value a farmer added to raw tubers, the better the income obtained from the market.

1.5 Recommendations to Government, Stakeholders, and Small-Scale Agripreneurs

1.5.1 Establishment of a value-addition center for small-scale entrepreneurs

The government could collaborate with private partners to establish a centralized value addition center for small-scale entrepreneurs. The center would provide processing equipment, training, and support services such as branding, marketing, and quality assurance to encourage entrepreneurs to engage in legume value addition.

1.5.2 Access to finance

The government can initiate policies and programs that provide small-scale entrepreneurs with easy access to affordable finance for equipment purchase, capacity building, and working capital. This can be done through microfinance schemes, government-funded grants, and soft loans, and grants from development institutions.

1.5.3 Technical and vocational education training

The government could collaborate with vocational training institutes to offer technical and vocational education training courses on legume value addition and effective marketing of value-added products. This will help entrepreneurs to acquire the necessary skills and knowledge to engage effectively in the value chain.

1.5.4 Provision of infrastructure

The government can invest in supplying reliable infrastructure such as roads, energy, and water for efficient transportation of products, processing activities, and storage facilities. The construction of modern storage facilities like silos will encourage storage of surplus production and boost the income of small-scale entrepreneurs.

1.5.5 Promote research and development

The government and development partners can support or fund participatory research initiatives to explore emerging trends, best practices, and new eco-friendly technologies in legume value addition. Such initiatives will help to increase production and profitability and improve the quality of processed products.

1.5.6 Incentives

The government can provide incentives for graduating entrepreneurs who are engaged in traditional value addition of legumes such as access to export markets, tax holidays, reduced licensing fees, and more favorable tariffs on imports and exports. By implementing these policy recommendations, Zimbabwe's government can leverage traditional value addition practices for legumes to encourage small-scale entrepreneurship growth and increase income and standard of living for its citizens in the Warren Park district.

1.5.7 Create a business plan

Creating a business plan will help traditional legume entrepreneurs to allocate resources. Resource management helps to make decisions on what should be bought and to allocate resources before spending starts. If there is a need to apply for credit, for example loans, a business plan is essential especially in banks who only meet with entrepreneurs to discuss financing when they have a business plan to present (Prijadi *et al.*, 2020).

1.5.8 Quality products and services

Providing quality products and reliable services in the market can help increase the reputation of the business for consistently delivering excellent customer experience (Klink *et al.*, 2021). This will help to attract more customers to the business and in gaining loyal customers, hence increasing the profitability.

1.6 Recommendations on Areas of Further Research

In this quantitative research study, the major limitation was that the effects of traditional value addition of grain legumes were limited to only one geographic area – the Harare Metropolitan Province. Therefore, recommendations for future research studies are to explore the effects in other geographic locations to gain a greater perspective which will potentially help legume entrepreneurs in Zimbabwe. There is also a need to carry out a feasibility study on the potential for upscaling the processing of legumes with traditional methods, taking into account supply and demand, and other factors. Future research could be on comparing the cost effectiveness of traditional value addition with modern processing methods used in other countries and how Zimbabwe can incorporate these lessons.

References

Adebanwi, Wale. (2024) *How to Become a Big Man in Africa: Subalternity, Elites, and Ethnic Politics in Contemporary Nigeria*. Indiana University Press.

Adeola, O., Gyimah, P., Appiah, K.O. and Lussier, R.N. (2021) Can critical success factors of small businesses in emerging markets advance UN sustainable development goals? *World Journal of Entrepreneurship, Management and Sustainable Development* 17(1), 85–105. DOI: 10.1108/WJEMSD-09-2019-0072.

Aderibigbe, O.R., Ezekiel, O.O., Owolade, S.O., Korese, J.K., Sturm, B. *et al.* (2022) Exploring the potentials of underutilized grain amaranth (*Amaranthus* spp.) along the value chain for food and nutrition security: A review. *Critical Reviews in Food Science and Nutrition* 62(3), 656–669. DOI: 10.1080/10408398.2020.1825323.

Affrifah, N.S., Uebersax, M.A. and Amin, S. (2023) Nutritional significance, value-added applications, and consumer perceptions of food legumes: A review. *Legume Science* 5(4), e192. DOI: 10.1002/leg3.192.

Agregán, R., NihalGuzel, M.G., SnehPuniaBangar, G., Kumar, M. and Lorenzo, J.M. (2023) The effects of processing technologies on nutritional and anti-nutritional properties of pseudocereals and minor cereal. *Food and Bioprocess Technology* 16(5), 961–986.

Albuquerque, T., AntóniaNunes, M., Bessada, S.M., Costa, H.S. and Oliveira, M.B.P. (2020) Biologically active and health promoting food components of nuts, oilseeds, fruits, vegetables, cereals, and legumes. In: *Chemical Analysis of Food*. Academic Press, pp. 609–656.

Alekseeva, A.Y., Groenenboom, A.E., Smid, E.J. and Schoustra, S.E. (2021) Eco-evolutionary dynamics in microbial communities from spontaneous fermented foods. *International Journal of Environmental Research and Public Health* 18(19), 10093.

Amin, S. and Borchgrevink, C.P. (2021) A culinology® perspective of dry beans and other pulses. In: Siddiq, M. and Uebersax, M.A. (eds) *Dry Beans and Pulses: Production, Processing, and Nutrition*. pp. 453–480.

Branca, G., Cacchiarelli, L., D'Amico, V., Dakishoni, L., Lupafya, E. *et al.* (2021) Cereal-legume value chain analysis: A case of smallholder production in selected areas of Malawi. *Agriculture* 11(12), 1217. DOI: 10.3390/agriculture11121217.

Brown, C., EszterKovács, I.H., Villamayor-Tomas, S., AmaiaAlbizua, A.G., IoannaGrammatikopoulou, D.M. *et al.* (2021) Simplistic understandings of farmer motivations could undermine the environmental potential of the common agricultural policy. *Land Use Policy* 101, 105136. DOI: 10.1016/j.landusepol.2020.105136.

Canaan, J.M.M., Brasil, G.S.A.P., de Barros, N.R., Mussagy, C.U. and Guerra, N.B. (2022) Soybean processing wastes and their potential in the generation of high value added products. *Food Chemistry* 373, 131476. DOI: 10.1016/j.foodchem.2021.131476.

Chang, K.-H., Hsiung, T.-Y. and Chang, T.-Y. (2022) Multi-commodity distribution under uncertainty in disaster response phase: Model, solution method, and an empirical study. *European Journal of Operational Research* 303(2), 857–876. DOI: 10.1016/j.ejor.2022.02.055.

Desire, M.F., Blessing, M., Elijah, N., Ronald, M., KamotaAgather, Z. *et al.* (2021) Exploring food fortification potential of neglected legume and oil seed crops for improving food and nutrition security among smallholder farming communities: A systematic review. *Journal of Agriculture and Food Research* 3, 100117. DOI: 10.1016/j.jafr.2021.100117.

De Vuyst, L., Comasio, A. and Van Kerrebroeck, S. (2023) Sourdough production: Fermentation strategies, microbial ecology, and use of non-flour ingredients. *Critical Reviews in Food Science and Nutrition* 63(15), 2447–2479. DOI: 10.1080/10408398.2021.1976100.

Dhull, S.B., Bamal, P., Kumar, M., Bangar, P. *et al.* (2023) Fenugreek (*Trigonella foenum graecum*) gum: A functional ingredient with promising properties and applications in food and pharmaceuticals—a review. *Legume Science* 5, e176. DOI: 10.1002/leg3.176.

Famuwagun, A.A., Fasogbon, B.M., Taiwo, K.A., Sobukola, O.P., Akinlade, F.A. *et al.* (2024) Selected traditional legume products of south-western Nigeria. In: *Nutritional and Health Aspects of Food in Western Africa*. Academic Press, pp. 211–227.

Hassan, S.Z., ShakeelSadiqJajja, M., Asif, M. and Foster, G. (2021) Bringing more value to small farmers: A study of potato farmers in Pakistan. *Management Decision* 59(4), 829–857. DOI: 10.1108/MD-12-2018-1392.

Hendra, R. and Hill, A. (2019) Rethinking response rates: New evidence of little relationship between survey response rates and nonresponse bias. *Evaluation Review* 43(5), 307–330. DOI: 10.1177/0193841X18807719.

Hughes, J., Pearson, E. and Grafenauer, S. (2022) Legumes—a comprehensive exploration of global food-based dietary guidelines and consumption. *Nutrients* 14(15), 3080. DOI: 10.3390/nu14153080.

Kebede, E. (2020) Grain legumes production and productivity in Ethiopian smallholder agricultural system, contribution to livelihoods and the way forward. *Cogent Food & Agriculture* 6(1), 1722353. DOI: 10.1080/23311932.2020.1722353.

Kennedy, G., Wang, Z., Maundu, P. and Hunter, D. (2022) The role of traditional knowledge and food biodiversity to transform modern food systems. *Trends in Food Science & Technology* 130, 32–41. DOI: 10.1016/j.tifs.2022.09.011.

Kirtley, J. and O'Mahony, S. (2023) What is a pivot? Explaining when and how entrepreneurial firms decide to make strategic change and pivot. *Strategic Management Journal* 44(1), 197–230. DOI: 10.1002/smj.3131.

Klink, R.R., Zhang, J.Q. and Athaide, G.A. (2021) Measuring customer experience management and its impact on financial performance. *European Journal of Marketing* 55(3), 840–867. DOI: 10.1108/EJM-07-2019-0592.

Kpossilande, C.E., Honfoga, B.G. and Ferre, T. (2020) Economic potentials of artisanal food processing microenterprises in West Africa: case of "atta" production in Cotonou (Benin). *Agricultural and Food Economics* 8(1), 24.

Kwiatkowski, G., Hjalager, A.-M., Ossowska, L., Janiszewska, D. and Kurdyś-Kujawska, A. (2024) The entrepreneurial orientation of exhibitors and vendors at food festivals. *Journal of Policy Research in Tourism, Leisure and Events* 16(no.2), 207–226. DOI: 10.1080/19407963.2021.1999967.

Lorato, T., Tadesse, T., Mamo, Y.A. and Getinet, B. (2023) The urban informal sector as a means of livelihood improvement among youth: Evidence from Hawassa city, Ethiopia. *Cogent Economics & Finance* 11(1), 2185346. DOI: 10.1080/23322039.2023.2185346.

Lowe, A., Nama, Y., Bryer, A., NihelChabrak, C.D., Jeacle, I. *et al.* (2020) Problematizing profit and profitability: Discussions. *Accounting, Auditing & Accountability Journal* (4), 753–793. DOI: 10.1108/AAAJ-10-2019-4223.

Maphosa, V. and Maphosa, M. (2020) E-waste management in Sub-Saharan Africa: A systematic literature review. *Cogent Business & Management* 7(1), 1814503. DOI: 10.1080/23311975.2020.1814503.

Materia, V.C., Linnemann, A.R., Smid, E.J. and Schoustra, S.E. (2021) Contribution of traditional fermented foods to food systems transformation: Value addition and inclusive entrepreneurship. *Food Security* 13(5), 1163–1177.

Mazenda, A., Obi, A. and Masiya, T. (2022) The contribution of the small-scale agricultural sector into South Africa's food value addition agenda. In: *The Palgrave Handbook of Africa's Economic Sectors*. Springer International Publishing, Cham, pp. 117–134.

Mbanaso, U.M., Abrahams, L. and Okafor, K.C. (2023) Research philosophy, design and methodology. In: *Research Techniques for Computer Science, Information Systems and Cybersecurity*. Springer Nature Switzerland, Cham, pp. 81–113.

Mendonca, A., Thomas-Popo, E. and Gordon, A. (2020) Microbiological considerations in food safety and quality systems implementation. In: *Food Safety and Quality Systems in Developing Countries*. Academic Press, pp. 185–260.

Miklas, P.N., Kelly, J.D. and Cichy, K.A. (2021) Dry bean breeding and production technologies. In: Siddiq, M. and Uebersax, M.A. (eds) *Dry Beans and Pulses: Production, Processing, and Nutrition*. pp. 29–56.

Mohsin, A.K.M., Lei, H., Hossain, S.F.A., HasanuzzamanTushar, M.C. and Hossain, M.E. (2020) Operation and coordination mechanism of closed-loop supply chain considering corporate social responsibility behavior consciousness. *Cogent Business & Management* 7(1), 1845937. DOI: 10.1080/23311975.2020.1845937.

Mokoena, M.P., Omatola, C.A. and Olaniran, A.O. (2021) Applications of lactic acid bacteria and their bacteriocins against food spoilage microorganisms and foodborne pathogens. *Molecules* 26(22), 7055. DOI: 10.3390/molecules26227055.

Moonga, H.B., Schoustra, S.E., Van den Heuvel, J., Linnemann, A.R., Samad, M.S. *et al.* (2020) Composition and diversity of natural bacterial communities in mabisi, a traditionally fermented milk. *Frontiers in Microbiology* 11, 1816. DOI: 10.3389/fmicb.2020.01816.

Mwadzingeni, L., Afari-Sefa, V., Shimelis, H., SognigbéN'Danikou, S., Depenbusch, L. *et al.* (2021) Unpacking the value of traditional African vegetables for food and nutrition security. *Food Security* 13(5), 1215–1226. DOI: 10.1016/j.tifs.2021.03.059.

Nanda, P.K., Das, A.K., PremanshuDandapat, P., SamiranBandyopadhyay, A.L.D., Lorenzo, J.M. *et al.* (2021) Nutritional aspects, flavour profile and health benefits of crab meat based novel food products and valorisation of processing waste to wealth: A review. *Trends in Food Science & Technology* 112, 252–267. DOI: 10.1016/j.tifs.2021.03.059.

Nwaila, G.T., Frimmel, H.E., Zhang, S.E., Bourdeau, J.E., Tolmay, L.C. *et al.* (2022) The minerals industry in the era of digital transition: An energy-efficient and environmentally conscious approach. *Resources Policy* 78, 102851. DOI: 10.1016/j.resourpol.2022.102851.

Odeku, O.A., Ogunniyi, Q.A., Ogbole, O.O. and Fettke, J. (2024) Forgotten gems: Exploring the untapped benefits of underutilized legumes in agriculture, nutrition, and environmental sustainability. *Plants* 13(9), 1208. DOI: 10.3390/plants13091208.

Okech, R.N. and Timothy, D.J. (2023) Culinary traditions and heritage foods in Africa. In: *Cultural Heritage and Tourism in Africa*. Routledge, pp. 154–169.

Olateju, A.O., Aminu, A.W., Shehu, A. and Danmola, R.A. (2017) The determinants of women entrepreneurs' access to micro credit programme: A case study of Cowries Microfinance Bank (CMB) Lagos, Nigeria. *International Journal of Development and Sustainability* 6(8), 783–791.

Prijadi, R., PermataWulandari, P.M.D., Pinagara, F.A. and Novita, M. (2020) Financing needs of micro-enterprises along their evolution. *International Journal of Ethics and Systems* 36(2), 263–284. DOI: 10.1108/IJOES-05-2018-0071.

Rampa, J.E., HerviusAskling, H., Lang, P., Zens, K.D., NejlaGültekin, Z.S. *et al.* (2020) Immunogenicity and safety of the tick-borne encephalitis vaccination (2009–2019): A systematic review. *Travel Medicine and Infectious Disease* 37, 101876.

Siddiq, M., Uebersax, M.A. and Siddiq, F. (2021) Global production, trade, processing and nutritional profile of dry beans and other pulses. In: Siddiq, M. and Uebersax, M.A. (eds) *Dry Beans and Pulses: Production, Processing, and Nutrition*. pp. 1–28.

Song, X., EsraCapanoglu, J.S.-G., Chen, F. and Xiao, J. (2022) Different food processing technologies: A general background. In: *Retention of Bioactives in Food Processing*. Springer International Publishing, Cham, pp. 37–89.

Szutowska, Julia. (2020) Functional properties of lactic acid bacteria in fermented fruit and vegetable juices: A systematic literature review. *European Food Research and Technology* 246(3), 357–372.

Tanguy, A., Carrière, L. and Laforest, V. (2023) Low-tech approaches for sustainability: Key principles from the literature and practice. *Sustainability: Science, Practice and Policy* 19(1), 2170143. DOI: 10.1080/15487733.2023.2170143.

Yilmaz, B., Bangar, S.P., Echegaray, N., Suri, S., Tomasevic, I. *et al*. (2022) The impacts of lactiplantibacillusplantarum on the functional properties of fermented foods: A review of current knowledge. *Microorganisms* 10(4), 826. DOI: 10.3390/microorganisms10040826.

2 Rural Youth Empowerment through the Agricultural Entrepreneurship Program Model Youth Entrepreneurship and Employment Support Services (YESS): The Case of the Miduana Indigenous Community in Indonesia

Iwan Setiajie Anugrah[1]*, Lukman Nul Hakim[1,2], Carolina Carolina[1], Tri Margono[1], Husmiati Husmiati[1], Seno Basuki[1] and Dian Purwasantana[1]

[1]Research Centre for Social Welfare, Villages, and Connectivity – National Research and Innovation Agency (BRIN), Jakarta, Indonesia; [2]Faculty of Education, Universitas Islam Internasional Indonesia (UIII), Depok, Indonesia

2.1 Introduction

The increasing age of farmers is a significant issue, particularly in regions where the agricultural sector is one of the leading sectors in economic development. The phenomenon of aging farmers also affects the resilience of rural areas, considering that agriculture is the primary livelihood for most rural residents, making its sustainability essential.

The aging farmer phenomenon is not a problem faced by Indonesia alone, it also occurs on a global scale. Susilowati (2016a) cites several sources that describe the occurrence of aging farmers and the decline in the number of young agricultural workers in countries such as the USA (USDA, 2007), Australia (NFF, 2012), the European Union (European Commission, 2012), and Asian countries including Japan (Uchiyama, 2014), Vietnam (Dang, 2014), Korea (Ma, 2014), and others. Susilowati (2016a) explains that both Indonesia and the world in general have faced severe issues with the declining number of young workers in the agricultural sector since 2016.

Agricultural census data over the past two decades illustrate a shift in the agricultural labor structure, leading to the dominance of older farmers and a decrease in the proportion of young workers. One cause of this scarcity is the process of urbanization, which occurs alongside the shift from a village-based agricultural economy to an industrial and service sector economy in urban areas. This condition has driven rapid urbanization (Susilowati, 2016b), which still continues. In addition to the aging agricultural workforce, Indonesia also faces the problem of low educational levels among agricultural workers, with the majority (74.89%) having only completed elementary

*Corresponding author: iwan035@brin.go.id

DOI: 10.1079/9781836992646.0002

school; meanwhile, 58% of them are over 45 years old (Central Bureau of Statistics, 2023b). Besides the crisis in farmer regeneration, poverty in rural areas, where farmers live and work, remains high at 12.22%. Meanwhile, 48.86% of households that rely on agriculture as their primary source of income fall into the poor category (Central Bureau of Statistics, 2023b).

With the complex issues of aging farmers, low educational levels among farmers, and poverty, Indonesia needs to have a strategy to strengthen the farming community, focusing on rural youth. A strategic effort initiated jointly by the Ministry of Agriculture (Kementan, 2024) and the International Fund for Agricultural Development (IFAD) is the Youth Entrepreneurship and Employment Support Services (YESS) program for young farmer communities, which is also implemented in indigenous communities that face various unique challenges.

This chapter discusses a strategic endeavor implemented by Indonesia through the YESS program to motivate young people to master knowledge and skills in agricultural entrepreneurship, particularly in the Miduana traditional village area in West Java Province of Indonesia. Miduana is a village where the majority of the community still upholds traditional customs in their lives, including in agricultural activities.

2.2 Methodology

The research was conducted in the traditional village of Miduana, located in Balegede village, Naringgul district, Cianjur Regency, West Java, Indonesia. The research site was purposively selected based on the focus areas of the YESS-IFAD program activities, which were implemented in many traditional villages in West Java Province as pilot project location. The informants who provided data and information for this research include young farmers and their community groups, the traditional leaders and administrators of the Miduana traditional organization, the head of Balegede village and its officials, as well as other sources of information at the district and regency levels in Cianjur. The data were analyzed descriptively and qualitatively, referencing the concepts and implementation of the YESS-IFAD program as the main basis for the research and the content of this chapter.

2.3 Challenge for Agricultural Development in Indonesia

At the global level, the issue of aging farmers receives less attention than other issues, such as climate change, even though this problem is a severe demographic challenge that needs attention due to its implications for the sustainability of the agricultural sector (Susilowati, 2016a). The increase in the number of elderly farmers over 60 years and the simultaneous decline in young farmers are occurring in almost all parts of the world. Structural demographic changes in the agricultural labor sector are also happening in Asian countries, Europe, the Americas, Canada, and other continents. The declining interest in agricultural labor has become a common phenomenon, requiring serious attention from policy makers to save the agricultural sector (Susilowati, 2016b).

Various reasons for the declining interest of young labor in the agricultural sector primarily include the less prestigious image of the agricultural sector and its inability to provide adequate rewards. The phenomenon stems from the relatively small average land ownership (Susilowati, 2016b). Another reason is that the worldview and lifestyle of young workers have changed in the current postmodern society. For rural youth, the agricultural sector is losing its appeal because it is becoming less economically promising and young people's reluctance to farm is influenced by new subcultures emerging in the digital era (Suyanto, 2016). According to Suyanto (2016), their reluctance to work as farmers is not solely because the agricultural sector is no longer economically attractive but because their worldview and lifestyle have changed significantly, transcending their parents' geographic and cultural boundaries.

The declining percentage of young labor, along with the increasing age of elderly labor, implicitly indicates a mismatch between the types of job opportunities desired by young workers in rural areas and the job opportunities available (Susilowati, 2016b). Moreover, as the education level of rural youth increases, they become more selective in choosing jobs. Young people are experiencing a shift in perception along with the wave of modernization, making farming no longer their primary choice (Arvianti *et al.*, 2015, 2016, Arvianti *et al.*, 2019).

According to Susilowati (2016a), internal and external factors cause the agricultural sector to be increasingly abandoned by youth. Internal factors are those caused by conditions of the individual or the agricultural industry that fail to attract youth to work in agriculture. These factors include:

1. Limited land size and land ownership status.
2. The agricultural sector needs more social prestige, and is dirty and risky.
3. There needs to be a better match between the quality of education and the available job opportunities in the village.
4. The perception is that agriculture is high risk and needs more assurance of income levels, stability, and continuity.
5. Low wages and income in agriculture.
6. Lack of development of non-agricultural businesses and agricultural industries in the village.
7. Low succession of farm management to children.
8. There need to be specific incentive policies for young or beginner farmers.
9. Limited access to financing support services and agricultural extension.
10. Limited production infrastructure (water, electricity, roads, telecommunications).

Other factors influencing rural youth attraction to agriculture are higher incentives in non-agricultural sectors and the perception that non-agricultural work in urban areas is more prestigious than being a farmer. Young people prefer to migrate to the city, even if only to work as construction laborers or in other informal jobs. Those with higher education work in formal employment, such as becoming civil servants, or in the industrial, service, and other sectors (Susilowati, 2016b). Many young farmers also switch professions due to various factors such as limited access to capital and resources, lack of skills and entrepreneurship training, limited market opportunities, and cultural and social barriers (Giriatna *et al.*, 2019; Lubis *et al.*, 2020).

The crisis of young farmers and their accessibility to land became the basis for IFAD's implementation of the YESS program. According to IFAD's records (IFAD, 2018), the low proportion of formal employment, low income, and limited career advancement cause the agricultural sector to have the lowest youth participation rate: only 17% of those working in this sector are aged 20–29 years. In 2016, only one in four youths entered the agricultural labor force. According to the Ministry of Agriculture, in 2011, almost 80% of farmers in the country were aged 45 years and over, compared with an average age of 40 years only 3 years earlier. Although agricultural vocational schools produce around 10,000 trained youths annually, only a few enter the agricultural sector. IFAD interviews with key informants conducted during program design in rural areas of target provinces show that less than 25% of young people work in the agricultural sector.

Further, IFAD notes that access to land is a significant obstacle to youth due to shrinking land size (68% of farmers are smallholders operating on less than 1 ha), widespread landlessness, and increasing competition for land driven by massive agricultural land conversion every year. Youth have little opportunity to inherit land, and buying land is becoming increasingly difficult. The lack of access to financial support further limits their access to agricultural land.

It is estimated that to meet the rapidly growing domestic demand for food, the productivity of Indonesian farmers will need to increase by more than 60% in the next 15 years. Failure to attract the younger generation to this sector will affect future agricultural productivity and rural transformation. Based on the 2023 Agricultural Census, results show that most farmers in Indonesia as of 2023 have low educational attainment and are "aging." As of February 2023, the majority (74.89%) of agricultural laborers had completed only elementary education or less, and about 58% were aged 45 years and over. The data in Table 2.1 show that young agricultural workers who

Table 2.1. Agricultural workers by age group. From Central Bureau of Statistics (2023a).

	Age group (years)	Total (million people)	Percentage (%)
	<30	5.93	14.57
	30–44	11.16	27.43
	45–59	13.83	33.99
	>60	9.77	24.01
Total		40.69	100.00

Table 2.2. Changes in the proportion of farmers in different age groups within 10 years. From Central Bureau of Statistics (2023a).

	Age group (years)	Proportion of farmers in 2013 (%)	Proportion of farmers in 2023 (%)
	<15	0.01	0.00
	15–24	0.88	1.25
	25–34	11.97	10.25
	35–44	26.34	22.04
	45–54	28.03	27.09
	55–64	20.01	23.21
	≥65	12.76	16.16
Total		100.00	100.00

are less than 30 years of age account for only 5.93 million people (14.57%); those aged 30–44 years total 11.16 million people (27.43%); those aged 45–59 years total 13.83 million people (33.99%); and those aged 60 years and over total 9.77 million people (24.01%).

Another challenge is that the composition of farmers based on the distribution of agricultural household management shows a dynamic trend toward the management of farm enterprises by farmers mostly aged over 45 years. This situation has seen an increase in the number of older farmers managing agrarian enterprises. Data from 10 years ago show that this percentage was 60.80% in 2013 and that it increased to 66.46% in 2023, according to the 2023 Agricultural Census results (Table 2.2). The increasing number indicates a potential slowdown in achieving sustainable agricultural businesses expected to support food security and agricultural development. The

Table 2.3. Distribution of farmers by generation in 2023. From Central Bureau of Statistics (2023a).

	Generation classification	Proportion of farmers in 2023 (%)
	Gen X	42.39
	Baby boomers	27.61
	Pre-boomers	2.24
	Post Gen Z	0.0
	Gen Z	2.15
	Millennials	25.61
Total		100.00

regeneration of younger farm entrepreneurs who possess competence and an entrepreneurial spirit in the agricultural sector is becoming increasingly necessary.

The development of the agricultural sector is viewed through the participation of young farmers. Table 2.3 shows that, based on the distribution of farmers by generation, the current challenge is that agricultural activities are still predominantly carried out by Gen X, who are between the ages of 43 and 58 years. Although this age range is still considered productive, the process of farmer regeneration must begin.

The data also show that the involvement of the millennial generation in agricultural activities is quite promising, reaching 25.61%, with 2.15% of Gen Z also starting to get involved in agricultural activities. Nevertheless, the combined percentage of these two generations nearly matches the percentage of farmers from the baby boomer (27.61%) and pre-boomer (2.24%) generations, who are in the older age range of 59–77 years, or even older than 78 years after 2023.

Susilowati (2016b) explains that the young farmers' crisis in the agricultural sector, along with the dominance of older farmers, has consequences for the development of a sustainable farming sector, particularly regarding agricultural productivity, market competitiveness, and rural economic capacity. It is feared that this will impact and threaten future national food security and the sustainability of the agricultural sector.

2.4 Youth Entrepreneurship and Employment Support Services (YESS) Program

The Indonesian government, through the Ministry of Agriculture, continues to respond to the issue of an aging farming population by implementing a farmer regeneration model. This model is initiated through various programs to encourage the emergence of young champion farmers, particularly in rural areas, who are committed to directly supporting sustainable agricultural development. Programs such as the Young Farmer Program, the Millennial Farmer Program, youth entrepreneurship, and the promotion of young farmers are being massively implemented to support the development of the agricultural sector.

One of the programs intensively carried out by the Ministry of Agriculture of the Republic of Indonesia, in collaboration with IFAD, is the implementation of the YESS program. This is designed to produce young rural entrepreneurs and generate competent young workers in the agricultural sector. The Indonesian Ministry of Agriculture and IFAD have agreed to support and implement the development of youth entrepreneurship and employment in the agricultural sector through the YESS program.

Both parties are committed to producing resilient, advanced, independent, and modern millennial entrepreneurs through the YESS program. This program will create opportunities for young people, especially in rural areas, to develop economic ventures in the agricultural sector. From this program, agricultural millennial entrepreneurs will emerge who are capable of supporting Indonesia's agricultural development program (BPPSDMP, 2020).

Fig. 2.1 shows that the YESS program serves as an alternative policy to address the challenges currently and in the future faced by the agricultural sector in Indonesia. According to the YESS concept established by IFAD (2018), the YESS program aims to create opportunities for rural youth to build their economic livelihoods through entrepreneurship or employment in rural areas.

Furthermore, the YESS program also incentivizes youth involvement in rural transformation, potentially increasing employment and entrepreneurship opportunities for young people. To effectively engage the millennial generation, YESS needs to focus on innovation, leverage technology, modernizing employment and agricultural practices, and creating incentives for young people to pursue entrepreneurship in rural areas.

The YESS program is expected to become a model for youth entrepreneurship and employment development in the farming sector in IFAD member countries. The planning and implementation of YESS activities involve various government institutions, business actors, related social institutions, and other stakeholders from the central, provincial, and district levels.

The YESS program has been implemented from 2019 to 2025. As a pilot project, the activities of the YESS program have been carried out in four main provinces: West Java, East Java, South Sulawesi, and South Kalimantan. The program has been further developed by the Agricultural Extension and Human Resource Development Agency (BPPSDMP), Ministry of Agriculture, Indosnesia. Based on the program's targets, from 2019 to 2025, the implementation of the YESS program is expected to reach 320,000 rural youth.

The achievement of the program's targets is supported by four main activities: enhancing the capacity of rural youth in agriculture, developing young rural entrepreneurs, facilitating access to capital, and building a conducive business environment. Meanwhile, the leading indicators of the success of the YESS program include enhancing the capacity of rural youth in agriculture, increasing income for 33,500 people, developing agricultural businesses involving 50,600 people, enabling 100,000 people (including 4300 young migrant households) to use financial services, and providing financial education opportunities for 120,000 young people.

The approach and strategy of the YESS program are designed comprehensively and focus on three main pillars (Hartono, 2018; Giriatna *et al.*, 2019; Lubis *et al.*, 2020):

1. Capacity development. The YESS program provides training and mentorship to young farmers to develop the skills and knowledge necessary for them to become successful

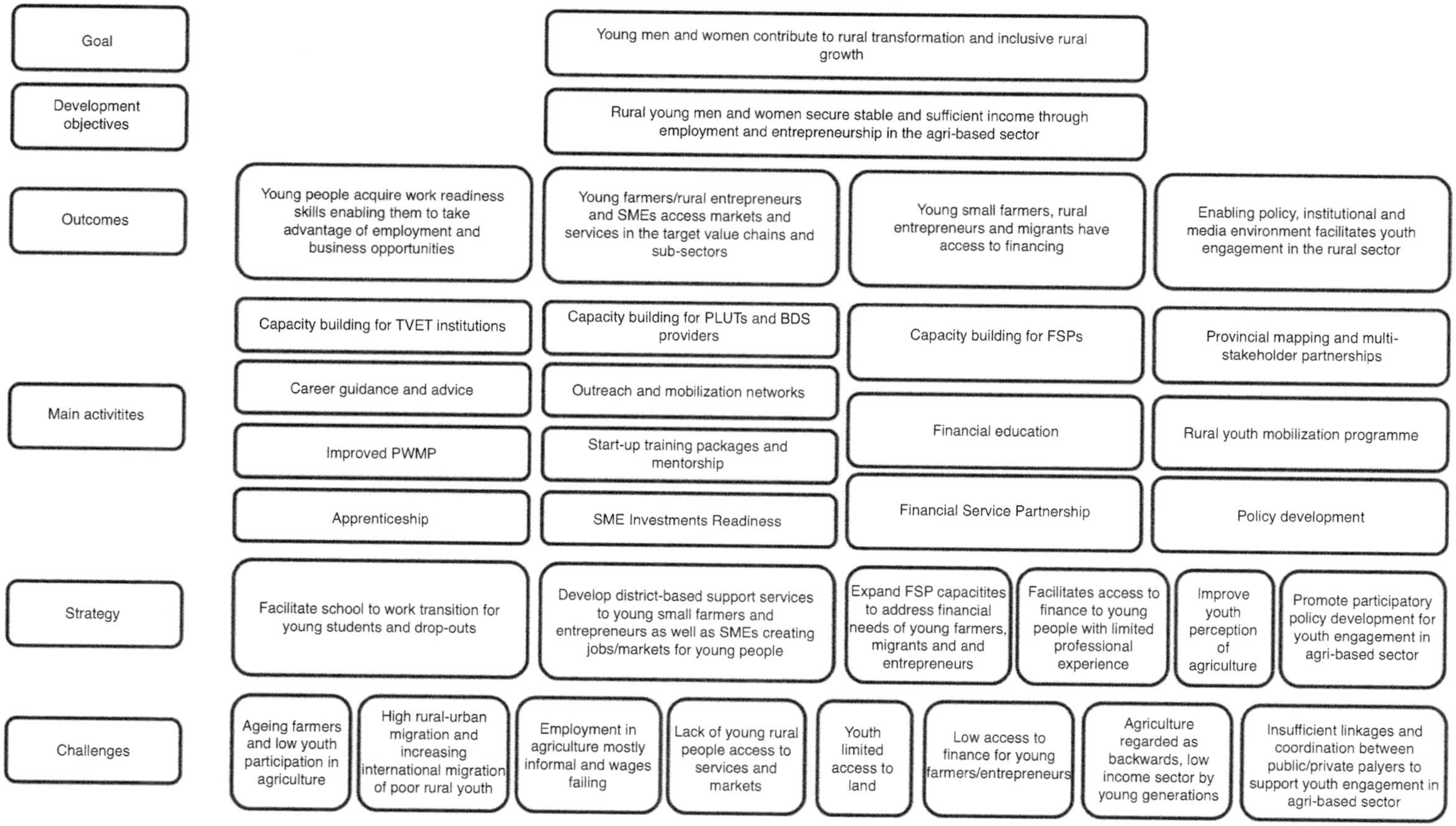

Fig. 2.1. YESS theory of change. BDS, business development services; FSP, financial service provider; PLUT, Pusat Layanan Usaha Terpadu (Integrated Business Service Center, a program of the Indonesian Ministry of Cooperatives and Small and Medium Enterprises); PWMP, Pengembangan Wirausaha Muda Pertanian (Young Agricultural Entrepreneurship Development Program, a program of the Ministry of Agriculture); SME, small and medium enterprises; TVET, technical and vocational education and training. From IFAD (2018).

entrepreneurs. This training covers various topics such as financial management, marketing, product development, and entrepreneurship.

2. Market access. The YESS program assists young farmers in gaining access to profitable markets for their products. The purpose is achieved through various means such as connecting young farmers with potential buyers, facilitating participation in trade shows and events, and helping young farmers develop effective marketing strategies.
3. Fostering collaboration and partnerships. The YESS program encourages collaboration and partnerships between young farmers, the private sector, and the government. These partnerships can help young farmers gain access to capital, resources, and markets and enhance their competitiveness.

The implementation mechanism of the YESS program is carried out through several steps, incorporating selection and training of young farmers and provision of business development support.

The YESS program has been implemented in four provinces and 19 regencies in West Java, East Java, South Kalimantan, and South Sulawesi (Fig. 2.2). By the end of 2023, the YESS program's collaboration was directed toward empowering young farmer communities based on indigenous societies. IFAD allowed this program to be implemented within indigenous communities. The involvement of indigenous communities in the YESS program is expected to improve their economic livelihood, preserve the nature and environment that have been their home and source of livelihood, involve them in the development process, and ensure they experience the benefits of development as citizens while contributing to it, both directly and indirectly.

Indigenous communities are considered an essential focus for the program because their unique background has long drawn government attention, particularly from the Ministry of Agriculture through the YESS program, in empowering farmers who have traditionally adhered to customary farming practices. The program aims to guide them towards broader agricultural practices, introducing technology and governance transformations without eliminating customary norms. This process begins with young farmers who are also core members of their indigenous communities.

In addition to empowerment through capacity building, identified young farmers or aspiring entrepreneurs (penerima manfaat or

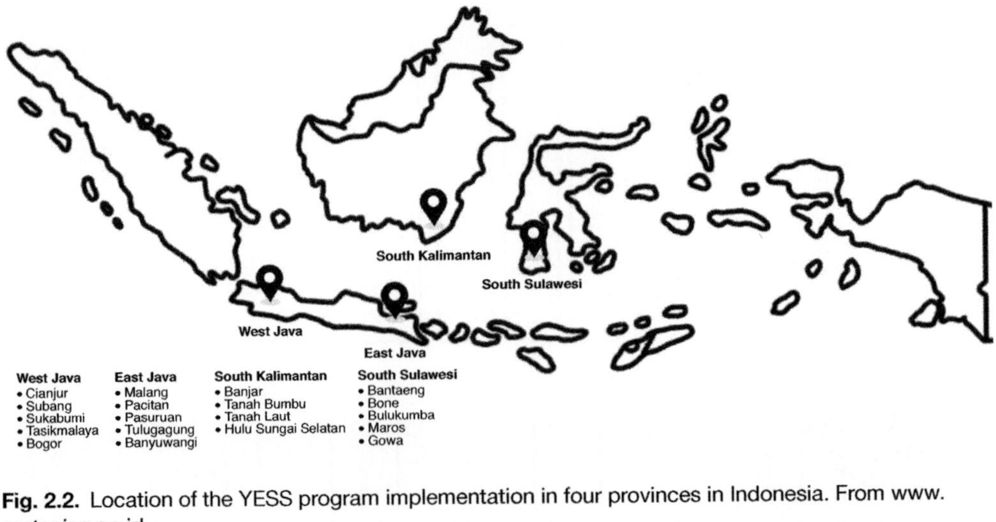

Fig. 2.2. Location of the YESS program implementation in four provinces in Indonesia. From www.pertanian.go.id.

PM) are also proposed for a competitive grant (hibah kompetitif or HK) selection process through the submission of agricultural business proposals. The opportunity to compete is provided as a continuation of their participation in various training initiatives under the YESS program.

As the program developed, competitive grant participants in the 2023 YESS program not only targeted young farmers within the YESS program areas but also included specific target groups. These specific targets include indigenous tribes, persons with disabilities, the development of local potential within YESS areas, and aspiring agricultural entrepreneurs. Free, prior, informed consent (FPIC), social, environmental, and climate assessment procedures (SECAP), and gender equality, disability, and social inclusion (GEDSI) are the three approaches applied by IFAD in the YESS program to address these specific target groups.

The FPIC model is used in the activities of the YESS program based on indigenous communities. It is a process free from pressure and intimidation ("free"), conducted before any company activities that might negatively impact them take place ("prior"), and based on sufficient knowledge of the company's planned activities and their impact on them ("informed"). Consequently, they can express agreement or disagreement to continue the company's planned activities ("consent").

In the initial phase of the pilot project, one of the locations for the YESS program's indigenous community-based activities was set in Kampung Adat Miduana, Balegede Village, Naringgul Subdistrict, Cianjur Regency, which also served as the research team's activity site. The YESS program based on indigenous communities was launched by the Ministry of Agriculture of the Republic of Indonesia to address the challenges confronted by young farmers in Indonesia, particularly in the Miduana indigenous community. This program was inspired by various studies and research showing that young farmers have great potential to contribute to agricultural and rural development (World Bank, 2017; Hartono, 2018). However, young farmers in Indonesia often face numerous obstacles to reaching their full potential, such as limited access to capital and resources, a lack of entrepreneurial skills and training, limited market opportunities, and cultural and social barriers (Giriatna *et al.*, 2019; Lubis *et al.*, 2020). The program is implemented with an approach tailored to the specific needs of young farmers in Miduana, considering local culture and customs (Lubis *et al.*, 2020).

2.5 Miduana Indigenous Community and Agricultural Resource Potential

2.5.1 Potential of agricultural and socio-cultural resources

The traditional village of Miduana is located in the administrative area of Balegede village, Naringgul district, Cianjur Regency, West Java. It has diverse natural resources and agricultural commodities, in the southern part of Cianjur Regency. The location and topography of the area are hilly, with clean air, nestled among vast tea plantations surrounded by preserved tropical forests. Agricultural activities are the daily routine and primary source of livelihood. There are also agricultural products such as coffee, palm sugar, granulated palm sugar, cardamom, and local rice for family consumption.

The availability of natural resources makes most of the Miduana indigenous community's livelihood dependent on farming. Most farmers are older individuals who apply conventional technology innovations based on customs and traditional rules passed down through generations as part of the 'customary rules' in agricultural management. However, the involvement of younger farmers in developing agriculture in the Miduana indigenous community has already begun. Young people have started using internet applications and mobile phones/gadgets for online marketing (e-commerce, Shopee) of palm sugar and dried coffee beans, with considerable sales potential, providing broader market opportunities. This cultural transformation in farming also presents a dilemma for young farmers in the Miduana community, as they balance embracing innovation with maintaining traditions.

Table 2.4. A design model for developing the potential management of the Miduana indigenous village through an off-taker system. The framework positions indigenous youth collectives as key actors, enhancing their capacity to access and expand into broader product markets. (Author's own analysis, 2023).

Age group (years)	Number of males	Number of females	Total	Percentage (%)
≤1	2	3	5	0.4
1–4	21	17	38	3.02
5–14	108	102	210	16.68
15–39	267	245	512	40.67
40–64	202	189	391	31.05
≥65	55	48	103	8.18
Total	655	604	1259	100.00

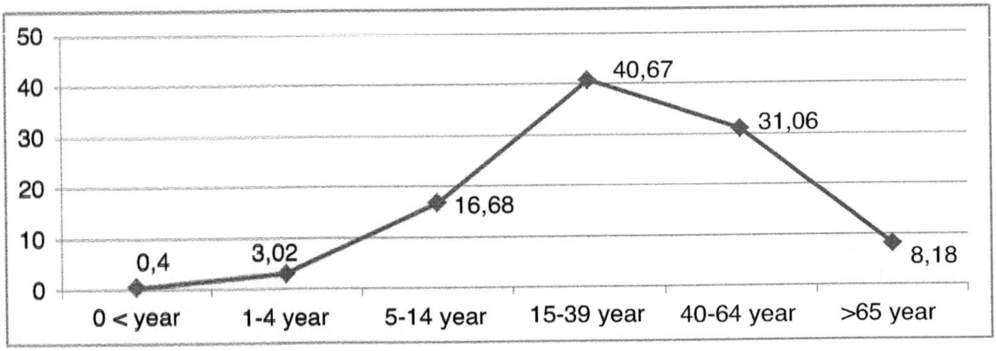

Fig. 2.3. Miduana population by age group in 2023.

2.5.2 Miduana human resources

Based on the structure and population of the Miduana traditional village in Table 2.4 and Fig. 2.3, the village is dominated by those aged 15–39 years (40.67%) and those aged 40–64 years (31.05%). The total population of the village was 1259 people in 2023: 655 men and 604 women.

The potential of the young population, particularly those in the productive age groups of 15–39 years and 40–64 years, is considered an asset as a driving force for rural development based on the local potential of Miduana, which has been traditionally managed. The enthusiasm of the young community in developing agricultural and local tourism businesses has been actualized through organization of the youth group in Miduana. The activities carried out by this community still need to be expanded to include the local/traditional environment and a variety of business activities.

The research shows that young farmers in Miduana previously faced various technical challenges that hindered them from reaching their full potential (Anugrah *et al.*, 2023; Hakim *et al.*, 2023a, 2023b; Husmiati *et al.*, 2023). These challenges included limited access to new technological innovations in agriculture, which needed to be aligned with traditional customs regarding planting methods and timing, crop selection, and technology. Additionally, the lack of capital and limited access to land they could cultivate posed significant obstacles to managing agricultural activities in their village. As a result, some young people from Miduana became migrants, working as industrial laborers, domestic helpers, shop employees, and in other non-agricultural jobs in the city.

2.6 Implementation of the YESS Program in Miduana Traditional Village

2.6.1 YESS program process initiation and stages

The main approaches and strategies implemented in the YESS program include prioritizing participatory engagement and diversity among young farmers in the Miduana indigenous community (Kusdiyanti and Karsidi, 2020). Young people are motivated to participate actively in decision making processes, ensuring their voices are heard, and encouraging inclusivity among various groups. The activities are done through regular consultations, meetings, and workshops where young people can express their ideas, opinions, and concerns. The YESS program also recognizes that understanding and managing the drivers of entrepreneurship is critical to the success of young farmers in Miduana.

In general, the challenges for these young farmers are relatively complex and prevent them from contributing to community development based on the agricultural resource potential of their environment (Anugrah *et al.*, 2023; Hakim *et al.*, 2023a, 2023b; Husmiati *et al.*, 2023). Given these conditions, the YESS program strategy was designed with an approach tailored to the specific needs of young farmers. The YESS program implementation mechanism is carried out through several steps:

1. Selection and training of young farmers.
2. Providing business development support.
3. Facilitation of market relations.
4. Monitoring and evaluation.

The search system to find millennial farmers who will be the program's target is carried out by looking at the activities and capacities of young prospective farmers. Those who meet the requirements are then submitted to the YESS program for further assessment. Then, after being selected, they are included in training directed at: (i) business model training (BMT); (ii) start-up; (iii) preparation of BMT proposal; (iv) internship (now overseas); (v) competitive grant (HK); and (vi) implementation cycle, including training.

The steps and stages carried out by the district implementation team in applying the YESS program at the field level were carried out using the following process (Anugrah *et al.*, 2023; Hakim *et al.*, 2023a, 2023b; Husmiati *et al.*, 2023):

1. Introducing the program (YESS).
2. Inputting data on the people who want to join.
3. Looking for prospective beneficiaries, preparing training, and competitive grant proposals (HK). Assistance is also carried out/accompanied by facilitators, mobilizers, and officers from the Balai Penyuluhan Pertanian (BPP or Agency for Agriculture Extension Service). Then, the internship stage is used or responded to by the Miduana indigenous community.
4. Interventions for beneficiary recipients, especially those related to grants, should be updated through monthly income indicators and reports from interventions. Recipients are expected to participate in 3 months of training every quarter.
5. Assisting in creating commodity clusters (chillies and sheep).

The results and impact of the YESS program in the initial stages of implementation show that it has partially shown its effectiveness in increasing the capacity and livelihoods of young farmers in Miduana. Anugrah et al. (2023) reported that this program has produced several positive impacts, such as increasing entrepreneurial skills and knowledge among young farmers, increasing access to capital and resources, better market opportunities, increasing income and livelihoods, and empowering young farmers to contribute to community development.

2.6.2 "Middle way" strategy to implement the YESS program based on local wisdom

This program is designed to improve young people's skills, access to education, and economic opportunities. YESS is a program that seeks to build prosperity based on justice and equitable development. A careful strategy is needed for this program to be successful and sustainable, paying attention to local wisdom and following the "middle way" approach. This approach ensures program sustainability and accommodates cultural diversity and local values, which are crucial to successful implementation. The middle ground strategy for implementing the YESS program is to focus on local wisdom by placing this as the primary foundation, along with active community participation, culture-based education, and regional economic development.

Local wisdom, the primary foundation, refers to the knowledge, values, and understanding that have existed in local communities for centuries. Including local wisdom as the main foundation of the YESS program is a strategic step because it creates a strong connection between the program and the local community. The sustainability of the program depends on the extent to which the program can be integrated into the local social, cultural, and economic context. One of the first steps that can be taken is to conduct in-depth research on local wisdom in the target area. Understanding local values, traditions, and ways of thinking of local people will help design programs that are more relevant and accepted by the community. Additionally, involving local stakeholders in planning and implementation can strengthen community support and participation (Hakim *et al.*, 2023a).

Training material in alignment with the YESS model should also be carried out using a bottom-up approach. Community participation is necessary by involving local stakeholders such as village leaders, traditional leaders, and youth leaders in the planning and implementation process. This approach will create a sense of ownership and ensure the program meets community needs and expectations. The YESS program seeks to encourage the creation of an entrepreneurial spirit among village youth. For this reason, it is essential to have local economic development efforts that can promote micro and small businesses based on local resources. This move creates economic opportunities while respecting and leveraging existing resources and expertise in the region region (Anugrah *et al.*, 2023).

2.6.3 Miduana empowerment model and the needs of young farmers

The generation of young farmers must be motivated to become the driving force of agriculture in Indonesia, which can be achieved through the YESS strategy, changing the behavior of young farmers so they remain committed to developing the agricultural sector. Extension institutions with appropriate programs, integrated with technical services and using innovative technology, are needed (Anugrah *et al.*, 2023). The next step is to increase institutional management support by completing extension infrastructure, developing standard operational procedures to guide measurable work achievements, and creating programs that provide added value for young farmers (Pardiana and Anantayu, 2024).

Based on resource potential, a local-based capacity-building model for the Miduana traditional village management pattern can be implemented to develop several activities that can improve the community's economy. This model can be used to increase various activities (Fig. 2.4): (i) agricultural/commodity potential capacity; (ii) potential capacity for agro/nature tourism education; and (iii) potential capacity for cultural and traditional tourism education.

The model for managing the potential of the Miduana traditional village is based on these three potential capacities, encouraging a significant role for, function of, and duties of offtakers, whose members all come from the youth community. They act as the overall managers of activities as well as being the leaders carrying out business collaboration with traders/dealers, badan usaha milik desa (BUMDES or village-owned business units), and cooperatives in order to maximize Miduana's potential products. This includes organizational cooperation with and assistance from Miduana traditional leaders and council, Balegede village government, and related ministries/institutions and departments, as well as looking for markets and arranging marketing models for Miduana's potential. This can be through online networks by utilizing internet media as well as through the offline marketing model. Miduana's potential tourist destination activities can provide opportunities for direct product promotion and marketing (Anugrah *et al.*, 2023).

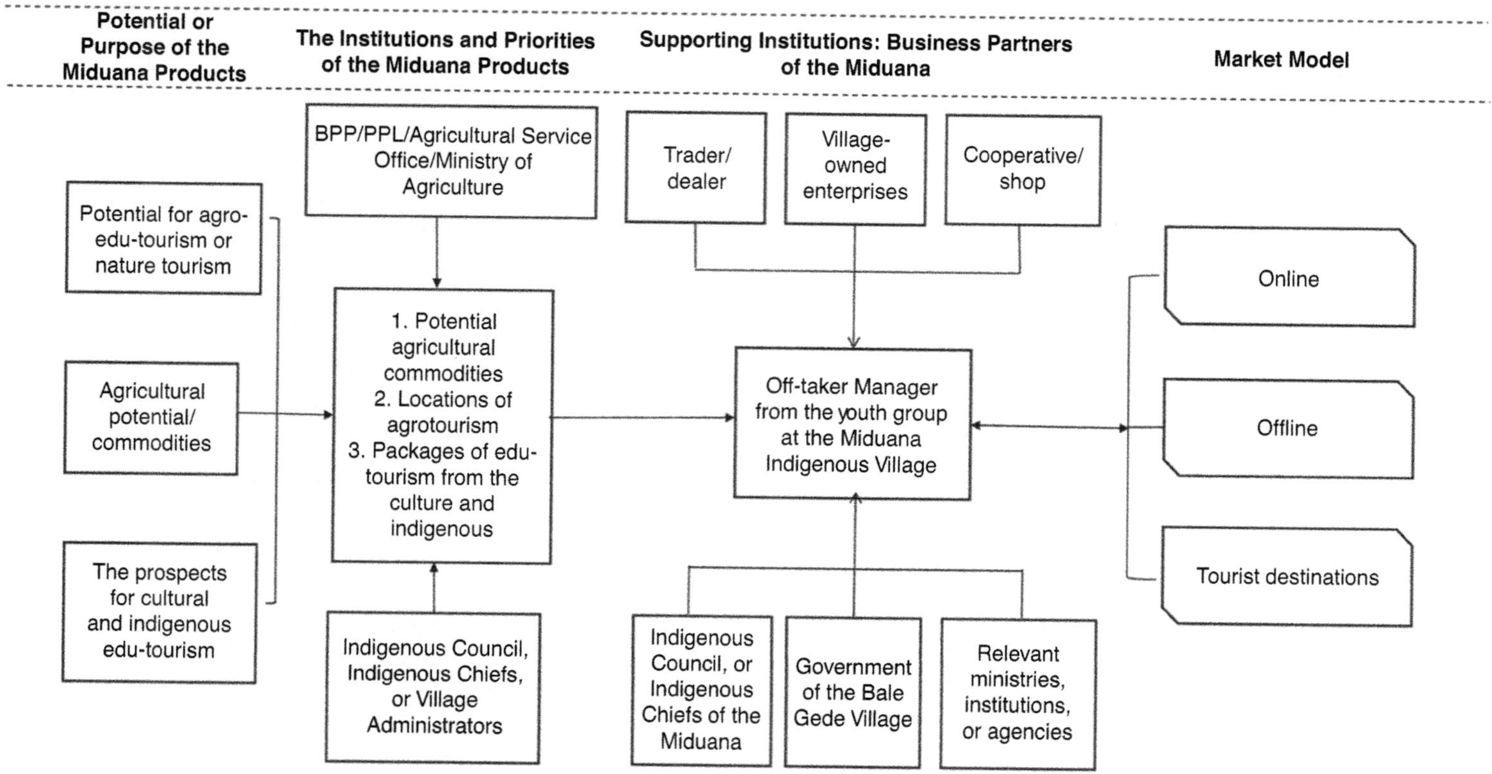

Fig. 2.4. A design model for developing the potential management of the Miduana indigenous village through an off-taker system. The framework positions indigenous youth collectives as key actors, enhancing their capacity to access and expand into broader product markets. (Author's own analysis, 2023).

The function of the role and duties of the off-taker internally is to compile a list of potential commodities – including their processed products, area (planting and harvest), estimated production quantities, and harvest time – as well as processed products so that everything is recorded and becomes a reliable source of information. The information benefits the market, consumers, and potential buyers because it is completed with the types and quantities needed by the market/consumers/buyers.

The same thing is also done for the potential for agro-tourism and edu-tourism for Miduana customs and culture. The function of the off-taker externally is looking for markets, relationships, and empowerment collaboration, as well as being a point of entry and exit for all organized activities. Farmers and commodity business actors in traditional village communities who are members of this organization are not permitted to run their trade/business directly – they must have one door through an off-taker. Vice versa, every buyer/consumer/buyer, if they need to order a commodity or product, must go through the off-taker, and, later, the off-taker will distribute the order to fulfill the quota to members of the organization or even to partners if the organization is short of stock. The transaction process (buying and selling) to the expedition/delivery agreement process for commodities and others are all carried out with/through the off-taker activity for direct product promotion and marketing (Anugrah *et al.*, 2023).

In the internal organizational mechanisms, efforts to enhance capacity related to agricultural commodity potential can be undertaken by stakeholders through the selection of potential agricultural commodities expected to be continuously developed in the traditional area. They will then request assistance to get special training. Cooperation and involvement between traders/merchants, BUMDES, and cooperatives/shops must be established to reach a mutual agreement to advance agriculture in the Miduana traditional village area. The outcomes of this agreement can then serve as a basis for regional policy recommendations regarding managing the agricultural ecosystem and marketing its products. In this cooperation, the off-taker acts as the leader in fulfilling the products the market/consumers need based on demand/orders by distributing quotas or seeking product assistance from trader partners, BUMDES, or cooperatives if there is a supply shortage. Conversely, if traders, BUMDES, or cooperatives face a supply shortage, it can be fulfilled by the off-taker (Anugrah *et al.*, 2023).

Supriyadi (2023) found that in the Malang area, the YESS program has successfully increased the motivation of youth to participate in agriculture. The same pattern can be applied in Miduana in which the off-taker carries out the primary role within the internal management organization, identifying potential and commercial commodities regarding planting location, farmers who cultivate it, the area planted/harvested, production, and products produced. Apart from identifying obstacles, problems, and training needs, off-takers continue to coordinate with the village leadership.

The choice of training materials should be listed in advance by the off-taker according to the needs and problems being faced by the community. The training implementation process can be carried out offline or online. The counseling can also be carried out directly in the field. This design is coordinated through one-door planning (in and out) to implementation with the off-taker and their team (Fig. 2.5).

In order to enhance the potential of sugar palm and red sugar production, the Bale Gede village government was able to collaborate with the Department of Agriculture/Ministry of Agriculture to create an ecosystem for the cultivation and management of products as well as post-harvest processes, together with sugar palm farmers and off-taker managers from the youth groups in the Miduana traditional village. Subsequently, sugar palm farmers are expected to receive training both on-farm and off-farm, as well as other related training such as for processing, packaging, and product marketing systems. In addition to acquiring agricultural skills, farmers are also expected to receive training related to financial services, agricultural insurance, and efforts to secure product partnerships (Anugrah *et al.*, 2023).

Product partnerships can be secured with start-ups as business opportunities for millennial

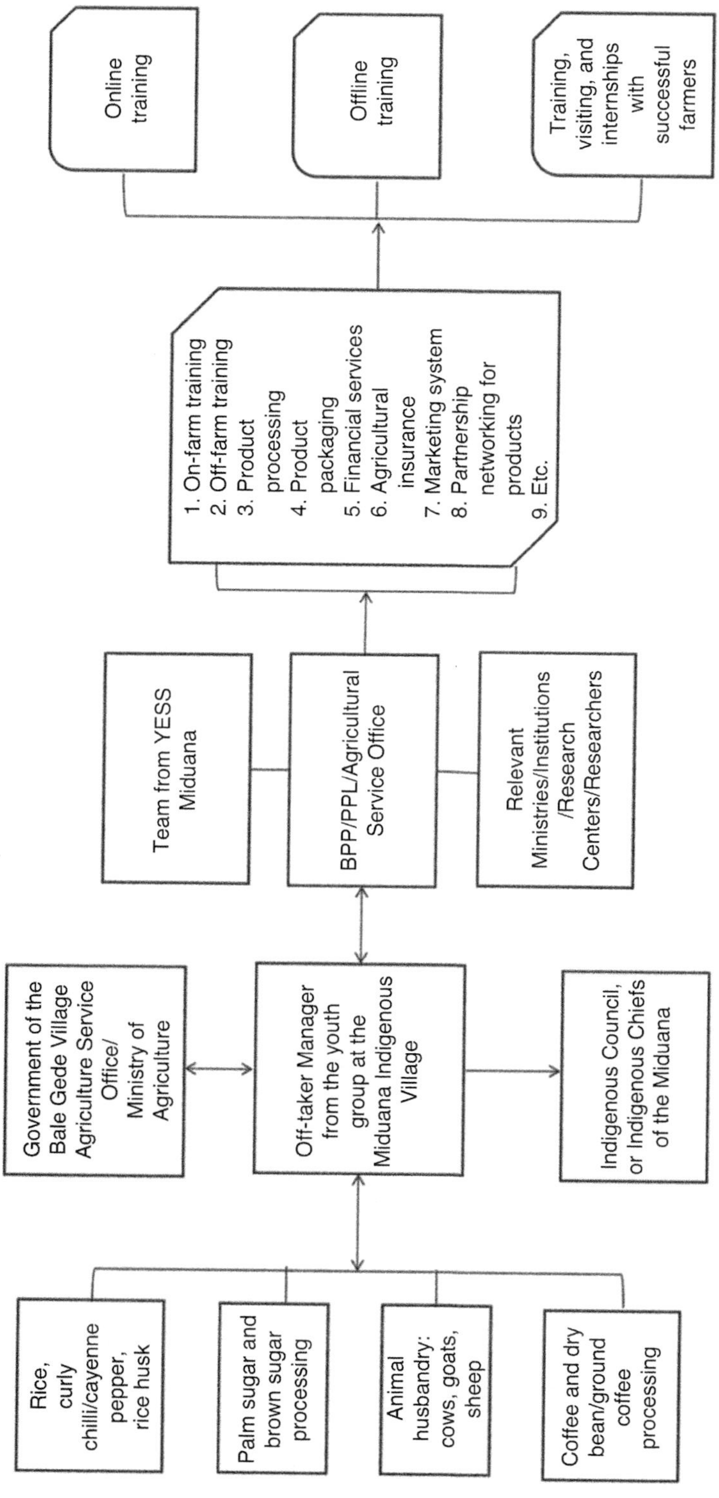

Fig. 2.5. Training model for agricultural commodity management via the off-taker scheme by BPP and extension services. (Author's own analysis, 2023).

farmers. Agriculture as practised among millennials is synonymous with technology that can be utilized to optimize production. Technology gives rise to various types of start-ups that enable brilliant ideas from the younger generation developing innovative solutions to solving various agricultural problems (Rachmawati and Gunawan, 2020). The younger generation in agriculture has the potential to become the driving force behind mutually beneficial cooperation between millennial farmers and local farmers in agricultural cultivation, such as in marketing partnerships, whether through conventional marketing by shortening supply chains or by developing online marketing system applications (Lakitan, 2019). The use of technology, institutions, and human resources in agriculture is often called the agricultural innovation system (AIS), where innovation requires comprehensive handling from marketing, supporting policies, financial incentives, and access to capital (Syahyuti *et al.*, 2018). AIS aligns with what millennial farmers are doing, which is utilizing technology to create start-ups in the agricultural sector.

2.6.4 YESS program synergy follow-up plan

The follow-up implementation of the planned guidelines for capacity building in the Miduana traditional village can be seen in Table 2.5. This shows that curly chillies are a commodity being cultivated by many people in the Miduana traditional village area. This commodity has high economic potential and is included in the YESS program. For market purposes, fresh chilli products are usually sold by farmers to the primary market. Chillies cultivated on-farm in this area are easily infected by several pests (whitefly, stem wilt, fruit aphid), so the market price for sales fluctuates. At the same time, farmers' farming capital is limited. The training farmers need is harvest management, processing into dried chillies, and product packaging. The information and training materials required are related to post-harvest handling, processing, packaging, market opportunities, and networks.

The sugar palm plant is a commodity that is grown a lot in the village and has become a source of livelihood for some of the community. The sugar palm tapping business can encourage the growth of household industries/micro, small and medium enterprises in the region utilizing red palm sugar and palm sugar products. The coffee plant is a plantation crop that the Miduana community has long cultivated; it has high economic value, and demand for coffee beans in this region continues to increase. This plant is cultivated through the shaded coffee program. Animal farming is also quite popular in this area because it is supported by favorable weather and continuous forage availability.

2.7 Conclusion

Sustainable agricultural development in Indonesia still needs to overcome challenges of agricultural human resources, land resources, and farmers' poverty in rural areas. The increase in the number of "old" farmers is inversely proportional to the contribution of young farmers' participation in managing agricultural businesses, which has become an essential phenomenon in formulating current and future agricultural sector development policies.

The YESS program, a collaboration between the Ministry of Agriculture and IFAD, is an alternative effort expected to be part of the answer to the stagnation in farmer regeneration that is currently occurring in Indonesia. Facilitating the younger generation with a digitally based agricultural entrepreneurship model to support the agricultural modernization transformation process is expected to attract young farmers committed to developing Indonesian agriculture.

Overall, the concept and implementation of the YESS program in the Miduana indigenous community has succeeded in having an impact on changes in insight and openness of information for young people to manage their agricultural potential and business activities.

Table 2.5. Training plan for Miduana commodities: potential, market goals and capacity needs.

Commodities and potential	Product/product market objectives	Constraints/problems	Training required
Chilli • Widely grown, high market price commodity • Included in YESS regional program • High economic value	Fresh red chillis sold to Pasar Induk	On-farm activities. Pest outbreaks (whitefly, stem wilt, fruit aphid) Fluctuations in sales market prices, high farming costs, and limited capital; During harvest time, price usually drops	Chilli cultivation procedures (good agricultural practices), handling of harvested products using alternative processing of dried chillies/ dried ground chillies, etc.
	Dry chillis in modern market	Price fluctuation; abundant product during harvest season	Harvest management and dry chilli processing and packaging
Aren • Widely grown; becomes main source of income for many farmers • Encourage home industry/micro, small, and medium enterprises	Solid palm sugar	Product quality, packaging, product variety Supporting facilities for post-harvest product processing technology Safety equipment utilized to support nira (sugar palm) sap	Processing technology of nira (sugar palm) sap to produce palm sugar and brown sugar Safety technique in nira (sugar palm) sap
	Palm sugar	Product quality, packaging, product variety Supporting facilities for post-harvest product processing technology	Processing technology of nira (sugar palm) sap to produce palm sugar and brown sugar
Coffee • A plant that has long been known although demand for it has increased lately • Farmers grow coffee in shade • High economic value	Coffee beans	Cultivation process, harvest, and post-harvest management process, product quality	Cultivation and product handling
	Coffee powder	Processing; facilities, and tools for the processing process	Cultivation, product handling, and further processing
Animals • Suitable environment and climate; forage available all year long • Included in YESS regional program	Breeding	Livestock cultivation, maintenance, livestock health, feed management, sales system	Methods and systems for raising livestock, feed management, maintaining livestock health, waste management/utilization
	Fattening	Livestock cultivation, maintenance, livestock health, feed management, sales system	Breeding methods and systems, feed management, livestock health maintenance, waste management/utilization, sales systems

References

Anugrah, I.S., Margono, T. and Dewi, A.K. (2023) Pemberdayaan masyarakat adat miduana melalui mekanisme peningkatan kapasitas dan program pembangunan pertanian berkelanjutan. *Laporan Hasil Kegiatan Tim Kajian Miduana - Bagian 2. Jakarta*. Unpublished.

Arvianti, E.Y., Ansah, A. and Prasetyo, A. (2015) Minat pemuda tani terhadaptransformasi sektor pertanian di kabupaten ponorogo. *Jurnal Buana Sains* 15(2), 181–188. DOI: 10.33366/bs.v15i2.377.

Arvianti, E.Y., Asnah, A. and Prasetyo, A. (2016) Tingkat konsistensi pemuda tani terhadap transformasi sektor pertanian di kabupaten ponorogo. *Jurnal Agriekonomika* 5(1), 16–30. DOI: 10.21107/agriekonomika.v5i1.1035.

Arvianti, EY., Masyhuri, W.LR. and Darwanto, D.H. (2019) Gambaran krisis petani muda di Indonesia. Agriekonomika. *Jurnal Sosial Ekonomi Dan Kebijakan Pertanian* 168–180. Available at: http://journal.trunojoyo.ac.id/agriekonomika

Balegede Village Monograph (2023) Integrated Subdistrict Electronic Report (LEKAT), Balegede Village, Naringgul District, Cianjur Regency.

BPPSDMP (2020) Badan Penyuluhan dan Pengembangan Sumber Daya Manusia Pertanian, or Agricultural Human Resources Extension and Development Agency. Available at: https://bppsdmp.pertanian.go.id

Central Bureau of Statistics (2023a) Diseminasi Hasil Sensus Pertanian ST-2023. Available at: https://doi.org/10.31949/jsa.v2i1.4772

Central Bureau of Statistics (2023b) Statistical year book of Indonesia. In: *Statistical Year Book of Indonesia. Central Bureau of Statistics*. 0126-2912

Dang, B.Q. (2014) Technology consultation anbackup for young generation's entry into farmin in Vietnam. In: *Proceedings of the 2014 FFTC-RDA International Seminar on Enhanced Entry of Young Generation into Farming*, Food and FertilizeTechnology Center for the Asian and PacificRegion, pp. 63–79.

European Commission (2012) The 2012 Ageing Report. Economic and Budgetary Projections for the 27 EU Member States (2010–2060). Brussels: Directorate-General for Economic and Financial Affairs of the European Commission. Available at: http://ec.europa.eu/economy_finance/publica tions/Europe an_economy/2012/pdf/ee-2012-2_en.pdf

Giriatna, Y., Lubis, A.M. and Syukur, M. (2019) Faktor-faktor yang mempengaruhi minat berwirausaha pada petani muda di desa cijawura, kecamatan karangpawitan, kabupaten garu. *Jurnal Agrikultur Dan Bisnis* 6(2), 113–122.

Hakim, L.N., Anugrah, I.S., and Carolina (2023a) Laporan kegiatan verifikasi lapangan Program Youth Entrepreneurship and Employment Support Services (YESS) pada masyarakat adat Miduana, Cianjur, Jawa Barat. Jakarta. Unpublished.

Hakim, L.N., Purwasantana, D. and Syifani, A.N. (2023b) Model pendekatan dan strategi program YESS pada masyarakat adat Miduana, Cianjur — Jawa Barat. Laporan Hasil Kegiatan Tim Kajian Miduana - Bagian 1. Jakarta. Unpublished.

Hartono, E. (2018) Pemberdayaan petani muda melalui Program Youth Entrepreneurship and Employment Support Services (YESS). *Jurnal Kajian Sosiologi* 12(1), 1–18.

Husmiati, Basuki, S. and Arofah, A. (2023) Rekomendasi internal untuk Program Youth Entrepreneurship and Employment Support Services (YESS). Laporan Hasil Kegiatan Tim Kajian Miduana - Bagian 3. Jakarta. Unpublished.

IFAD (2018) Youth Entrepreneurship and Employment Support Services Programme (YESS): Design completion report/main report and appendices.. 2000001202. Asia and the Pacific Division Programme Management Department.

Kementan (2024) YESS Programme. Available at: https://yesskementan.org/profile

Kusdiyanti, R.F. and Karsidi, R. (2020) Farmer groups (kelompok tani) capability towards food self-sufficiency by applying 'Special Efforts Program for Increasing Rice, Corn, and Soybean Production' in Juwangi District, Boyolali Regency.. In: *IOP Conference Series: Earth and Environmental Science*, Yogyakarta. Available at: https://doi.org/10.1088/1755-1315/423/1/012025

Lakitan, B. (2019) Double track strategy for agriculture development in Indonesia: Facilitating millennial generation and improving livelihood of smallholder farmers. In: Herlinda, S. (ed.) *Prosiding Seminar Nasional Lahan Suboptimal 2019*. Unsri Press, Palembang, pp. 1–8.

Lubis, A.M., Syukur, M. and Giriatna, Y. (2020) Strategi pemberdayaan petani muda melalui program youth entrepreneurship and employment support services (YESS) di kabupaten garut. *Jurnal Agrikultur Dan Bisnis* 7(1), 43–52.

Ma, S.J. (2014) How to encourage young generation to engage in farming: Korea's case. In: *Proceedings of the 2014 FFTC-RDA International Seminar on Enhanced Entry of Young Generation into Farming*, Food and Fertilizer Technology Center for the Asian and Pacific Region, pp. 147–162.

NFF (2012) FF Farm Fact: 2012. Available at: http://www.nff.org.au/farmfacts.html

Pardiana, W., Anantayu, S., Suwarto, and Suminah (2024) Strengthening institutional counseling for regeneration of young farmers in cianjur regency. In: *Proceeding International Conference on Multidisciplinary Studies -ICOMSI* (Vol. 1).

Rachmawati, R.R. and Gunawan, E. (2020) Peranan petani milenial mendukung ekspor hasil pertanian di Indonesia. *Forum Penelitian Agro Ekonomi* 38(1), 67–87. DOI: 10.21082/fae.v38n1.2020.67-87.

Supriyadi, D. (2023) Training and capacity building for eligible SMEs – Youth Entrepreneurship and Employment Support Services Programme (YESS Programme) untuk wilayah Malang, Jawa Timur. *Jurnal Ilmiah Abdimas* 4(1), 123–130.

Susilowati, S.H. (2016a) Fenomena penuaan petani dan berkurangnya tenaga kerja muda serta implikasinya bagi kebijakan pembangunan pertanian. *Forum Penelit Agro Ekon* 34(1), 35–55.

Susilowati, S.H. (2016b) Kebijakan Insentif Untuk Petani Muda: Pembelajaran Dari Berbagai Negara Dan Implikasinya Bagi Kebijakan Di Indonesia. *Jurnal Forum Penelitian Agro Ekonomi* 34(2), 103–112.

Suyanto, B. (2016) Kenapa generasi muda enggan bertani? Memahami subkultur dan gaya hidup anak muda dari perspektif cultural studies. In: *[Artikel] Bahan Pertemuan Upaya Meningkatkan Minat Generasi Muda Terhadap Pertanian." Pusat Perpustakaan Dan Diseminasi Informasi*. Bogor, Indonesia.

Syahyuti, A.B., Sumedi, E.S., Sunarsih, J.F.S. and Dabukke, F.B.M. (2018) Monograf internal] "Temuan-temuan pokok dan rekomendasi kebijakan pembangunan pertanian dari hasil-hasil penelitian pusat sosial ekonomi dan kebijakan pertanian (PSEKP) tahun 2016. In: *Pusat Sosial Ekonomi Dan Kebijakan Pertanian*. Bogor.

Uchiyama, T. (2014) Recent trends in young people's entry into farming in Japan: An international perspective. FFTC-RDA 2014. In: *International Seminar on Enhanced Entry of Young Generation into Farming*. Food and Fertilizer Technology Center for the Asian and Pacific Region, Taipei (TW), pp. 1–16. Available at: https://ap.fftc.org.tw/article/756

USDA (2007) Census of Agriculture. New York. Available at: https://www.agcensus.usda.gov/Publications/2007

World Bank (2017) *Empowering Young Farmers in Indonesia: A Policy Brief*. World Bank, Washington, DC.

3 Sustainable Agribusiness, Food Security, and Women's Empowerment: Rural Women at the Heart of the Virtuous Cycle

Hauwah K.K. AbdulKareem[1,2*], Loveth Chiduma Ode Omenka[2,3], Joe E. Assoua[4,5], Solomon O. Okunade[2,6], Eric Jr. Bomdzele[7,8,9,10], Olanrewaju O. Omosehin[11], Samuel O. Binuomote[12,13], Sarah Enwa[14], Evans Osabuohien[2,15] and Uduak S. Akpan[16]

[1]Department of Economics and Development Studies, Kwara State University, Malete, Nigeria; [2]DePECOS Institutions and Development Research Centre (DIaDeRC), Ota, Nigeria; [3]Department of Economics, Accounting and Finance, Bells University of Technology, Ota, Nigeria; [4]Department of Agricultural Economics and Agribusiness, Faculty of Agriculture and Veterinary Medicine, University of Buea, Buea, Cameroon; [5]Eastern and Southern African Management Institute and Trade Policy Training Centre in Africa (ESAMI-TRAPCA), Arusha, Tanzania; [6]Department of Economics, Chrisland University, Abeokuta, Nigeria; [7]Alliance Bioversity and CIAT, Cali, Colombia; [8]Department of Agricultural Economics and Agribusiness, Faculty of Agriculture and Veterinary Medicine, University of Buea, Buea, Cameroon; [9]Sustainable Outreach Foundation (SOF), Buea, Cameroon; [10]World Bank Group, Cotonou, Benin; [11]Department of Agricultural Economics, University of Ibadan, Ibadan, Nigeria; [12]Department of Agricultural Economics, Ladoke Akintola University of Technology, Ogbomoso, Nigeria; [13]Department of Agricultural Economics and Management, University of Eswatini, Luyengo Campus, Eswatini; [14]Department of Agricultural Economics, Delta State University, Abraka, Nigeria; [15]Department of Economics and Development Studies, Covenant University, Ota, Nigeria; [16]Department of Development Studies, SOAS University of London, London, UK

3.1 Introduction

Agriculture, agribusiness, and food security are closely associated and interdependent. The agricultural sector, which comprises agribusiness, is Africa's biggest economic sector and is driven by several factors including demographic, economic, institutional, and socio-cultural drivers (HLPE, 2017). These factors influence intra-household gender and power dynamics and

*Corresponding author: hauwah.kamaluddeen@kwasu.edu.ng

DOI: 10.1079/9781836992646.0003

shape food security outcomes in food systems. In developing regions like the sub-Saharan African countries, half of their gross domestic product (GDP) is contributed by agriculture, which employs 60–80% of the entire labor force (IFC, 2023). Agribusiness on the other hand is reported to constitute one-fifth of the total GDP in these regions (World Bank, 2013). From the market opportunities that agriculture provides has evolved the term agribusiness. Agribusiness has emerged as a vast range of businesses in the sector, from farming, processing, marketing, and distribution, to sales (Otaokpukpu *et al.*, 2023). Agribusiness not only provides an avenue for those involved to earn income but also contributes to food security due to improved agricultural practices and access to food markets. More specifically, agribusiness is fundamental to the four pillars of food security: food availability, accessibility, utilization, and stability (DTN, 2022). Agriculture together with agribusiness are crucial in addressing the multifaceted challenges confronting the African continent (Haggblade, 2011; Castañeda *et al.*, 2018; Assoua *et al.*, 2024) and in the attainment of sustainable development goals SDG2, SDG3, SDG5, and SDG8.

According to World Bank estimates, agriculture and agribusiness account for about 45% of the sub-Saharan economy, and by 2030 both sectors are estimated to be worth US$ 1 trillion. Recognizing the growing role and importance of agribusiness for food security, African governments in collaboration with regional and international development partners have implemented several strategies and initiatives at both the regional and continental levels including the African Agribusiness and Agro-Industries Development Initiative (3ADI), and the Strategic Framework for Pillar 2 on markets and agribusiness (Yumkella, 2011).

For rural women, the challenge of food insecurity is a case of double jeopardy. This is because rural communities and females bear a larger burden of food insecurity. According to the Food and Agriculture Organization of the United Nations (FAO), food insecurity affected over 33% of those dwelling in rural areas, while only 26% of those living in urban areas suffered from food insecurity in 2022 (FAO *et al.*, 2023). Even though the food security gender gap is reducing, the gap between males and females is still quite significant at about 3%, with 25% of males being food insecure while the figure is 27.8% for females. It is rather ironic as women account for 50% of the world's food production and make up 43% of labor in the agri-food sector (Kehinde *et al.*, 2021). Also, rural women encounter other challenges such as limited access to productive assets and resources like finance and land and earning less than men for the same amount of work done (Zereyesus, 2017). This stifles their productivity, contribution, and significance in agribusiness (Ikhaghu, 2022). These limitations hinder women's potential in agribusiness and food security, and they require effective strategies to promote women's empowerment. This will allow rural women to harness their potential and increase their contributions in the agricultural sector in particular, and sustainable economic development at large.

This chapter has three main objectives. The first is to synthesize the growing body of literature on agriculture and agribusiness to elicit the important role of rural women in agribusiness and food security. The second is to explore how these can lead to female empowerment, and lastly, to construct a use case to examine how the linkages and interrelationships among the constructs lead to a virtuous cycle.

This chapter seeks to make a variety of novel contributions to the existing body of agribusiness literature in several ways. First, it provides a synthesis of these various aspects which presents a holistic view and offers fresh insights from an interdisciplinary perspective to address the menace of food insecurity and gender inequality and disempowerment in agriculture. Second, and from the development perspective, exploring the interconnectedness and linkages among agribusiness, food security, and women's empowerment in the same contextual framework is novel and will provide a comprehensive analysis of how development in one area affects the other areas. This will guide policy makers, development practitioners, and relevant stakeholders on effective strategies and interventions that will have a triple impact on agribusiness, food security, and women's empowerment. Third, looking at gender dimensions, is the exposition of the specific contributions of rural women in agribusiness and food security along the agricultural value chain and pillars of food security, which have not enjoyed adequate attention in mainstream literature. This chapter will therefore be of great policy import in designing targeted policies and programs that will address gender inequality in agriculture and also support and amplify the contributions of rural women in order to harness their potential for sustainable development.

The organization of this chapter is given as follows: the next section presents the conceptual framework and theoretical foundation of the study and section 3.3 presents the methodology adopted for the study. The roles and contributions of rural women in agribusiness and food security are then described, while expounding the challenges and barriers confronting women. Section 3.5 explores pathways of empowering rural women in agribusiness and food security, establishing a virtuous cycle of agribusiness, food security, and female empowerment. Section 3.6 presents a use case of the virtuous cycle, and the final section draws conclusions and policy implications.

3.2 Conceptual and Theoretical Frameworks

3.2.1 Conceptual framework

With millions of people impacted, food security continues to be a central, urgent issue in Africa (FAO, 1997). Over a billion people in the world are engaged in agriculture, amounting to more than a quarter of the labor force. Agricultural employment is declining in many nations due to technological developments, yet it is still essential for food security, livelihoods, and rural development. However, in order to eradicate poverty, lessen inequality, and put an end to hunger, a sustainable agriculture sector with decent work is necessary (ILO, n.d.).

Women constitute a large portion of the agricultural workforce in developing nations and play critical roles in the world's agri-food systems. They carry out essential activities in the household and financial spheres of society. Neglecting this significant and influential portion of the population makes it difficult to advance rural and national development (Damisa and Yohanna, 2007; Swati and Bhumika, 2024). As shown in Fig. 3.1, enhancing women's involvement in agriculture by means of investments and business ventures is expected to yield two benefits – improved food security and nutrition as well as increased economic expansion (Bunch and Mehra, 2008). However, because of societal norms, inadequate resource access and decision-making authority, and small- and medium-sized business ownership, women only own one-third of these businesses in emerging economies. There is, however, a rising demand for female labor in agricultural activities such as post-harvesting which involves the preservation, processing or value addition, packaging, and delivery of agricultural products.

Research supports the fact that sustainable agribusiness is a viable means of increasing disposable income, particularly in rural communities where agriculture is the main source of livelihood and, especially, for women who constitute the majority of the agricultural labor force (Asogwa *et al.*, 2020). Sustainable agribusiness has also brought about job opportunities that link producers of agricultural products (farmers) with the consumers who demand their products since agribusiness entails the preservation and transformation of crops and other farm produce for consumption, whether at the intermediate or final stage.

Furthermore, sustainable agricultural practices provide profits that may be used to invest in land, credit, and technology resources that are essential for women to participate in agriculture. Modern agricultural techniques are frequently taught in sustainable agribusiness, which can be especially helpful for women by improving their productivity and skill sets. Crop rotation and organic farming are examples of sustainable agriculture techniques that have an impact on food security by raising agricultural productivity, which

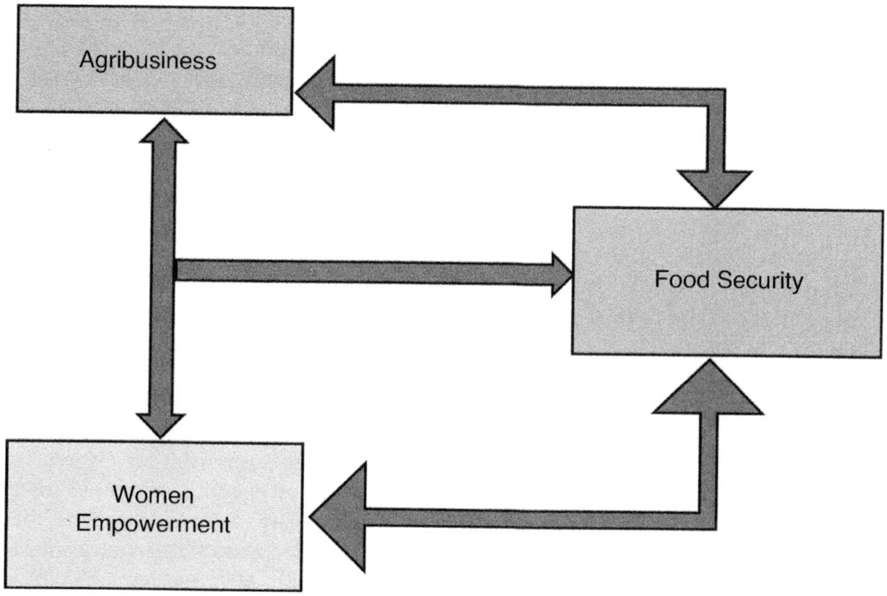

Fig. 3.1. Agribusiness, food security, and rural women's empowerment framework.

in turn results in better yields and a steadier supply of food. It encourages the growth of a range of crops, improving nutrition and its diversities. By decreasing environmental deterioration, these sustainable measures also guarantee agricultural land's long-term visibility and secure food production for coming generations. van Westen *et al.* (2019) findings support the significant impact of agribusiness on food security; however, the study is inconsistent with the findings of Iervolino (2023).

Given the dominant roles of women in agribusiness, it is crucial that rural women are empowered to not only reach their full potential in the field but also to maximize the potential gains of agribusiness in the economy. Empowering them will improve their economic and social status, and afford them increased income from their business ventures which will translate to improved well-being of the women and their families. This will facilitate inclusion, gender-sensitive outcomes, and contribute to gender equality. When women have access to training and productive resources, they become empowered, and this will significantly boost agricultural productivity and output through increased financial investment in agribusiness, enhanced knowledge, improved techniques, and sustainable practices. With these, food supply becomes more abundant, stable, and dependable, which in turn facilitates food and nutrition security.

Farmers are motivated to use sustainable farming practices that improve agriculture production when there is a steadily increasing demand for these agricultural products. A stable food supply enables farmers to make investments in sustainable farming methods and lowers economic doubt. People who have better access to food tend to be healthier, which makes them better equipped to support sustainable agriculture while also allowing women and their children to engage more fully in the workforce and educational pursuits. Food security helps women increase their earnings, thereby improving their economic empowerment and leadership capacity as well as the ability to promote laws and policies that encourage more empowerment.

Assisting women to progress from subsistence farming to agribusiness needs to be prioritized as a fundamental goal of the development agenda and food security intervention. Transitioning from subsistence farming to agribusiness, particularly in conjunction with processing and value addition, paves the path for highly significant and sustainable improvements in living conditions. With sustainable agribusiness, increased agricultural output and profits will be realized which will afford women access to valuable resources and other income-generating opportunities. These will in turn enhance food security in the household as well as improve health and educational benefits in the entire household since women own the care-giving and household managing responsibilities.

3.2.2 Theoretical framework

The theory of multifunctional agriculture (MFA) forms the theoretical foundation of this chapter due to its significance and suitability for the subject matter. The MFA theory posits that agriculture should be regarded beyond its traditional roles of crop production and animal rearing to its social and environmental benefits (Tohidyan Far and Rezaei-Moghaddam, 2019; Gaitán, 2020). Some of these spillover gains from the multifunctionality of agriculture in these spheres include: (i) employment generation which will bring about economic gains; (ii) festivals related to agriculture such as the new Yam Festival in Nigeria which promotes social cohesion; and (iii) adoption of green technologies and farming techniques like crop rotation which improves soil quality and facilitates environmental sustainability. The theory, therefore, advocates for agricultural production to be integrated with other sectors of the economy (Zhichkin *et al.*, 2022) while agricultural policies take cognizance of their potential spillover and trade-offs, such as urbanization and sustainability. Agriculture should yield several commodity and non-commodity products that are interrelated in a way so their supply is interdependent (Cahill, 2001). Outshoots of agriculture such as food and nutrition security, biodiversity, animal welfare, land and environmental quality, and cultural values are regarded as part of the multifunctionality feature of the agricultural sector (Cahill, 2001; Zhichkin *et al.*, 2022).

Some commonly cited multifunctionalities of agriculture include rural development through agricultural employment; biodiversity and improved landscape via animal product production on mountain pastures and terraced fields; and food security from domestic production, among others. However, some criticism of the theory is its non-universality or applicability to all agricultural producers or circumstances (Zhichkin *et al.*, 2022). Instances include private farm households in Russia and the environment-degrading tendency of agriculture, such as pollution and soil erosion (Cahill, 2001). This theory is, however, suitable for the present chapter as food security, agribusiness, and rural development that encompasses empowerment are regarded as part of the social and environmental outcomes of the multifunctionality of agriculture. This theory is considered appropriate in this study context given that not only can MFA improve agricultural productivity, but it can also address socio-economic challenges like food insecurity and female empowerment (Gaitán, 2020).

3.3 Methodology

This chapter conducts a review of the extant literature to synthesize a conceptual analysis of the roles of agribusiness, food security, and empowerment while bringing to the fore the criticality of rural women and discussing policy options to promote women's empowerment through agriculture towards sustainable development. As the preferred methodological technique, in-depth literature review provides a comprehensive understanding of the currently available knowledge and allows for a critical analysis of the subject in addition to gap identification and the establishment of potential linkages and interrelationships among the various constructs of the study. The thematic review was conducted on the basis of the different themes of the study vis-à-vis rural women's role in agribusiness and food security, and women's empowerment. This helped these authors to identify and synthesize trends and patterns in literature, provide a critical analysis, and uncover the critical relationships necessary for achieving the set objectives of the study.

3.4 Rural Women in Agribusiness and Food Security

3.4.1 Role of rural women in agribusiness

Agribusiness, which is made up of production, value addition, and marketing of farm produce, is a crucial sector for many economies, especially the developing countries (Arthur *et al.*, 2024). Studies show that rural women's contributions are essential in the various stages and activities of the agricultural sector, significantly enhancing food security, household incomes, and community well-being (Savari *et al.*, 2020; Elum and Digitemie, 2023). The multifaceted role and contributions of rural women in agribusiness include the following.

1. *Agricultural production.* For rural communities, women are at the forefront of production on the farm, participating in various operations which include but are not limited to planting, nurturing, and harvesting of crops, rearing of livestock, as well as post-harvest value addition (Antriyandarti *et al.*, 2024). They often manage smallholder farms and engage in subsistence farming including the maintenance of vegetable gardens, which is critical for household food security. Women comprise over 40% of the agricultural labor force in developing countries (SOFA Team and Doss, 2011; IITA, 2024). Their involvement in crop production, horticulture, and animal husbandry is vital for maintaining food supplies and supporting local economies.
2. *Agricultural processing.* Beyond primary production, rural women participate actively with regard to the processing of agricultural products, adding value through traditional methods and modern techniques. This includes processing grains into flour, dairy into cheese and yoghurt, and fruits into preserves and juices (Gaffney *et al.*, 2019; Antriyandarti *et al.*, 2024).
3. *Value addition.* Processing not only increases the product life and storage capacity as well marketability of agricultural commodities, it also enhances household incomes and reduces post-harvest losses (Arthur *et al.*, 2024). Women's knowledge of traditional processing methods, combined with modern practices, contributes significantly to the agribusiness value chain.
4. *Agricultural entrepreneurship.* Rural women's involvement in the marketing of agricultural commodities is on the rise, selling their produce in local, regional, and even international markets (Schoneveld, 2022). They engage in entrepreneurial ventures, establishing small agribusinesses that produce and market a variety of farm products. These enterprises range from small-scale food processing units to handicraft production using agricultural by-products (Gaffney *et al.*, 2019; Ikuemonisan *et al.*, 2022; Arthur *et al.*, 2024). Women's participation in agribusiness entrepreneurship therefore promotes economic diversification, job creation, and sustainable rural development.

3.4.2 Challenges confronting rural women in agribusiness

Despite their substantial contributions, a myriad of challenges often confronts rural women which thwart their full potential in agribusiness, including restricted access to assets and productive resources such as land, education, skills and training, and markets (Adeyanju *et al.*, 2023; Elum and Digitemie, 2023; Antriyandarti *et al.*, 2024).

First, inadequate access to resources constitutes a major challenge as women in rural areas are generally handicapped in acquiring essential resources and assets like land (Elum and Digitemie, 2023), credit, and inputs like seeds and fertilizers. Usually, there are disparities in gender distribution as regards land ownership and rights of inheritance; this significantly restricts their ability to secure land for agricultural activities. Additionally, financial institutions have a lower probability of providing credit to women, limiting the ability of rural women to invest in and expand their agribusinesses (World Bank, 2017; Savari *et al.*, 2020). Second, many rural women have restricted access to education and agricultural training, impeding in their ability to adopt modern farming techniques and improve productivity (Elum and Digitemie, 2023). The fundamental importance of education in empowering rural women with better decision-making abilities, adopting innovative practices, and managing their agribusinesses effectively is very critical. The gender gap in education and training opportunities remains a significant barrier to women's advancement in agribusiness (Adeyanju *et al.*, 2023).

In addition, basic facilities are insufficient, including unavailability of transportation, storehouses, and market information, which disproportionately affects rural women. According to Ikuemonisan *et al.* (2022), without reliable roads and transportation networks, women struggle to bring their products to market, which often results in farm produce wastage as well as income losses. Inadequate storage facilities also contribute to food spoilage and waste, further exacerbating economic vulnerabilities (Savari *et al.*, 2020). Furthermore, traditional gender roles and cultural norms as a result of ethno-communal limitations often restrict the involvement of women in agribusiness activities (Schoneveld, 2022). In several rural areas, there is a strong expectation for women to place a higher priority on household and care-giving roles above businesses. These social constraints restrict their mobility, time, and opportunities to engage in productive work, limiting their contributions to agribusiness (Agarwal, 2018; Adeyanju *et al.*, 2023).

In conclusion, the contribution of rural women in agribusiness is essential for the accomplishment of sustainable and inclusive food systems transformation. Recognizing and addressing the challenges they face is essential for unlocking their full potential and ensuring inclusive growth. By improving resource accessibility, enhancing education, promoting training, investing in infrastructure, and promoting social and cultural change, rural women can be empowered to lead successful agribusinesses. Empowerment of rural women benefits not their households alone, it impacts their communities positively and also enhances broader development of the rural economy and food security.

3.4.3 Rural women and food security

Rural women play a preponderant role in all four dimensions of food security, namely food availability, food accessibility, food utilization, and food stability. This is occasioned by their multifaced contributions in the production of food, in agricultural entrepreneurship, as gatekeepers in food and nutrition security, and as managers of stable food supply (UN, 2012).

1. *Food production.* Sustainable food production is positioned as the most important pillar for food security. Women comprise almost 43% of the total agricultural workforce in developing economies (SOFA Team and Doss, 2011; IITA, 2024). Agriculturally active rural women account for a sizable number in the global population, which continues to experience rapid growth, and they account for the production of 60–80% of the food in many developing nations. They are mainly resonsible for around 50% of the total food production (Ulker, 2013; OECD, 2022). Women are known to be players of a key role in food systems in different capacities as producers, laborers, processors, and marketers, including food preservation (Njuki *et al.*, 2022). The usual daily chores of rural women include crop cultivation, vegetable farming, livestock rearing, and food production among others (FAO, 2021). Rural woman most of the time not only work on household farmlands, but they also become workers on the farmland owned by others, especially during planting and harvesting seasons when they work to generate extra, or the main, income for family sustainability. Thus, rural women's contributions to agriculture are invaluable and cannot be overemphasized.
2. *Food accessibility*. Women are generally key to the distribution and accessibility of food, with most rural women being small-scale farmers who are the main food providers (Owren, 2017). According to Quisumbing *et al.* (1996), having access to available food does not only depend on income but also on

who earns the income. Evidence from the literature establishes that a substantial percentage of women spend a larger portion of their earnings on food, which improves the welfare of the entire family compared with the spending of men's income. Case studies in various parts of the globe support the positive relationship between female income and food security in the household. Specifically, rural women's income has been found to increase calorie consumption in Kenya and Rwanda, food budget in Cote d'Ivoire, food expenditure in Guatemala, and overall food access at the household level (Quisumbing *et al.*, 1996; FAO, 2023).

3. *Food utilization.* All around the globe, rural women have been widely recognized as critical stakeholders in human nutrition given their role as producers of food at the level of the household (Ghosh *et al.*, 2021). With women having the primary responsibility of care-giving given their role as home managers, they ensure the nutritionary quality and dietary needs and consumption of their family members. Rural women spend 75–80% of their income on food utilization and preparing food for their household, which plays a significant role in dietary diversity, food quality, and food safety, especially when this is combined with their knowledge of local food systems and traditional recipes FAO (2021). Further studies acknowledge the positive correlation women's empowerment has with food security and household nutrition (Kalansooriya and Chandrakumara, 2014; Galiè *et al.*, 2019; Essilfie *et al.*, 2020; Wei *et al.*, 2021). Studies reveal that when women are empowered, it significantly improves the food security and nutrition of rural households (Verhart *et al.*, 2015).
4. *Food stability.* The role rural women play in the promotion of a stable food supply is becoming increasingly acknowledged, particularly in the context of climate change. Rural women are significantly involved in adaptive and sustainable agricultural practices, preserving biodiversity, and managing natural resources to improve the stability of food systems (World Bank, 2014). For instance, the women-led oyster farming introduced in Saloum Delta, Senegal helped in addressing environmental degradation, hence promoting food system stability in Senegal (Meighan, 2024). In addition, the involvement of rural women in cooperative associations, savings and/or credit groups, and other informal support groups ensures the continuous supply and accessibility of food in times of shocks, which promotes food stability. This is even more so as studies have indicated that women reinvest their incomes in their household needs, which promotes resilience in the household in times of food insecurity (World Bank, 2014).

The most crucial part of agriculture is done by women (Sraboni *et al.*, 2014; UNDP, 2016), and their tripartite roles as producers of food, income earners, and home keepers make them of great value and indispensable in the quest towards food security.

3.4.4 Gender inequality, rural women's empowerment, and household food security

However, despite the great and multiple levels of women's contribution in agriculture and agribusiness, they earn less when compared with the earnings of their male counterparts principally as they do not possess equal access to assets and productive resources (Srivastava and Srivastava, 2010; FAO, 2011; Subathra, 2015; Chanana-Nag and Aggarwal, 2020). It is noteworthy that while women's contribution to food production is significant and undeniable, it is not possible for their contribution to production to be quantified. In addition, women face certain problems arising from gender inequality, threatening the sustainability of rural women's roles and contributions to developing countries' agriculture.

In developing countries, females are often relegated, restricted, and often subsumed in norms and practices (Beyene, 2015). Within the allocation and distribution of productive inputs and resources, they are often discriminated against. Hence, they are mostly overlooked in critical times of decision making whether in the household, community, employment, agriculture, leadership, or distribution of resources like credit land and livestock, credits, etc. (Galiè *et al.*, 2015; MOWCY, 2019). Consequently, women's power to create a livelihood from participating in agricultural operations is greatly restricted by their lack of possession, use, and control of resources (Quisumbing and Maluccio, 2003). The percentages of unpaid, seasonal, and part-time jobs are generally higher among women, while available data show that women receive lesser wages for the same number of jobs as their male counterparts (Chanana-Nag and Aggarwal, 2020). Some studies have evaluated the inequalities in accessing agricultural resources in sub-Saharan countries with the conclusion that women are generally disadvantaged (Kabeer, 2016; Chanana-Nag and Aggarwal, 2020).

According to IFAD (2010), the role of rural women in maintaining household food security is still largely unrecognized when it comes to policy design, execution, and resource allocation, especially in developing nations. The voices and concerns of rural women are not generally heard, and, at times, suppressed at the country and global levels. Women's groups are often confined to the local level. This translates into an unfortunate but dramatic mismatch between rural women's voices and decision-making roles and their immense contribution to the agricultural value chain and livelihoods (IFAD, 2010).

Empowerment is the "process of acquisition of the capacity for self-determination, a situation where an individual is living a life that he or she has reason to value" (Galiè *et al.*, 2017). Many studies have established that women's empowerment is positively associated with food security and better nutrition, which enhances household welfare and children's growth (Smith *et al.*, 2003; Essilfie *et al.*, 2020; Hossain *et al.*, 2021; Wei *et al.*, 2021). The literature suggests that when women are empowered, it significantly improves the food and nutrition security of rural households (Verhart *et al.*, 2015). Researchers and policy makers alike have come up with different techniques to make agriculture a viable vehicle to empower women. Part of these methods have focused on education to build capacities (Ibrahim and Alkire, 2007; Negin *et al.*, 2009), while another dimension of empowerment which is important but has not received the needed attention is women's participation in the process of decision making. In addition, empirical evidence shows that enhancing women's access and control over resources will ultimately exert a positive influence on indicators of development (Sraboni *et al.*, 2014).

3.5 Agribusiness, Food Security, and Rural Women's Empowerment

3.5.1 Empowering rural women through agribusiness

Agribusiness is crucial in empowering rural women as it provides them with economic opportunities and enhances their involvement in the agricultural value chain (Asogwa *et al.*, 2020). Women's involvement in agribusiness can significantly contribute to a reduction in the prevalence of poverty and food insecurity and in propelling sustainable development (Akinbami *et al.*, 2019; bin Duwa bin Khoja, 2021). Women also impact agricultural production, with estimates of female involvement ranging from 60 to 80% in Nigeria (Asogwa *et al.*, 2020). However, women face various challenges in this sector, including socio-cultural restrictions, poverty, and lack of education (Asogwa *et al.*, 2020). To overcome the challenges, improving resource accessibility, market opportunities, and knowledge and skills development become essential (bin Duwa bin Khoja, 2021).

Singh *et al.* (2022) opined that empowering women through agribusiness can increase income generation, self-sufficiency, and sustainable growth (Manyise and Dentoni, 2021). Diiro *et al.* (2018) posited that women's empowerment in the agricultural sphere is positively associated with increased maize productivity, particularly on female-managed plots, while Mariyono *et al.* (2020) found that women's participation in vegetable-based agribusiness increases income generation and rural modernization. Educational status, job creation, and extension services are key determinants of female entrepreneurship development in rural areas (Ijatuyi *et al.*, 2022). Also, women's entrepreneurial activities, social networks, and agribusiness innovation contribute to poverty alleviation in rural communities (Osei and Zhuang, 2020). Similarly, Mulema *et al.* (2021) emphasized that it is crucial to invest in various community capitals to empower rural women in agriculture, with human, social, and financial capital as key entry points.

In the same vein, microcredit programs combined with extension services and training can upscale agribusinesses, increase incomes, and prevent credit capture by men (Kimengsi *et al.*, 2020). Studies on women's empowerment in the agricultural domain have advanced, using measurement innovations and evidence-based interventions (Quisumbing *et al.*, 2023). Microfinance services such as credit access, skill acquisition, and development, and saving practices (Abebe and Kegne, 2023), play critical roles in women's entrepreneurship orientation as they positively influence their outcomes. For example, in rural Zimbabwe, the main contributors to entrepreneurship development among women are financial, environmental, psychological, and sociological factors that contribute positively to sustainable development (Mashapure *et al.*, 2023).

Riley (2024) consolidated in his study that adopting digital accounts for loan disbursement as against physical cash can usher in more profit for female entrepreneurs and business increase in Uganda. Ghimire *et al.* (2024) showed that the socio-economic status of rural women in terms of power relations and decision making, credit access, and savings in Nepal can significantly be improved if joint land ownership is permitted. Fetter and Usmani (2024) highlighted the possibilities of increasing employment opportunities for both men and women in India as a result of combined forces of rural electrification and economic boom in affected villages.

In addressing the structural constraints of gender roles in agricultural food systems, arguments have been posed by Lecoutere *et al.* (2024) for a multidimensional way of nipping these barriers in the bud at individual, local, sectorial, national, regional, and international levels. The review put forward suggestions that rural women could be empowered in agribusiness through the provision of credit, skill acquisition, and development which then foster quality social clusters.

3.5.2 Empowerment of rural women and food security

The nexus between the empowerment of women in rural communities and achieving food security has been widely discussed by researchers and policy makers. Women's contribution to agricultural production in sub-Saharan Africa is significant, with estimates suggesting they produce up to 80% of the region's food (Lufuke *et al.*, 2022). However, they often produce under severe resource constraints including lack of land ownership, access to credit, farm inputs, etc. Access to agricultural resources among Kenyan women farmers was found to be reduced compared with men, which depressed productivity and hence household food security (Ingutia and Sumelius, 2022).

Empowering rural women is suggested to wield an incremental impact on food security in the household (Adenidji and Özçatalbaş, 2022). Moreover, the literature reveals that empowering Ghanaian women through education and decision making had positive effects on food security (Lufuke *et al.*, 2022). In addition, the empowerment of South African women through various means like agency, capital, and social and cultural aspects seemed to have lowered their vulnerability to food insecurity (Lufuke *et al.*, 2022). However, the association between the empowerment of women and the achievement of food security is nuanced. There have been some surprising results, mostly when it comes to financial empowerment. In South Africa, women's empowerment in the financial capital domain was significantly positively associated with household food insecurity (Lufuke *et al.*, 2022). This implies that care should be taken when designing and implementing female empowerment programs.

Over the years, Africa's economic transformation and the implementation of development policies have influenced the dual role that women play in the production of goods both as economic agents (farmers), and in reproductive duties (domestic unit providing labor). Studies reveal that women have more problems in engaging in independent agriculture and that less benefits accrue with them from production proceeds (Whitehead, 1991). This paradigm shift in intrahousehold power dynamics and farm decision making has repercussions on the food security status of households and in safeguarding children's welfare. Also, gender inequality concerning decision making and resource access poses a barrier to food security status. A study in eastern Africa showed the vicious cycle of women overtasked with food production activities and lack of time or power to make independent decisions which have negative consequences for health and social outcomes (Hyder *et al.*, 2005). This further illustrates the nexus among women's empowerment, food security, and household welfare.

Evidence suggests that programs or interventions focusing on empowering rural women's groups impact food security positively. Agricultural cooperatives, for example, have been reported to improve technical efficiency among smallholder farmers and to also improve food security in the home (Adenidji and Özçatalbaş, 2022; Ingutia and Sumelius, 2022). In addition, other techniques such as drip irrigation systems have been able to increase crop production while tackling gender issues (Ingutia and Sumelius, 2022). This chapter definitely agrees with the need to empower rural women and thus facilitate food security in sub-Saharan countries but it also highlights the complexity of this relationship and the need for context-specific, multifaceted approaches to empowerment that address the various constraints faced by rural women.

3.6 Use Case: The Virtuous Cycle of Sustainable Agribusiness, Food Security, and Women's Empowerment

To further illustrate the interconnectedness and process of sustainable agribusiness, food security, and women's empowerment, a use case was developed to expound on the preconditions, flows, interactions, and outcomes needed to achieve the desired tripartite goals. In Fig. 3.2, the outer linkages (flow) begin from the precondition of providing rural women with improved access to productive assets, services, and

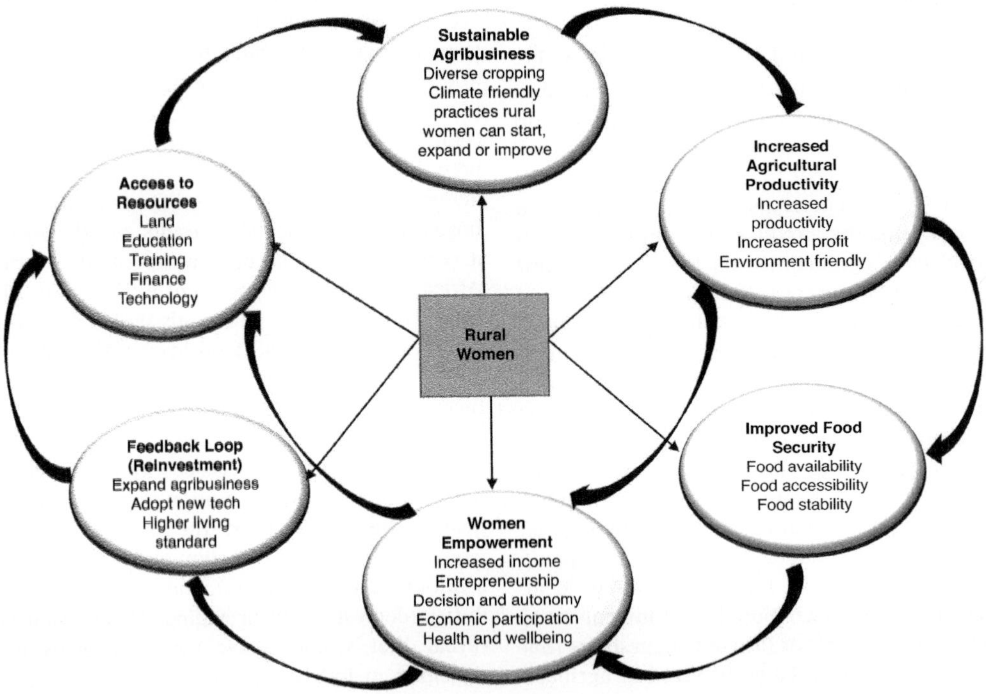

Fig. 3.2. Use case for the virtuous cycle of sustainable agribusiness, food security, and women's empowerment.

resources such as land, education, and training on sustainable agricultural technologies and techniques like water conservation methods and use of organic fertilizers. Other interventions could be the provision of financial services through microcredit to rural women to start an agribusiness like livestock rearing or processing since microfinance has been established as boosting business start-ups and growth (AbdulKareem, 2023).

With these resources and increased income, rural women can start, expand, or improve agribusiness through the adoption of green agricultural practices and processes, climate-resilient approaches, diverse cropping systems, and value addition to agricultural products. Implementing these leads to increased agricultural productivity, higher profits, and environmentally friendly agriculture which means higher crop yields, better quality agricultural products, practices, and processes, and ultimately, improved overall output and productivity of the agricultural sector. The increased agricultural output and productivity will enhance food security through improved availability and accessibility as well as stability from the improved and more reliable food supply, which reduces the risk of food shortages.

Furthermore, with available and accessible food, rural women can start up small-scale business in food processing which will earn them income and empower them economically. It also creates opportunities for agricultural entrepreneurship in terms of food production, processing, and distribution. This facilitates the empowerment of rural women in the areas of autonomy, economic participation, and decision making. The psychological and socio-political empowerment of rural women can also be enhanced through this process from increased social mobility, self-worth and dignity, and community leadership (Kamaluddeen, 2019) in sustainable agricultural practices.

Lastly, food security boosts the health and well-being of both rural women as well as their families. Improved nutritional status and reduced vulnerability of the family means rural women will have more time for self-development like education and skills acquisition, which will boost gender equality and reduce the risk of experiencing domestic and gender-based violence. Consequently, female empowerment will lead to a feedback mechanism where rural women reinvest in agribusiness by either expanding the business or adopting new technologies, which puts them in a better stead to access additional productive resources, which leads to a virtuous cycle.

The inner linkage (flow), which is an alternative pathway, indicates that increased output and improved productivity in agriculture have the potential of improving the socio-economic status of rural women through increased profit, enhanced market access, and entrepreneurship opportunities. From women's empowerment to improved accessibility to productive resources, women will have better access to training, finances, land, and other assets to engage productively in sustainable agribusiness and the virtuous cycle continues.

3.7 Conclusion and Policy Recommendations

3.7.1 Conclusion

This chapter examines the important roles of rural women in agribusiness and food security, explores how these can lead to female empowerment and, lastly, constructs a use case to examine how the linkages and interrelationships among the three constructs lead to a virtuous cycle. Anchored on the theory of multifunctional agriculture, an in-depth literature review has been adopted as the methodological approach to synthesize conceptual analysis of the roles of agribusiness, food security, and women's empowerment. Consequently, the chapter develops a virtuous cycle of agribusiness, food security, and women's empowerment while exposing their interconnectedness and linkages. It is established that rural women play central roles in agribusiness and food security.

Findings show that rural women engage in the entire chain of agribusiness from production, preservation, processing, value addition, and marketing to distribution. In food security, rural women are essential in managing food production in the home, maintaining vegetable gardens, and rearing livestock and are responsible for household diet and nutrition. Both agribusiness and food security were discovered to contribute to women's empowerment in various dimensions – economic, social, psychological, and political. In turn, with women's empowerment, where rural women have better accessibility and control over productive resources, agribusiness and food security can be boosted since women make up 52% of the agri-food sector while accounting for half of the labor on sub-Saharan farms. This not only creates a virtuous cycle but underlines the criticality of rural women in the cycle. Furthermore, this chapter also underscores the critical roles of rural women in achieving the SDGs, particularly as regards zero poverty (SDG1) and ending hunger (SDG2), gender equality and empowerment (SDG5), and reduced inequality (SDG10).

3.7.2 Policy recommendations

3.7.2.1 Land ownership

The first policy recommendation is improving land access for women. This can be achieved through policies on land reforms that will prioritize land ownership for women in agribusiness. Such agricultural intervention can include allocating farmlands to rural women in agriculture or on a farm-operate-and-transfer (FOT) basis in instances where outright transfers cannot be made. In this case, rural women farm on lands on profit-sharing terms after which the land eventually becomes theirs after a specified period. This is crucial as studies such as Djurfeldt *et al.* (2020) have established that land tenure for women improves agricultural productivity and women's empowerment.

3.7.2.2 Financial access

As limited access to finance has been recognized as a major hurdle for rural women in agribusiness, it is important for government intervention programs to include a finance component. These include tailored financial products and services appropriate to the needs of rural women with less stringent conditions on interest rates, collateral, documentation, and repayment plans (Kamaluddeen, 2019). Credit schemes can also be designed specifically for rural women in order to improve their involvement and productivity in agribusiness (Malapit and Quisumbing, 2021). Financial literacy needs to be an important entry point for financial grants and loans extended to rural women for them to have the financial acumen to effectively manage funds without diverting their use, so as to attain the intended target of boosting agribusiness, food security, and rural women's empowerment.

3.7.2.3 Green agricultural technology

Rural women also need to be provided with green technologies applicable to the different areas of agribusiness in which they are involved in order to promote sustainability, food security, and empowerment. For instance, solar irrigation systems and organic fertilizers can be provided for those in production while providing fuel-efficient cooking stoves, solar-powered dryers, and storage units for those in processing and value addition activities. In addition, those in sales and marketing can be provided with environmentally friendly packaging and digital market information systems. These green technologies have been established in literature to boost productivity in agribusiness and to empower women (Dherani *et al.*, 2008; Abubakar *et al.*, 2021; Aker *et al.*, 2021; Burney *et al.*, 2021; James and James, 2021; Ramesh *et al.*, 2021). Rural women can be granted access to these technologies through agricultural intervention programs, women's empowerment initiatives, government subsidies, and cooperatives.

3.7.2.4 Capacity building

The role of education and training cannot be overemphasized, especially where rural women are concerned. For agribusiness in particular, rural women need to be offered continuous education, training, and retraining on the use and importance of various green agricultural technologies (such as those mentioned above) that can enhance their output, productivity, and income. The capacity of rural women can also be developed through vocational training in various aspects of agribusiness and its management to improve the rate of success for start-ups, to improve business growth, and to increase profit (Datt and Ravallion, 2020) which will, on the whole, drive productivity, food security, and women's empowerment.

3.7.2.5 Gender mainstreaming

Given the significant role of rural women in agribusiness and food security as expounded in this chapter, it is pertinent to mainstream gender into all agricultural policies, programs, and interventions that not only cater to the needs of women but also address the challenges they encounter in the sector. The World Bank (2021) and United Nations (UN Women, 2020) have been strong proponents of this with the latter advocating for gender mainstreaming even at monitoring and evaluation stages to capture gender-sensitive impacts at the grassroots level.

3.7.2.6 Addressing gender norms

Ultimately, the gender norms hindering women from having a level playing field in terms of equal access to opportunities, resources, and markets in agribusiness need to be addressed. Continuous sensitization programs, town hall meetings, policy frameworks, and community engagements by the government, non-governmental organizations and international development organizations should be organized to address discriminatory and derogatory cultural norms and myths, while emphasizing the critical roles of empowering women, especially for promoting agribusiness and food security.

References

AbdulKareem, H.K.K. (2023) Financial inclusion and poverty reduction among the operators in the small and medium scale enterprises of lagos state. *Ilorin Journal of Economic Policy* 10(1), 15–31. Available at: https://ijep.org/Abdulkareem2023.pdf

Abebe, A. and Kegne, M. (2023) The role of microfinance institutions on women's entrepreneurship development. *Journal of Innovation and Entrepreneurship* 12(1), 17. DOI: 10.1186/s13731-023-00285-0.

Abubakar, M.S., Elhassan, A.M. and Uddin, M.S. (2021) Assessment of solar drying technology as a tool for women empowerment in Africa. *Renewable Energy* 174, 56–63.

Adenidji, E.M.C. and Özçatalbaş, O. (2022) A view of Sub-Saharan Africa from the perspective of food security and gender. In: *Food Security in Africa*. IntechOpen. DOI: 10.5772/intechopen.106392.

Adeyanju, D., Mburu, J., Gituro, W., Chumo, C., Mignouna, D. *et al.* (2023) Harnessing the job creation capacity of young rural agripreneurs: A quasi-experimental study of the ENABLE program in Africa. *Social Sciences & Humanities Open* 8, 100791. DOI: 10.1016/j.ssaho.2023.100791.

Agarwal, B. (2018) Gender equality, food security and the sustainable development goals. *Current Opinion in Environmental Sustainability* 34, 26–32. DOI: 10.1016/j.cosust.2018.07.002.

Aker, J.C., Ghosh, I. and Burrell, J. (2021) Digital inclusion in rural areas: A framework for analyzing challenges and opportunities for rural e-commerce. *World Development* 141, 105405.

Akinbami, C.A.O., Olawoye, J.E., Adesina, F.A. and Nelson, V. (2019) Exploring potential climate-related entrepreneurship opportunities and challenges for rural Nigerian women. *Journal of Global Entrepreneurship Research* 9(1), 1–28. DOI: 10.1186/s40497-018-0141-3.

Antriyandarti, E., Suprihatin, D.N., Pangesti, A.W. and Samputra, P.L. (2024) The dual role of women in food security and agriculture in responding to climate change: Empirical evidence from rural java. *Environmental Challenges* 14, 100852. DOI: 10.1016/j.envc.2024.100852.

Arthur, K.K., Bannor, R.K., Masih, J., Oppong-Kyeremeh, H. and Appiahene, P. (2024) Digital innovations: Implications for African agribusinesses. *Smart Agricultural Technology* 7, 100407. DOI: 10.1016/j.atech.2024.100407.

Asogwa, I.S., Omah, E.C. and Asogwa, M.A. (2020) Empowering women through agribusiness: A key to reducing poverty and food insecurity in Nigeria. *Sapientia Foundation Journal of Education, Sciences and Gender Studies (SFJESGS)* 2(3), 55–67.

Assoua, J.E., Bomdzele, E., Jr, Omosehin, O., Binuomote, S.O., Osabuohien, E. *et al.* (2024) Integrating climate change, food safety, nutrition, and health outcomes in food systems policies and programs in Africa. *World Food Policy* 10(1), 86–110.

Beyene, H. (2015) Final report national assessment: Ethiopia gender equality and the knowledge society. Women in Global Science and Technology. Available at: https://www.owsd.net/sites/default/files/NationalAssessmentonGenderandSTI-Ethiopia_0.pdf

bin Duwa bin Khoja, D.A. (2021) Participation of rural women in agricultural activities. *International Journal of Modern Agriculture and Environment* 1(2), 1–25. DOI: 10.21608/ijmae.2023.215954.1014.

Bunch, S. and Mehra, R. (2008) Women help solve hunger: Why is the world still waiting? International Center for Research on Women (ICRW). Available at: https://www.icrw.org/wp-content/uploads/2016/10/Women-Help-Solve-Hunger.pdf

Burney, J., Naylor, R. and Postel, S. (2021) The case for solar-powered irrigation in Sub-Saharan Africa. *Nature Sustainability* 4(2), 101–106.

Cahill, C. (2001) The multifunctionality of agriculture: What does it mean? *Eurochoices* 1, 36–41. DOI: 10.1111/j.1746-692X.2001.tb00073.x.

Castañeda, A., Doan, D., Newhouse, D., Nguyen, M.C., Uematsu, H. *et al.* (2018) A new profile of the global poor. *World Development* 101, 250–267. DOI: 10.1016/j.worlddev.2017.08.002.

Chanana-Nag, N. and Aggarwal, P.K. (2020) Woman in agriculture, and climate risks: Hotspots for development. *Climatic Change* 158(1–2), 13–27. DOI: 10.1007/s10584-018-2233-z.

Damisa, M.A. and Yohanna, M. (2007) Role of rural women in farm management decision making process: Ordered probit analysis. *World Journal of Agricultural Sciences* 3(4), 543–546.

Datt, G. and Ravallion, M. (2020) Poverty, growth, and safety nets in the village economy: A rural employment guarantee case study. *Journal of Development Economics* 142, 102373.

Dherani, M., Pope, D., Mascarenhas, M., Smith, K.R., Weber, M. *et al.* (2008) Indoor air pollution from unprocessed solid fuel use and pneumonia risk in children aged under five years: A systematic review and meta-analysis. *Bulletin of the World Health Organization* 86(5), 390–398C. DOI: 10.2471/BLT.07.044529.

Diiro, G.M., Seymour, G., Kassie, M., Muricho, G. and Muriithi, B.W. (2018) Women's empowerment in agriculture and agricultural productivity: Evidence from rural maize farmer households in western Kenya. *PloS One* 13(5), e0197995. DOI: 10.1371/journal.pone.0197995.

Djurfeldt, A.A., Holmén, H., Jirström, M. and Larsson, R. (2020) Gender and the African agricultural development path: Reflections on the literature. *Geoforum* 109, 1–7.

Sofa Team and Doss, C. (2011) The role of women in agriculture (ESA Working Paper no.11-02). Food and Agriculture Organization of the United Nations. Available at: https://www.fao.org/4/am307e/am307e00.pdf

DTN (Data Transmission Network) (2022) Agribusiness' role in food security. Available at: https://www.dtn.com/agribusiness-role-in-food-security (accessed 12 May 2024).

Elum, Z.A. and Digitemie, T. (2023) Assessment of food security status of rural women in Bayelsa State, Nigeria. *Scientific African* 21, e01878. DOI: 10.1016/j.sciaf.2023.e01878.

Essilfie, G., Sebu, J., Annim, S.K. and Asmah, E.E. (2020) Women's empowerment and household food security in Ghana. *International Journal of Social Economics* 48(2), 279–296. DOI: 10.1108/IJSE-05-2020-0328.

FAO (Food and agriculture organization) (1997) FAO focus on women and food security. *Food and Agriculture Organization of the United Nations*, Rome. Available at: http://www.fao.org/focus/e/womenlsustin-e.htm

Food and Agriculture Organization (2011) The state of food and agriculture 2010–11: Women in agriculture: Closing the gender gap for development. FAO, Rome. Available at: http://www.fao.org/docrep/013/i2050e/i2050e00.htm

FAO (Food and Agriculture Organization) (2021) *National Gender Profile of Agriculture and Rural Livelihoods – Ukraine, Revised*. Budapest. DOI: 10.4060/cb4501en.

FAO (Food and Agriculture Organization) (2023) *The Status of Women in Agrifood Systems*. Rome: Food and Agriculture Organization of the United Nations. DOI: 10.4060/cc5343en.

FAO, IFAD, UNICEF, WFP, and WHO (2023) *The State of Food Security and Nutrition in the World 2023: Urbanization, Agrifood Systems Transformation and Healthy Diets Across the Rural–Urban Continuum*. DOI: 10.4060/cc3017ee.

Fetter, T.R. and Usmani, F. (2024) Fracking, farmers, and rural electrification in India. *Journal of Development Economics* 170, 103308. DOI: 10.1016/j.jdeveco.2024.103308.

Gaffney, J., Challender, M., Califf, K. and Harden, K. (2019) Building bridges between agribusiness innovation and smallholder farmers: A review. *Global Food Security* 20, 60–65. DOI: 10.1016/j.gfs.2018.12.008.

Gaitán, L. (2020) Sustainable agriculture: Principles and concepts. In: *Sustainable Agriculture: Current Trends and Innovations*. Springer, Cham, pp. 1–11.

Galiè, A., Mulema, A., Benard, M.A.M., Onzere, S.N. and Colverson, K.E. (2015) Exploring gender perceptions of resource ownership and their implications for food security among rural livestock owners in Tanzania, Ethiopia, and Nicaragua. *Agriculture & Food Security* 4, 1–14.

Galiè, A., Jiggins, J., Struik, P.C., Grando, S. and Ceccarelli, S. (2017) Women's empowerment through seed improvement and seed governance: Evidence from participatory barley breeding in pre-war Syria. *NJAS-Wageningen Journal of Life Sciences* 81, 1–8. DOI: 10.1016/j.njas.2017.01.002.

Galiè, A., Teufel, N., Girard, A.W., Baltenweck, I., Dominguez-Salas, P. *et al.* (2019) Women's empowerment, food security and nutrition of pastoral communities in Tanzania. *Global Food Security* 23, 125–134. DOI: 10.1016/j.gfs.2019.04.005.

Ghimire, P.R., Devkota, N., Maraseni, T., Khanal, G., Deuja, J. *et al.* (2024) Does joint land ownership empower rural women socio-economically? Evidence from Eastern Nepal. *Land Use Policy* 138, 107052. DOI: 10.1016/j.landusepol.2024.107052.

Ghosh, S., Sen, L.C., Mali, S.K., Islam, M.M. and Bakchi, J. (2021) The role of rural women in household food security and nutrition management in Bangladesh. *Asian Journal of Women's Studies* 27(3), 441–459. DOI: 10.1080/12259276.2021.1970351.

Haggblade, S. (2011) Modernizing African agribusiness: Reflections for the future. *Journal of Agribusiness in Developing and Emerging Economies* 1(1), 10–30. DOI: 10.1108/20440831111131532.

HLPE (High Level Panel of Experts) (2017) Nutrition and food systems. A report by the high level panel of experts on food security and nutrition of the committee on world food security. Rome.

Hossain, M., Asadullah, M.N. and Kambhampati, U. (2021) Review of Economics of the Household,1–29. Available at: https://link.springer.com/article/10.1007/s11150-021-09546-x

Hyder, A.A., Maman, S., Nyoni, J.E., Khasiani, S.A., Teoh, N. *et al.* (2005) The pervasive triad of food security, gender inequity and women's health: Exploratory research from sub-saharan Africa. *African Health Sciences* 5(4), 328–334..

Ibrahim, S. and Alkire, S.. (2007) Agency and empowerment: A proposal for internationally comparable indicators. *Oxford Development Studies* 35(4), 379–403.

Iervolino, Bruna. (2023) Agribusiness possible threats to food security of a kalunga community. *Thesis. Examensarbete Vid Institutionen För Geovetenskaper*. Available at: https://www.diva-portal.org/smash/record.jsf?pid=diva2%3A1761092&dswid=-4554

IFAD (2010) *Report of the Third Global Meeting of the Farmers' Forum*. International Fund for Agricultural Development (IFAD), Rome.

IFC (International Finance Corporation) (2023) Enhancing food security and safety. Available at: https://www.ifc.org/en/what-we-do/sector-expertise/agribusiness-forestry/enhancing-food-security-and-safety#:~:text=Agricultural%20development%20has%20tremendous%20potential,80%20percent%20of%20total%20employment

IITA (International Institute of Tropical Agriculture) (2024, January 8) Transforming Nigerian female farmers' lives through soybean and cowpea production to alleviate poverty. CGIAR. Available at: https://www.cgiar.org/news-events/news/transforming-nigerian-female-farmers-lives-through-soybean-and-cowpea-production-to-alleviate-poverty

Ijatuyi, E.J., Oladele, O.I., Abiolu, O.A. and Omotayo, A.O. (2022) Socioeconomic determinants of female development in the rural North West province of South Africa. *Sustainability* 14(1), 547. DOI: 10.3390/su14010547.

Ikhaghu, A. (2022) Tackling challenges faced by women in agro-businesses. In *The Guardian*, 2022. Available at: https://guardian.ng/features/tackling-challenges-faced-by-women-in-agro-businesses/ (accessed 12 May 2024).

Ikuemonisan, E.S., Abass, A.B., Feleke, S. and Ajibefun, I. (2022) Influence of agricultural degree programme environment on career in agribusiness among college students in Nigeria. *Journal of Agriculture and Food Research* 7, 100256. DOI: 10.1016/j.jafr.2021.100256.

ILO (International Labour Organization) (n.d.) Agriculture, aquaculture, plantations. In: *International Labour Organization*. Available at: https://www.ilo.org/topics-and-sectors/industries-and-sectors/agriculture-aquaculture-plantations

Ingutia, R. and Sumelius, J. (2022) Determinants of food security status with reference to women farmers in rural Kenya. *Scientific African* 15, e01114. DOI: 10.1016/j.sciaf.2022.e01114.

James, S.J. and James, C. (2021) The food cold-chain and climate change. *Food Research International* 99, 28–38. DOI: 10.1016/j.foodres.2021.01.013.

Kabeer, N. (2016) Gender equality, economic growth, and women's agency: The 'endless variety' and 'monotonous similarity' of patriarchal constraints. *Feminist Economics* 22(1), 295–321. DOI: 10.1080/13545701.2016.1147096.

Kalansooriya, C.W. and Chandrakumara, D.P.S. (2014) Women's role in household food security in rural Sri Lanka. *International Journal of Multidisciplinary Studies* 1(1), 1–10. DOI: 10.4038/ijms.v1i1.31.

Kamaluddeen, H.K. (2019) Effect of microfinance banks on women empowerment: Evidence from Kwara State, Nigeria, Doctoral dissertation, Kwara State University (Nigeria).

Kehinde, M.O., Shittu, A.M., Adeyonu, A.G. and Ogunnaike, M.G. (2021) Women empowerment, Land Tenure and Property Rights, and household food security among smallholders in Nigeria. *Agriculture & Food Security* 10(1), 25. DOI: 10.1186/s40066-021-00297-7.

Kimengsi, J.N., Balgah, R.A., Buchenrieder, G., Silberberger, M. and PridedinorahBatosor, H. (2020) An empirical analysis of credit-financed agribusiness investments and income poverty dynamics of rural women in Cameroon. *Community Development* 51(1), 72–89. DOI: 10.1080/15575330.2020.1716031.

Lecoutere, E., Achandi, E.L., Ampaire, E.L., Fischer, G., Gumucio, T. *et al.* (2024) Fostering an enabling environment for equality and empowerment in agri-food systems: An assessment at multiple scales. *Global Food Security* 40, 100735. DOI: 10.1016/j.gfs.2023.100735.

Lufuke, M., Bai, Y., Fan, S. and Tian, X. (2022) Women's empowerment, food security, and nutrition transition in Africa. *International Journal of Environmental Research and Public Health* 20(1), 254. DOI: 10.3390/ijerph20010254.

Malapit, H.J.L. and Quisumbing, A.R. (2021) What dimensions of women's empowerment in agriculture matter for nutrition in Ghana. *Food Policy* 101, 102034. DOI: 10.1016/j.foodpol.2021.102034.

Manyise, T. and Dentoni, D. (2021) Value chain partnerships and farmer entrepreneurship as balancing ecosystem services: Implications for agri-food systems resilience. *Ecosystem Services* 49, 101279. DOI: 10.1016/j.ecoser.2021.101279.

Mariyono, J., Abdurrachman, H., Suswati, E., DwitaSusilawati, A., Sujarwo, M. *et al.* (2020) Rural modernisation through intensive vegetable farming agribusiness in Indonesia. *Rural Society* 29(2), 116–133..

Mashapure, R., Nyagadza, B., Chikazhe, L., Mazuruse, G. and Hove, P. (2023) Women entrepreneurship development and sustainable rural livelihoods in Zimbabwe. *Arab Gulf Journal of Scientific Research* 41(4), 557–584. DOI: 10.1108/AGJSR-07-2022-0112.

Meighan, J. (2024) Food systems and gender: The groundbreaking role of rural women. *Journal of International Affairs*, Available at: https://jia.sipa.columbia.edu/news/food-systems-and-gender-groundbreaking-role-rural-women (accessed 30 August 2024).

MOWCY (Ministry of Women, Children and Youth) (2019) *Gender Equality, Women's Empowerment and Child Wellbeing in Ethiopia*. UNICEF Ethiopia.

Mulema, A.A., Boonabaana, B., Debevec, L., Nigussie, L., Alemu, M. *et al.* (2021) Spiraling up and down: Mapping women's empowerment through agricultural interventions using the community capitals framework in rural Ethiopia. *Community Development* 52(1), 113–130. DOI: 10.1080/15575330.2020.1838589.

Negin, J., Remans, R., Karuti, S. and Fanzo, J.C. (2009) Integrating a broader notion of food security and gender empowerment into the African green revolution. *Food Security* 1(3), 351–360. DOI: 10.1007/s12571-009-0025-z.

Njuki, J., Eissler, S., Malapit, H., Meinzen-Dick, R.S., Bryan, E. *et al.* (2022) A review of evidence on gender equality, women's empowerment, and food systems. *Global Food Security* 33, 100622. DOI: 10.1016/j.gfs.2022.100622.

OECD (2022) Fact sheet: Food security and gender. United States Agency for International Development. Available at: https://www.oecd.org/dac/gender-development/46460857.pdf (accessed 9 June 2022).

Osei, C.D. and Zhuang, J. (2020) Rural poverty alleviation strategies and social capital link: The mediation role of women entrepreneurship and social innovation. *SAGE Open* 10(2), 2158244020925504. DOI: 10.1177/2158244020925504.

Otaokpukpu, N.J., Nwankwo, L.A. and Okonkwo, C.J. (2023) Youth involvement in agribusiness for sustainable economic development and national security: A conceptual review. *Forshen Hub International Journal of Entrepreneurial and Cooperative Studies* 6, 5. Available at: www.forshenhub.com

Owren, C. (2017) Women and the environment. In: Aragon, J. and Miller, M. (eds) *Global Women's Issues: Women in the World Today, Extended Version*. Bureau of International Information Programs, U.S. Department of State, pp. 171–175. Available at: https://opentextbc.ca/womenintheworld/chapter/chapter-11-women-and-the-environment/

Quisumbing, A., Cole, S., Elias, M., Faas, S., Galiè, A. *et al*. (2023) Measuring women's empowerment in agriculture: Innovations and evidence. *Global Food Security* 38, 100707. DOI: 10.1016/j.gfs.2023.100707.

Quisumbing, A.R. and Maluccio, J.A. (2003) Resources at marriage and intra-household allocation: Evidence from Bangladesh, Ethiopia, Indonesia, and South Africa. *Oxford Bulletin of Economics and Statistics* 65(3), 283–327. DOI: 10.1111/1468-0084.t01-1-00052.

Quisumbing, A.R., Brown, L.R., Feldstein, H.S., Haddad, L. and Peña, C. (1996) Women: The key to food security. *Food and Nutrition Bulletin* 17(1), 1–2. DOI: 10.1177/156482659601700116.

Ramesh, P., Panwar, N.R., Singh, A.B. and Ramana, S. (2021) Organic farming: Its relevance to the Indian context. *Indian Journal of Agronomy* 66(1), 1–15. DOI: 10.5236/ijafr.2021.01.001.

Riley, E. (2024) Resisting social pressure in the household using mobile money: Experimental evidence on microenterprise investment in Uganda. *American Economic Review* 114(5), 1415–1447. DOI: 10.1257/aer.20220717.

Savari, M., Sheykhi, H. and Amghani, M.S. (2020) The role of educational channels in the motivating of rural women to improve household food security. *One Health* 10, 100150. DOI: 10.1016/j.onehlt.2020.100150.

Schoneveld, G.C. (2022) Transforming food systems through inclusive agribusiness. *World Development* 158, 105970. DOI: 10.1016/j.worlddev.2022.105970.

Singh, S., Rana, A., Sharma, N. and Kumar, M. (2022) A review on women agri-entrepreneurship: Roles and opportunities in agriculture for sustainable growth in India. *Humanities* 10(2), 56–67. DOI: 10.34293/sijash.v10i2.5117.

Smith, L.C., Ramakrishnan, U., Ndiaye, A., Haddad, L. and Martorell, R. (2003) The importance of women's status for child nutrition in developing countries. Research Report no.131. Washington D.C.

Sraboni, E., Malapit, H.J., Quisumbing, A.R. and Ahmed, A.U. (2014) Women's empowerment in agriculture: What role for food security in Bangladesh. *World Dev* 61, 11–52. DOI: 10.1016/j.worlddev.2014.03.025.

Srivastava, N. and Srivastava, R. (2010) Women, work, and employment outcomes in rural India. *Economic and Political Weekly* XLV(28), 49–63.

Subathra, C. (2015) Scenario of cultivators and agricultural laborers—A study. *International Journal of Research* 2(10), 1133–1139.

Swati, M. and Bhumika, T.V. (2024) Women agripreneurs need gender-responsive interventions. *Farming First*. Available at: https://farmingfirst.org/2024/03/women-agripreneurs-need-gender-responsive-interventions

Tohidyan Far, S. and Rezaei-Moghaddam, K. (2019) Multifunctional agriculture: An approach for entrepreneurship development of agricultural sector. *Journal of Global Entrepreneurship Research*. DOI: 10.1186/s40497-019-0148-4.

Ulker, U.E. (2013) Factors affecting women's entrepreneurship in rural areas and determination of current situation in tekirdağ province. Master's thesis, Namık Kemal University, Graduate School of Natural and Applied Sciences, Department of Agricultural Economics.

UN (United Nations) (2012) Overview: Food security. Women Watch. Available at: https://www.un.org/womenwatch/feature/ruralwomen/overview-food-security.html

UNDP (United Nations Development Programme) (2016) Gender, climate change and food security. UNDP. Available at: www.undp.org/gender

UN Women. (2020) *Progress of the World's Women 2020: Women and Sustainable Development*. UN Women.

van Westen, A.G., Mangnus, E., Wangu, J. and Worku, S.G. (2019) Inclusive agribusiness models in the Global South: The impact on local food security. *Current Opinion in Environmental Sustainability* 41, 64–68. DOI: 10.1016/j.cosust.2019.11.003.

Verhart, N., Wijngaart, A.V.D., Dhamankar, M. and Danielsen, K. (2015) Bringing agriculture and nutrition together using a gender lens. KIT Working Paper 6. Amsterdam.

Wei, W., Sarker, T., Roy, R., Sarkar, A. and Rabbany, M.G. (2021) Women's empowerment and their experience to food security in rural Bangladesh. *Sociology of Health & Illness* 43(4), 971–994. DOI: 10.1111/1467-9566.13273.

Whitehead, A. (1991) *Rural Women and Food Production in Sub-Saharan Africa*. Oxford University Press eBooks. DOI: 10.1093/acprof:oso/9780198286356.003.0011.

World Bank. (2013) *Growing Africa: Unlocking the Potential of Agribusiness*. World Bank.

World Bank. (2014) *Gender in Agriculture Sourcebook*. World Bank.

World Bank. (2017) *World Development Report 2017: Governance and the Law*. World Bank.

World Bank. (2021) Rural women and food production in Sub-Saharan Africa. World Bank. Available at: www.ifad.org/farmer/2010/doc/concepe.pdf

Yumkella, K.K. (2011) *Agribusiness for Africa's Prosperity*. UNIDO.

Zereyesus, Y.A. (2017) Women's empowerment in agriculture and household-level health in northern Ghana: A capability approach. *Journal of International Development* 29(7), 899–918. DOI: 10.1002/jid.3307.

Zhichkin, K.A., Nosov, V.V., Zhichkina, L.N. and Gubadullin, A.A. (2022) The theory of agriculture multifunctionality on the example of private households. *Agriculture* 12, 1870. DOI: 10.3390/agriculture12111870.

4 Agripreneurs: The Accelerators of Agricultural Development in India

Punita Malik[1]*, Renu Hooda[2] and Piera Centobelli[3]
[1]Department of Commerce, Government College for Women, Kulana, Jhajjar, India; [2]Department of Business Administration, Maharaja Surajmal Institute, New Delhi, India; [3]Department of Industrial Engineering, University of Naples Federico II, Naples, Italy

4.1 Introduction

Agriculture is a prominent sector in the world economy as it accounts for 3% of the global gross domestic product (GDP) and employs around 1 billion people. In India, 85% of the rural population depends on agriculture for their livelihood. This sector engages about 140 million farmers throughout the country and its contribution to GDP accounts for 15%. For many years agriculture has been considered to be the backbone of the Indian economy as this sector provides food security to the nation, supplies various commodities like rice, coffee, and tea for exports, and produces raw materials for various consumer goods' industries. The Indian agriculture sector successfully withstood the Covid-19 pandemic and braced the nation during the time when major industries were shutting down.

However, the seasonal dependence of productivity and profitability of agriculture is a major impediment to its growth. This sector also bears the grave issues of farmer suicides and disguised unemployment. The lack of technological development, farm mechanization, and educational awareness contribute further towards the sluggish growth of this sector.

According to the Food and Agriculture Organization, the demand for food and agricultural produce is projected to increase by 50% between 2012 and 2050 owing to rapid urbanization and population growth. Moreover, the consequences of the recent Covid-19 pandemic on the three economic sectors showed the agriculture sector to be the most sustainable (Sreedhar and Siva Sri Mani Kanta, 2022). This generates the need to revitalize the agriculture sector by addressing the key concerns hindering its growth. Entrepreneurial initiatives provide a successful pathway for agricultural development in India. According to Dollinger (2003), "entrepreneurship in agriculture is the creation of innovative economic organization for the purpose of growth or gain under the conditions of risk and uncertainty in agriculture." This fusion of agriculture and entrepreneurship led to the germination of a new spirit of enterprise called "agripreneurship." In 1955, this concept was advocated by John Davis of Harvard University as "agribusiness." However, with the influx of innovation in agriculture, a new breed of entrepreneurs came to light called "agripreneurs." The Indian agriculture sector has undergone major transformations from

*Corresponding author: punitamalik93@gmail.com

 Agripreneurship, Digital Inclusion and Sustainability: Rural Innovation and Growth in the Global South (eds Dutta and Trivedi)
DOI: 10.1079/9781836992646.0004

conventional farming practices to incorporating more integrated farming practices. With the augment of technological advancements in farm machinery, supply chain logistics, and entrepreneurship-oriented policies; India is in the process of forging a highly sophisticated configuration of resources for the development of entrepreneurship in agriculture.

The agriculture sector has enormous potential to boost the national economy, and integrating it with entrepreneurship is an effective way to extract the potential of this sector. Agripreneurship can lead to more jobs and revenue in agriculture which can further eradicate poverty, provide a check on migration, educate and train rural people, and finally enhance the food security of the nation.

4.1.1 Agripreneurs

The word agripreneur is a combination of two words:

Agriculture + Entrepreneur = Agripreneur

Thus, Uplaonkar and Biradar (2015) simply define an agripreneur as "an entrepreneur whose main business is agriculture or agriculture-related." An agripreneur advances beyond the conventional farming practices by developing a new farming method, cultivating a new crop, or discovering a new market for agricultural produce. Hence, an agripreneur manifests the characteristics of an entrepreneur such as being a risk-taker, initiator, and innovator while dealing with the uncertain agricultural environment.

4.1.2 Agripreneurship

Agripreneurship is a profitable fusion of agriculture and entrepreneurship. It converts agricultural activity into an entrepreneurial activity. Chand (2019) defines agripreneurship as "the process of adopting [a] new method, process or technique in agriculture and its allied sectors for better output and economic earnings." Thus, agripreneurship gives a social perspective to business and exhibits a holistic and community-oriented approach. Agripreneurship involves the creation of an innovative economic structure that intends to gain from the uncertain and risky agricultural environment. This facilitates the creation of incremental wealth from agriculture that has been an undervalued economic sector for so many years.

4.1.3 Agrienterprise

Agrienterprise refers to a business venture pertaining to agriculture or its allied sectors. It generally involves a farmer or a group of farmers who integrate their entrepreneurial thinking into establishing a venture that gives a new direction to conventional agricultural practices. Thus, an agrienterprise is the output of the agripreneurship process. The agripreneurship process given by Usanga (2012) can be depicted as an action-oriented agripreneur who undertakes to gain an economic advantage from the agricultural business venture and this whole process can be encapsulated as shown in Fig. 4.1.

4.2 Agripreneurs

4.2.1 Types of agripreneurs

Agripreneurs can undertake different types of ventures in agriculture and its allied sectors. India has abundant natural resources for agricultural use with 15 different types of climates and various types of crops and irrigation styles

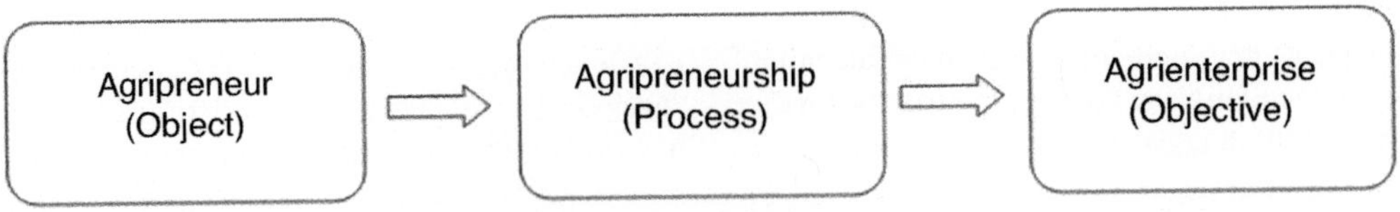

Fig. 4.1. Agripreneurship process. Adapted from Usanga (2012).

depending on the region. Thus, ample opportunities exist for entrepreneurs who can pursue farm-oriented activities to boost production or offer hiring services for farm equipment and operations. Building on the trend for organic products, agripreneurs can venture into biofertilizers, biopesticides, and vermicompost. Focusing on the solution-oriented approach, agripreneurs can provide digital solutions via apps or satellites for effectively meeting the needs of farming communities. Considering these various avenues, Uplaonkar and Biradar (2015) have provided four distinct categories of agripreneurs (Fig. 4.2).

1. *Farm-level producers.* The farm-level producers are the agripreneurs who as a family work on the field. All the family members together form one agrienterprise. Their main entrepreneurial function is to utilize the technology, resources, and market demand to gain from the agrienterprise.
2. *Service providers.* There is an array of services related to agricultural activities such as the retail and distribution of farm inputs, hiring of farm machinery like tractors and harvesters, irrigation amenities like seeding, yielding, harvesting, and warehousing. Thus, the agripreneurs exploiting these opportunities are categorized as service providers. Related opportunities also exist in the allied agricultural sectors like livestock husbandry, breeding, and disease diagnosis.
3. *Input producers.* Inputs required for agriculture stretch across a plethora of light and heavy equipment. The production of crop seeds, plant roots, pesticides, fertilizers, compost, agricultural tools, and accessories present a wide range of opportunities for agripreneurs not only in farming but other allied activities like fisheries, poultry, and sericulture. Therefore, agripreneurs involved in production of such inputs are known as input producers.
4. *Processing and marketing cooperatives.* An important aspect of the post-production phase in agriculture is the processing and marketing of the output. Agripreneurs in this category must have a well-organized and specialized support as this type of agripreneurship requires higher level of knowledge and investment. In India, the most successful agrienterprise in this category has been the Amul dairy cooperative.

4.2.2 Qualities of an agripreneur

Agripreneurship is exclusively dependent upon the agripreneur's ability to exploit the resources and manage risk in order to successfully realize benefits from agricultural activities. The key issues pertaining to the Indian agriculture sector are seasonal dependence of the output, unskilled workers, lack of education regarding agricultural operations, and obscure purchase of agricultural output. The entrepreneurial behavior of an agripreneur can pursue these issues as business opportunities and convert them into business ventures (Boye *et al.*, 2022). The success and profitability of these ventures depends largely upon the entrepreneurial qualities present in the agripreneur (Fig. 4.3). An agripreneur has to perform a lot of functions simultaneously and the presence of the following qualities ensures efficient management of those tasks leading the enterprise towards success (Garima *et al.*, 2023, April 24).

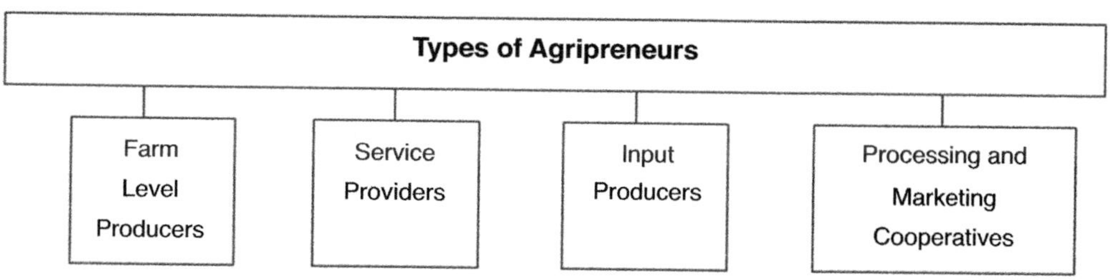

Fig. 4.2. Types of agripreneurs. Adapted from Uplaonkar and Biradar (2015).

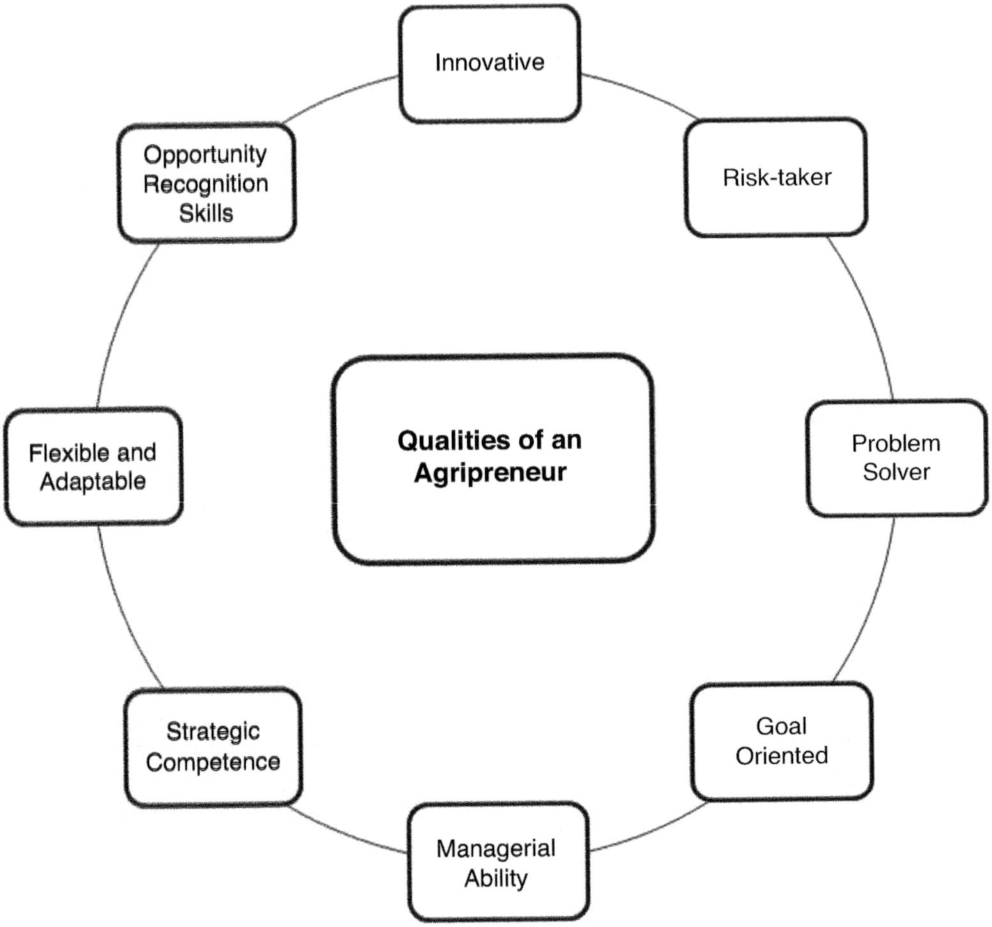

Fig. 4.3. Qualities of an agripreneur.

1. *Innovative.* This is the core of an entrepreneur. Agripreneurs should be creative enough to extract unique ideas and use their managerial and innovative skills to build them into reality.
2. *Risk-taker.* Every kind of business entails some degree of risk. Thus, an agripreneur must have the capability to take risk and realize their goals.
3. *Problem solver.* Agripreneurs should work with an approach to generate solutions to problems. A creative person inherently focuses more on the solution and less on the problem and this attitude helps in saving time and moving ahead.
4. *Goal oriented.* An agriprenuer must be driven by the purpose to attain the goal. This helps in staying focused and committed towards the main task at hand by overcoming distractions.
5. *Managerial ability.* An agripreneur should be a thinker and planner. They must possess leadership and organizational skills to stimulate team members towards better implementation of the plan and attainment of goals.
6. *Strategic competence.* Agriprenuers must have knowledge and information about the internal as well as external business environment. This would facilitate the formulation of better strategies as the agripreneur would be more equipped to tackle the uncertain external environment with the competencies and strengths of the internal environment.

7. *Flexible and adaptable.* An agripreneur must be open to changes and be an initiator of change. The entreprenurial approach should always be to do what is best for the business. Thus, an agripreneur must be flexible and adaptable according to the volatile business environment.
8. *Opportunity recognition skills.* An agripreneur must be an information-seeker. Awareness about the surrounding business environment like governmental policies, schemes, and national and international developments in agriculture would enable the agripreneur in recognizing the opportunities beforehand.

4.3 Agripreneurship

4.3.1 Need for agripreneurship

India is primarily called an agrarian economy because it is naturally possessed with an assortment of climates and soil types with about 20 agro-climatic regions. The tropical climate along with a rich biodiversity cover and a teeming livestock of cattle, buffaloes, goats, and sheep add to India's resources that have remained underutilized for years. Indian farmers have not been able to adapt to new developments and still rely on traditional methods of farming. Moreover, due to lack of education and awareness, they predominately focus on land tilling and crop harvesting, which involve physical labor and do not emphasize the selling of their produce, which would more effectively enhance their profitability. Indian society perceives agriculture as an occupation of rural and uneducated people that cannot produce any economic gains. There is a need to make this sector economically viable to increase the economic growth of the nation.

The Indian agriculture sector contributes about 18% to the GDP and provides employment to about 42% of the Indian population. In terms of global agricultural exports, India has ranked eighth, preceded by the European Union, USA, Brazil, China, Mexico, Canada, and Indonesia (WTO (World Trade Organization) World Trade Report 2023, 2023). India dominates the world in the production of rice, wheat, pulses, cotton, sugar, sugarcane, peanuts, tea, spices, fruits, and vegetables. Thus, the agriculture sector has remained a priority for the Indian government and is given due consideration in the yearly budget allocations (Adapa and Yarram, 2022). Recently, there has been a substantial shift in the outlook of agriculture with the widespread creation of infrastructure pertaining to transportation, warehousing, grading, processing, and packaging of agricultural produce. The allied sectors of agriculture are also being given due importance as the policy makers are emphasizing crop diversification in fruits, vegetables, medicinal and aromatic plants and spices, and plantation crops like coconut, cashews, and cocoa. India is the highest producer of milk, and third highest in fish and egg production. Despite such high rankings, India still lacks the basic infrastructural facilities like cold storage and refrigerated transportation.

Thus, the Indian agriculture sector is booming with business opportunities that need to be converted into ventures or enterprises for optimum utilization of such an organically rich sector. Hence, agripreneurship seems to be the chief remedy for enhancing agricultural production and profitability, reducing the burden on the agriculture sector by exploiting opportunities in the allied sectors, generating employment avenues, and developing rural areas.

4.3.2 Challenges in agripreneurship

Agripreneruship integrates entrepreneurship in agriculture. Since the agriculture sector is the most deficient sector in terms of resources, it is quite difficult to foster entrepreneurship in this sector (Sharma and Singh, 2006; Otache, 2017). The agriculture sector poses the following challenges for an agripreneur, as summarized in Fig. 4.4.

1. *Inadequate education and knowledge.* Most of the workers involved in farming or farm-based activities do not have any formal education relating to it. They work on the basis of inherited knowledge of cultivation. This is a major drawback in this sector as the workers do not have the right education and knowledge to extend their activities. This

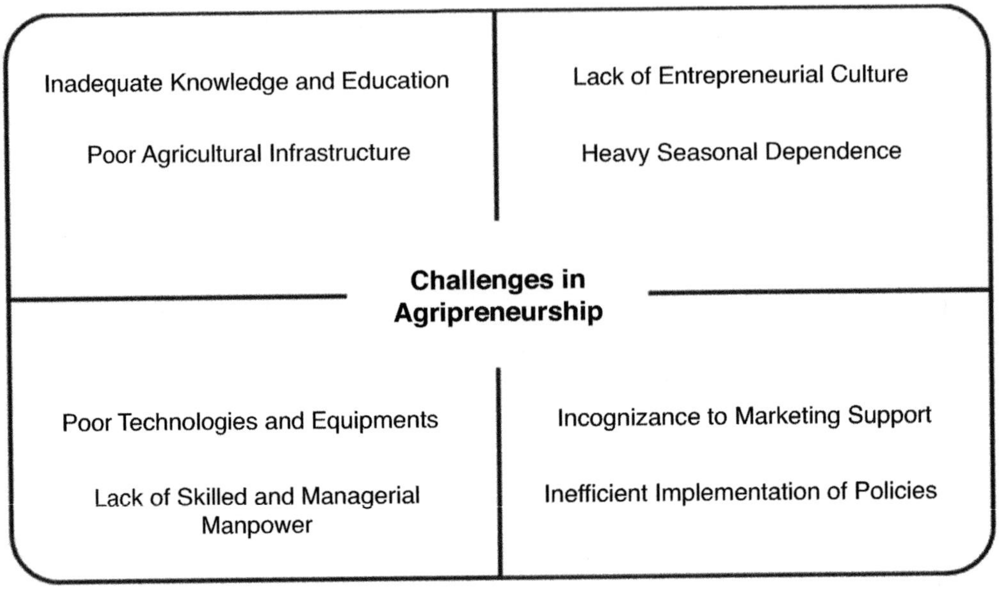

Fig. 4.4. Challenges in agripreneurship.

adds to their lack of awareness about new developments in the sector and thus they keep following the traditional methods and do not opt for the upgraded market place to sell their produce and rely on middlemen.

2. *Poor agricultural infrastructure.* Despite governmental policies and schemes, infrastructure in agriculture still lacks warehousing and transportation. In every season agricultural produce is seen lying in open spaces where it gets spoiled by rain or insects. This leads to huge losses not only to the farmers but also the government as it depreciates the value of that produce. Moreover, inadequate communication, power, and marketing networks obstruct the availability of infrastructure to remote areas.
3. *Lack of entrepreneurial culture.* Agriculture in India is mainly considered as an occupation for rural or uneducated, illiterate people. Most of the students who study agriculture do not pursue it as their main occupation and leave it for white collar jobs. Moreover, the attitude of Indians towards entrepreneurship is also not supportive as a lack of family support, low risk-taking ability, lack of finance, lack of awareness about government policies, and the tendency to stick to traditional ways of livelihood discourage the youth from entrepreneurship. Thus the pooled concept of agripreneurship will take time to flourish given Indian societal norms and culture.
4. *Heavy seasonal dependence.* The major drawback of the agriculture sector is the dependence of crop production on seasons. Any variation in the seasonal cycle disrupts the process of crop production and harvesting. Thus, there is a dire need for research and development in this area to make crops less prone to seasonal variations.
5. *Poor technologies and equipment.* Workers in agriculture and its allied sectors rely on the same traditional methods and are reluctant to use new techniques and equipment. This is mostly due to lack of funds or lack of awareness. However, the government has made considerable efforts in this direction by launching awareness programs and missions to make these tools and equipment available through rent or hire schemes as in the case of Namo Drone Didi. Workers or their employers are usually reluctant to change and show disinterest towards adopting such technologies.

6. *Lack of skilled and managerial manpower.* Another hindrance towards the adoption of new technologies in agriculture is the lack of skilled labor and managerial expertise. The workers who are operating on the farm are mostly uneducated and face difficulties in learning new things. There is also the problem of using equipment properly which can lead to machine breakdowns, which further makes it difficult to hire equipment. Another factor is a lack of managerial ability in the workers. Management encapsulates all the functions of planning, organizing, motivating, and controlling that are evidently necessary for managing agricultural operations and would help in tackling the problems of uncertainty in this sector. Moreover, improved managerial ability would help farmers in storing and retailing their produce more systematically. The lack of such key qualities in agricultural manpower is hampering its efficiency and growth.
7. *Incognizance to marketing support.* Traditionally, the government provided "mandis" or agricultural produce market committees (APMCs) to procure agricultural produce from farmers. This hindered the procurement process as it was not catering to pan-India, and farmers could sell only to their regional APMCs at whatever rate they offered. In 2016, the government formed an e-marketplace called e-NAM that unified the market for the whole country and brought transparency in prices. However, out of the estimated population of about 140 million farmers only 17.7 million had registered on the e-NAM portal by March 2024. This shows the lack of awareness and execution prevailing in the agriculture sector.
8. *Inefficient implementation of policies.* The resistance to change, lack of awareness among the farming community, and inefficient implementation of government policies and schemes are major impediments towards growth in the agriculture and allied sectors. Despite policies promoting horticulture, innovation, marketing, and entrepreneurship in the agriculture sector, the inefficiency of the government's executive branch and the adamant attitude of the agriculture community are hampering the development of the agriculture sector.

4.3.3 Government initiatives in agripreneurship development

The plethora of benefits that come from agripreneurship affect not only the nation but the farmers and rural community as well. After the agriculture sector witnessed transitions due to the green revolution in the 20th century, it is apparent that developments need to be initiated in the 21st century to keep the agriculture sector at par with the technological advancements being witnessed by other globalized economies. Thus, the Government of India introduced the National Agriculture Policy in 2000 to realize the untapped potential of the agriculture sector and to aid in growing the sector through equity and sustainability. This policy was operationalized through various initiatives like agri-infrastructural facilities, technology missions for modernization of agriculture, promotion of cooperatives for better productivity, crop and farmer insurance, and call centers for farmers; agriclinics and agribusiness centers were also established under this policy in 2001–2002.

Moreover, with the launch of the Make in India project in 2014, and the Atal Innovation Mission and Start-Up India initiatives in 2016, entrepreneurship became the main highlight of the Indian economy. The Ministry of Agriculture and Farmers Welfare, Government of India, in collaboration with agricultural institutions like the National Bank for Agricultural and Rural Development (NABARD) and National Institute of Agricultural Extension Management (MANAGE), launched various schemes and programs to promote agripreneurship development in the country. Such initiatives include the following.

1. *MANAGE.* This was an autonomous institute established in 1987 in Hyderabad by the Ministry of Agriculture and Farmers Welfare. This institute was opened with the objective of increasing the level of agricultural technology in order to modernize and reorient the agricultural system. After liberalization and globalization

in 1991, this institute was set up to bring the Indian agriculture sector to be at par with other globalized economies in the world. MANAGE provides education, training, consultancy, research, and information services pertaining to agriculture.

2. *Agriclinics*. Agriclinics were launched to provide expert advice in agriculture by forming a pool of agricultural experts available in the region. They may or may not possess any formal agriculture education but must have ample experience and knowledge that can help other farmers in improving agricultural productivity in the region. These experts in agriculture and allied activities like horticulture, poultry farming, and forestry provide advice regarding crop selection, market trends, pricing, insurance, maintaining sanitary conditions, credit availability, and so on.
3. *Agribusiness centers (ABCs)*. These centers cater to the agriculture and allied sectors by providing paid services. They provide loans up to INR 25 lakh for individual agricultural projects and up to INR 100 lakh for group agricultural projects which can be repaid in 5–10 years. ABCs provide a package of facilities like farm equipment for hire, consultancy services, warehousing, and retailing services.
4. *Mission for Integrated Development of Horticulture (MIDH)*. This is a centrally sponsored scheme set up in 2014 and aimed at growth in the horticulture sector. The horticulture sector covers cultivation of fruits, vegetables, root crops, mushrooms, spices, aromatic plants, flowers, and plantation crops like cashews, cocoa, and bamboo. This scheme aims to provide infrastructure facilities like the establishment of new gardens and orchards or the rejuvenation of old ones, as well as protected cultivation and pollination, support for organic farming, and post-harvest management and marketing assistance.
5. *National Bamboo Mission*. This mission focuses on the development of a complete value chain of bamboo production. It uses a cluster mode approach to link growers with consumers. Under this mission the government has established bamboo nurseries, bamboo plantations, and bamboo processing units along with product development and capacity building units for farmers and artisans.
6. *Sub-Mission on Agriculture Mechanization (SMAM)*. This mission, set up in 2014, aims at attaining a high level of farm mechanization and making it accessible to small and marginal farmers. Under this mission, custom hiring centers have been established as hubs for hi-tech and high-value farm equipment. These centers have made heavy farm equipment easily available for all kinds of farmers, on a for sale as well as a hire basis. These centers also provide demonstrations, capacity building, training, performance testing, and certification of farming machineries and equipment.
7. *Rainfed Area Development (RAD)*. This scheme, established in 2014, uses an area-based approach for performing agriculture and allied activities in cluster mode. Thus, according to the soil health and climatic condition of an area, clusters of workers are formed and an integrated farming system is used where different cropping practices are mixed together for better productivity like multi-cropping and rotational cropping.
8. *Rashtriya Krishi Vikas Yojana (RKVY)*. In this mission, the government provides autonomy to each state to cater to the local needs of farmers in that state. It intends to fill the resource gap in agriculture and allied sectors by providing pre- and post-harvest infrastructure development including marketing facilities. Under RKVY, 1524 start-ups have been selected since 2019 for funding by the state governments.
9. *Agriculture Infrastructure Fund (AIF)*. This fund was launched in 2020 under the Atmanirbhar Bharat Package as a medium- and long-term debt financing facility for investment in agriculture infrastructure to fill the existing resource gap in agriculture and allied sectors. It provides up to INR 1 lakh crore through interest subvention and credit guarantee support for infrastructure projects relating to post-harvest management and community farming. This fund intends to benefit farmers, agripreneurs, start-ups, cooperative societies, state agencies, self-help groups

(SHGs), farmer producer organizations (FPOs), APMCs, and such.

10. *National Beekeeping and Honey Mission (NBHM)*. This mission, set up in 2020, also comes under the ambit of Atmanirbhar Bharat Abhiyan, with the goal of achieving a "sweet revolution" by promoting and developing beekeeping. Under this mission, hi-tech honey testing labs have been established, a Madhukranti portal has been launched for registering beekeepers, cooperative societies, and firms involved in beekeeping, and 160 projects have been sanctioned.
11. *National Mission on Edible Oils – Oil Palm (MNEO)*. This mission was launched in 2021 to make the country self-reliant in edible oil. It promotes oil palm cultivation with a special focus on north-eastern states of India and the Andaman and Nicobar Islands.
12. *Sub-Mission on Seed and Planting Material (SMSP)*. Since 2023, this mission has focused on holistic development of the seed production chain, from seed germination to supply. It provides support to seed-producing organizations to enhance their capacity and quality. The SATHI portal was also launched, known as the Seed Authentication, Traceability and Holistic Inventory, to effectively monitor the seed chain. This mission also founded seed banks to tackle the unforeseen circumstances of natural calamities.
13. *Namo Drone Didi*. Under this scheme, set up in 2024, the central government provides drones to 15,000 selected women's SHGs as a rental service for the application of pesticides and fertilizers. This scheme also aims to provide sustainable livelihoods and business support to SHGs.
14. *Agri-Fund for Start-Ups and Rural Enterprises (Agri-SURE) and NABVENTURES*. This fund was established in 2024 to support start-ups in agriculture and allied sectors through debt and equity financing. It specifically targets high-risk and high-impact activities pertaining to the agriculture value chain to foster innovation and sustainability in the Indian agriculture sector. The fund is set up to encourage IT-based solutions for farming activities to enhance the agriculture value chain, creating rural infrastructure and supporting FPOs. The fund was set up with an initial corpus of INR 750 crore of which NABARD and the Ministry of Agriculture contributed INR 250 crore each and the rest funded by other institutions. This is a wholly owned subsidiary of NABARD, and acts as the fund manager for Agri-SURE.
15. *National e-Governance Plan in Agriculture (NeGPA) – Digital Agriculture*. NeGPA provides easily accessible information services to farmers for crop planning, health, input access, credit, insurance, market intelligence, and support for agri-tech industry and start-ups.
16. *Market Intervention Scheme and Price Support Scheme (MIS-PSS)*. The MIS focuses on the procurement of agricultural and horticultural commodities that are perishable in nature and do not come under the purview of the PSS of the government. Whereas the PSS focuses on the procurement of pulses, oilseeds, and copra the objective of MIS-PSS is to assist farmers in pricing, marketing, and retailing their produce according to the economic conditions.
17. *Paramparagat Krishi Vikas Yojana*. This scheme is meant to promote and develop organic farming culture in India. It uses the cluster mode approach where a group of 20 farmers or more are encouraged to farm without the use of agri-chemicals. It also provides financial incentives to such farmer groups through direct beneficiary transfer.
18. *National Food Security Mission (NFSM)*. This mission aims to increase the productivity of rice, wheat, pulses, and coarse cereals in some specially identified districts across 28 states and two union territories of India. It emphasizes increasing the fertility and production in these areas to improve the farm-level economy and instill confidence among farmers by assisting them in post-harvest value addition.
19. *Integrated Scheme for Agriculture Marketing (ISAM)*. Marketing of agricultural produce is a key hindrance faced by the farmers. ISAM assists state governments in creating and improving market structures, enhancing the access to market information, and capacity building. National Agricultural

Marketing Scheme known as e-NAM is an electronic trading portal covering pan India to provide a unified national market for agricultural produce.

20. *Mission Organic Value Chain Development for North Eastern Region (MOVCDNER).* This mission aims to develop commodity-specific organic value chains in the north-eastern region of India covering the states of Arunachal Pradesh, Assam, Manipur, Meghalaya, Mizoram, Nagaland, Sikkim, and Tripura. The mission focuses on the holistic development of the value chain from the initial stages of inputs and seeds to processing, marketing, and brand building initiatives.

4.3.4 Role of agripreneurship in the national economy

Agripreneurship not only benefits the agriculture sector but also uplifts the rural and urban economies by enhancing employment opportunities and increasing their income levels. Thus, the whole economy stands to benefit from the growth in even one sector. However, the benefits of growth in the agriculture sector due to agripreneurship are manifold (Bairwa *et al.*, 2014) as reflected in Fig. 4.5 and the following.

1. *Employment generation.* Entrepreneurs unravel business opportunities in agriculture sector through agripreneurship. This helps in establishing enterprises that require jobs pertaining to agricultural operations like irrigation, harvesting, etc. This not only creates more job opportunities in the agriculture sector but also helps in curbing the disguised unemployment prevailing in this sector. Thus, unproductive manpower under disguised unemployment can be optimally utilized through agripreneurship.
2. *Balanced regional development.* By tapping into the unexplored agriculture sector, agripreneurs are able to stimulate economic activity in remote and rural areas which

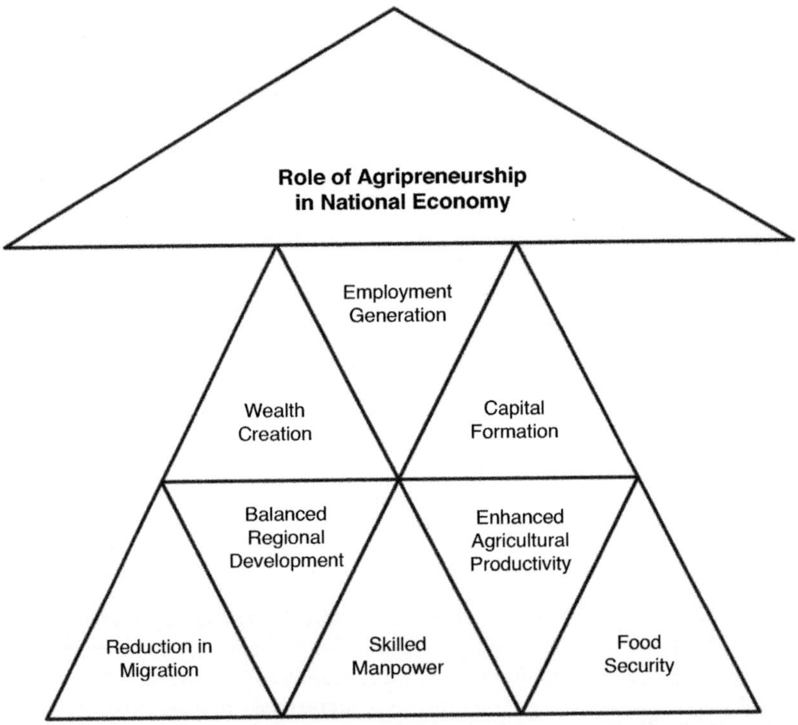

Fig. 4.5. Role of agripreneurship in the national economy.

had been lying idle for many years. This helps in removing regional disparities and increases the contribution of backward areas in national economic growth. Hence, agripreneurship promotes balanced regional development.

3. *Reduction in migration.* Due to the concentration of industries in urban areas and lack of employment opportunities in rural areas, manpower migrates to urban areas, creating a concentration of labor there. Agripreneurship taps into the potential of rural areas and creates job opportunities for the rural population in their homelands. This enables agricultural workers to find employment near their homes and reduces the need for migration to urban areas in search of work. This also helps in eradicating the burden on urban areas due to overpopulation.
4. *Skilled manpower.* The onset of agrienterprises has facilitated the development and training of manpower in rural areas. Agripreneurs are well informed about the kind of skills they require in workers. Thus, workers are educated and trained accordingly to cater to the requirements of agripreneurs. This enables workers to become skilled and gain experience which further improves their employability.
5. *Enhanced agricultural productivity.* Agripreneurship has enhanced the scope of agriculture through research and development. This has facilitated innovation in agriculture through foreign exposure and advanced technological developments by agripreneurs. This has raised agricultural productivity and enhanced the standard of agricultural exports from India in terms of quality and packaging.
6. *Food security.* Agripreneurs have facilitated development of the agriculture sector by providing efficiency in agricultural operations, creation of skilled labor, systematic marketing, and processing of agricultural produce. This has rendered an organized and systematic structure to the agriculture sector where supply and demand have been considerably armored against uncertainties. The advent of agripreneurship ensures sufficient agricultural output for national consumption and exports leading to food security in the nation.
7. *Wealth creation.* Wealth creation means generating income and gains for a long period of time. Agripreneurship has accelerated growth in the agriculture sector by encouraging new businesses in agriculture, generating employment, reducing poverty, and optimizing the utilization of resources that remained unexplored for years like horticulture, sericulture, etc. By diversifying the income of the agriculture sector through various business opportunities, agripreneurs have been able to expedite wealth creation in this sector.
8. *Capital Formation.* Economic activities like entrepreneurship require investments which can be attained by utilizing the unused savings of the public. Thus, agripreneurs induce banks or investors to supply them with funds to incorporate in agrienterprises. This increases the rate of saving and investment in the economy, leading to capital formation, which is a prerequisite for economic development.

4.4 Future Opportunities in Agripreneurship

Indian agriculture has witnessed huge diversification after the immersion of entrepreneurship in this sector. Numerous government initiatives in agripreneurship and the ability of agripreneurs have opened a whole new set of opportunities in agriculture, which now encompasses activities like sericulture, shrimp farming, floriculture, fisheries, etc. More opportunities can be identified at different phases of the agricultural value chain like input manufacturing or processing, providing services related to farming activities like harvesting and cultivation, or improving the logistics and marketing of farm produce (Nwibo *et al.*, 2016). Further, agripreneurs can establish forward or backward linkages with the manufacturing and service sectors like planting coffee beans for coffee powder or jute for jute bags and so on. Thus, the agriculture sector has ample opportunities for agripreneurs; keeping in mind the institutional support by the Indian government (Sindhu *et al.*, 2016), Fig. 4.6 depicts the opportunities worth exploring.

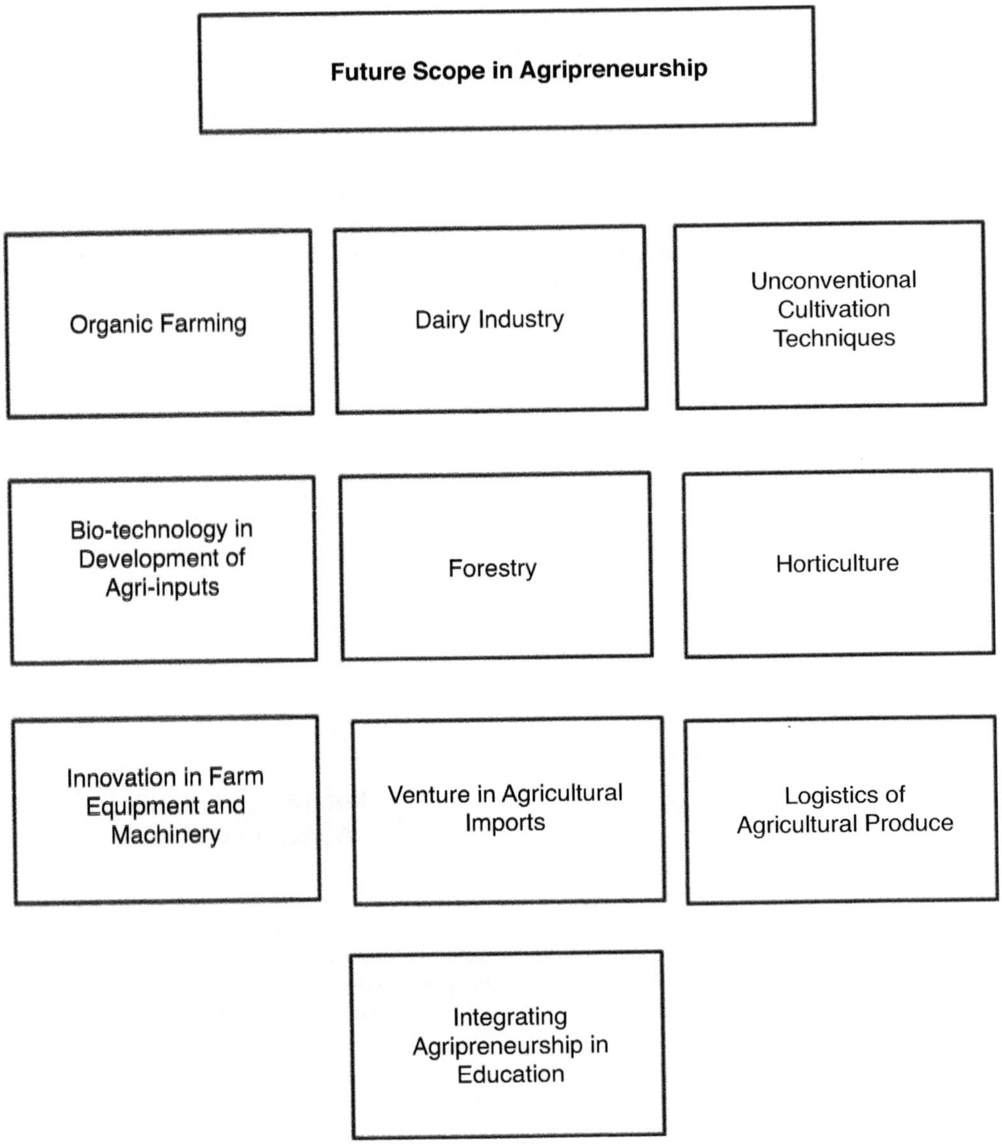

Fig. 4.6. Future scope in agripreneurship.

1. *Organic farming*. Organic farming is gaining popularity with the increasing consumer awareness of the negative effects of fertilizers and pesticides on the human body. Thus with government schemes like MOVCDNER and Paramparagat Krishi Vikas Yojana, agripreneurs can take advantage of such opportunities and convert them into viable business ventures.
2. *Dairy industry*. India is the highest producer of milk in the world, and dairy products are an integral part of the daily diet of an Indian household. Thus, ample opportunities can be discovered in the dairy industry by exploring the forward and backward linkages with the manufacturing or service sectors. Agripreneurs can provide ancillary services to dairy manufacturers

and cooperatives in the form of cattle care, sanitation in the dairies, cattle insurance, storage, marketing, or branding of the milk and milk products.

3. *Unconventional cultivation techniques*. There are various government schemes that promote and support unconventional cultivation methods in agriculture. Developing micro-irrigation techniques to reduce the dependence of agriculture on water or rainfall can be explored by taking the help of government schemes like the Micro Irrigation Fund and Per Drop More Crop (PDMC).
4. *Biotechnology for development of agri-inputs*. Agri-inputs include seeds, fertilizers, and pesticides apart from farming tools and machinery. Biotechnology can be used to develop a variety of seeds, fertilizers, pesticides, and other agri-inputs that can help in enhanced productivity, less chemically dependent crops, natural manure, and hybrid or genetically modified seeds adaptable to different Indian soil types and climatic conditions. Further schemes like SMSP can facilitate agripreneurs developing the seed production chain through innovations in the seed production and distribution stages.
5. *Forestry*. India has the tenth largest forest cover in the world and secured the third spot in the world for forest area gain. The government of India has been actively launching schemes for entrepreneurs to tap into this abundant natural resource of the country. The National Bamboo Mission emphasizes the development of a complete value chain for bamboo, thus opening opportunities for business. Moreover with the integration of the agroforestry scheme with the Rashtriya Kisan Vikas Yojana, the government aims to promote plantations on farmlands for diversifying the livelihood options of farmers.
6. *Horticulture*. Horticulture comprises the cultivation of fruits, vegetables, spices, flowers, aromatic plants, and plantation crops like cashew, cocoa, bamboo, etc. With the government's MIDH initiative, specific infrastructure development for horticulture is provided along with post-harvest management and marketing assistance. Thus, agripreneurs can exploit opportunities in horticulture as these are also major exports of India.
7. *Innovation in farm equipment and machinery*. Under the AgriSURE and Agriculture Infrastructure Fund, the government aims to make India self-reliant by filling the infrastructure gaps in agriculture. The scheme is meant for farmers, agripreneurs, start-ups, FPOs, etc. who can get funding of up to INR 2 crore for viable projects to build post-harvest management infrastructure and community farming assets.
8. *Logistics of agricultural produce*. The core deficiency of the Indian agriculture sector is the lack of storage facilities for agricultural produce. Thus, agripreneurs can tap into this stage of the agricultural value chain and provide farmers with warehousing facilities that will help them in avoiding the huge losses they suffer from the storage of produce in open spaces leaving it vulnerable to rain and insects. Moreover, with the advent of the food processing industry, cold storage and refrigerated transportation are the key deficiencies in the logistics of agricultural produce.
9. *Venture in agricultural imports*. The Indian government has been placing great emphasis on the Atmanirbhar Bharat Abhiyan since 2014. Since, vegetable oil is the largest imported agricultural commodity of India, especially palm oil, the Ministry of Agriculture and Farmers Welfare launched MNEO in 2021. It intends to promote cultivation of oil palm to reduce the reliance on imports and focuses on the north-eastern regions and Andaman and Nicobar Islands for plantation projects. Thus, agripreneurs can explore this as a business opportunity.
10. *Integrating agripreneurship in education*. There is a crucial need to change the attitude of Indians towards agriculture so that more educated and skilled people can consider it as a career option. This can be done by associating agriculture with entrepreneurship. Thus institutes providing training for entrepreneurship must include agripreneurship also. Moreover, the government has also established various training institutes in

agriculture like MANAGE in Hyderabad which should also include entrepreneurship in their curriculum. In addition to this, the Sub-Mission on Agriculture Extension emphasizes disseminating technology to farmers because in order to modernize the agriculture sector, the agriculture community also needs to be well versed with the new technologies.

4.5 Case Studies

This section gives an account of some agripreneurs who have successfully exploited the opportunities prevailing in the Indian agriculture sector and have benefited farmers and villages and subsequently improved economic growth. Table 4.1 shows the development of agripreneurship in India by depicting the government initiatives and agripreneurs who made use of those initiatives. The case studies below are just a few of the thousand inspiring agripreneurs in India (MANAGE (National Institute of Agricultural Extension Management), 2018).

4.5.1 Shanti Agro Tech, Andhra Pradesh: seed production and agri-consultancy

This company was founded by Mr Nageshwar Rao with a loan of INR 10 lakh from the State

Table 4.1. Status of agripreneurship development in India, with individual examples.

Type of agripreneurship	Agripreneurs in India	Government initiatives
Farm-level producers	NV Eco Farm: agritourism JK Agro: bee-keeping ShivShambhu Agro Search: bamboo cultivation RK Patel Agriclinics and Agribusiness Center: floriculture with marigold	National Beekeeping and Honey Mission (NBHM), 2020 National Bamboo Mission Mission for Integrated Development of Horticulture (MIDH), 2014
Service providers	Selvaperumal Paddy Harvester: custom hiring center Lakhmai Foundation Soil and Testing Laboratory: soil testing Rudrani Irrigation and Agro Services: agri-inputs installation Skyline Nutrition Pvt. Ltd: provides veterinary aids	Sub-Mission on Agriculture Mechanization (SMAM), 2014 Agriculture Infrastructure Fund (AIF), 2020
Input producers	Shanti Agro Tech: seed production NK Marine PCR Lab: shrimp breeding Kakade Agro Equipment: manufacturing agriculture machinery Ganpati Cattle Feed: animal feed producer Agriculture Development Foundation: eco-friendly pest trap Unique Hybrid Seeds Pvt. Ltd: seed production P&G Agro India: manufactures biofertilizers and micronutrients Big Farmer: farm inputs for apple cultivation	Sub-Mission on Seed and Planting Material (SMSP), 2023 Paramparagat Krishi Vikas Yojana
Processing and marketing cooperatives	Hindustan Bio-Coal: processing agriculture waste Shree Balaji Milk Dairy: selling dairy products MAX Agritech Pvt. Ltd: exporting gherkins Agrikart Co.: online marketplace	Rashtriya Krishi Vikas Yojana (RKVY), 2019

Bank of India and 36% subsidy support from NABARD. He established this company by setting up a composite agricultural development project in 6 acres of land. He has grown around 20 crop varieties and sells them to a network of six distributors across various districts of Andhra Pradesh. He employs 24 male and 10 female skilled workers and trades seed to 500 farmers across 15 villages. Nageshwar is a post-graduate in genetics and plant breeding and got training from the Participatory Rural Development Initiatives in Hyderabad under the Agri Clinics and Agri Business Centres (AC & ABC) Scheme. The company generates a turnover of INR 1.6 crore annually.

4.5.2 Selvaperumal Paddy Harvester, Tamil Nadu: custom hiring center

After getting training from the National Agro Foundation under the AC & ABC Scheme, Mr Selvaperumal decided to purchase a paddy combined harvester from his own savings and borrowing from relatives. He provides the harvester to 600–800 farmers across 45–50 villages to earn an annual turnover of INR 1.3 crore. Being the sole person running the custom hiring center, his monthly expenses include diesel expense of INR 90,000 and maintenance charges of INR 18,000. His monthly profit is around INR 115,000 during the harvesting season.

4.5.3 Hindustan Bio-Coal, Maharashtra: manufacturing biofuel to substitute coal

Manufacturing industries primarily use coal and charcoal, which cause a lot of pollution. With the advent of sustainability in business, industries are demanding alternatives or substitutes that would be less harmful for the environment. Mr Swapnil Satappa Patil discovered that using biomass briquettes for heating instead of coal decreased pollution levels. During his postgraduate degree in food science and technology, he visited a bio-briquette manufacturing unit and was inspired. Later, he got training from Krishna Valley Advanced Agriculture Foundation under the AC & ABC Scheme and a bank loan of INR 5 lakh and opened his own manufacturing unit. He employs four people and earns INR 35 lakh annually. He has collaborated with over 250 farmers across 12 villages for purchasing biomass waste like rice husk, ground nut shells, municipal solid waste, and other agricultural waste. Currently he supplies these biomass briquettes to three factories.

4.5.4 NK Marine PCR Lab, Tamil Nadu: shrimp breeding

Mrs K. Nirmala is a graduate in fishery sciences and got training at the Community Awareness Rural Educational Trust (CARE), Tamil Nadu. She opened the NK Marine PCR Lab to test shrimp eggs. She provides clinical services, recommends probiotics and chemicals for shrimp diseases, and assists 500 farmers in shrimp farming in the Nammakal district of Tamil Nadu. Her lab has become the backbone of that village. She employs 5 permanent and 25 unskilled laborers and earns INR 1 crore in a year.

4.5.5 Kakade Agro Equipment, Maharashtra: manufacturing farm machinery

Mr Subhash Kakade successfully tackled the main problem hindering farm mechanization in India. Being an agricultural engineer, he designed low-cost and user-friendly machinery to assist in farming operations as he believes that "when machines are used along with traditional methods in agricultural operations, then farm mechanization is partial and most acceptable." Thus, after attaining training from the Krishna Valley Advanced Agriculture Foundation, he set up his farm machinery manufacturing unit. He designed a sugarcane planter that ensures uniform distribution causing higher germination. Next he designed a power tiller to remove weeds and loosen soil that was so lightweight that women farmers could also operate it easily. With these inventions, he serves around 1000 farmers and employs five people and makes an annual turnover of INR 40 lakh.

4.5.6 Ganpati Cattle Feed, Maharashtra: producing animal feed

Mr Manoj Bharat Gore has a diploma in agriculture and after getting training in entrepreneurship from Shriram Gramin Sanshodhan Va Vikas Pratishthan, he got involved in manufacturing feeds for dairy cattle. He invested INR 15 lakh to establish a processing unit for animal feed and collaborated with 150 farmers for the procurement of green and dry fodder. He then contracted with nine retailers; his two highest selling products are mineral mixture and cattle cakes. He employs six people and creates an annual turnover of INR 50 lakh.

4.5.7 Lakhmai Foundation Soil and Testing Laboratory, Maharashtra: soil and water testing lab

After receiving training under the Krishi Vigyan Kendra, Mr Nikhil Chaware, a graduate in agriculture, was motivated to open a soil and water testing lab in his region. After attaining the entrepreneurial training under the AC & ABC Scheme, he opened the lab and caters to 4000 farmers across 23 villages and provides consultation for them to use compost, manures, and micronutrients to improve soil fertility. He employs four people and earns INR 5 lakh in turnover annually.

4.5.8 Agriculture Development Foundation, Haryana: eco-friendly pest trap

This foundation was started by Mr Sardar Singh who is an agricultural graduate and retiree from the State Department of Agriculture, Haryana. After attending the entrepreneurial training program at the Indian Society of Agri-Business Professionals, Karnal, he got inspired to commercially manufacture eco-friendly pest traps that can also lead to natural farming practices. He used pheromone attractants to lure insects onto the trap with a simple low-cost technique and made it into a financially viable business employing six people and generating a turnover of INR 1 crore by catering to 1000 farmers across 150 villages.

4.5.9 Shree Balaji Milk Dairy, Haryana: dairy and dairy products

Mr Mohanlal was a diploma holder in agriculture engineering and attended the entrepreneurial training at the Indian Society of Agri-Business Professionals, Karnal. On a visit during training to the National Dairy Research Institute, Karnal, he found out about scientific methods to add value to basic dairy products. After networking with 120 dairy farmers in the region, he collected 400 liters of milk and processed it further to produce curd, ghee, paneer, and chenna in order to add value. He still maintains that the key behind a successful business is the quality of the products and sufficient supply. He works with three people and earns INR 25 lakh in a year.

4.5.10 Unique Hybrid Seeds Pvt. Ltd, Uttar Pradesh: seed production and processing

Mr Harjeet Singh began with seed production on about 16 acres of land and sold his seeds to 100 farmers on a buy back basis. These further germinated into 1500 kg of seeds with genetic purity and were sold again. He has been awarded by NABARD for producing high-quality seeds. Harjeet is a postgraduate in seed technology and after being trained as an agripreneur from the Centre for Agricultural and Rural Development, Uttar Pradesh, he ventured into production of high-quality seeds of paddy, mustard, pulses, wheat, and vegetables. His core strategy is to produce the seed under standardized and well-organized conditions.

4.5.11 Rudrani Irrigation and Agro Services, Maharashtra: plasticulture

Mr Konark Thakur is a trained entrepreneur from Krishna Valley Advanced Agriculture Foundation under the AC & ABC Scheme. His shop provides a one-stop solution for all types of plasticulture works needed in agriculture like irrigation sets, greenhouses, shed nets, PVC pipes, soil and water testing kits, solar products, landscaping services, and other agricultural machinery. He serves 700 farmers across 100 villages and earns INR 75 lakh in turnover and provides employment to five skilled and ten unskilled workers.

4.5.12 MAX Agritech Pvt. Ltd, Tamil Nadu: gherkin production and export

This company was founded by Mr D. Inbasekaran who was in retail management and then joined the Voluntary Association for People Service (VAPS), Tamil Nadu, to realize his dream of creating an agrienterprise. Due to his trusted relationship with the farmers and well-organized production strategies he was able to convince 1500 farmers to associate with him in the production of gherkins, which were in great demand in foreign countries. The company employs 19 skilled and 100 unskilled workers and generates an annual turnover of INR 10 crore. Thus, he was able to establish a financially viable enterprise for the production and export of gherkins.

4.5.13 NV Eco Farm, Goa: agritourism

Mr Mahesh Patil is a postgraduate in agriculture and management and after taking a 2-month course with the Krishna Valley Agriculture Advance Foundation, he thought of combining agriculture with adventure. He developed a farm with modern as well as traditional facilities like the old banyan trees, 125 forest species, adventurous activities like zip-lining, a Burma bridge, trekking, rappelling, wall climbing, etc. Besides this they have bamboo houses, cottages, tent houses, and a spice farm for farm exposure. With ten people employed on the premises, NV Eco Farm makes an annual turnover of INR 20 lakh and covers about 300 farmers.

4.5.14 P&G Agro India, Maharashtra: manufacturing of biofertilizers and micronutrients

Mr Rajkumar Purad is involved in offering organic solutions for farming activities. His company manufactures biofertilizers and micronutrients for all types of soil and crops, selling to about 45 retailers. The firm covers 450 farmers across 25 villages and provides employment to eight people, rendering an annual turnover of INR 80 lakh. He also received training at the Krishna Valley Advanced Agriculture Foundation and started with INR 10 lakh to form a portfolio of 15 product types to cater to different soil types and crops.

4.5.15 Agrikart Co., Karnataka: online selling of organic food

This is an e-commerce company that sells organic food grains like cereals, pulses, millets, and spices. All the products are procured directly from 40 farmers across seven villages through contract farming. Under this kind of farming the agripreneur, Mr Rakesh Sindagi, provides all the technical information with farming inputs to the farmers, and the products are then certified as being organic and are marketed by the e-commerce company in its name. Thus, the products from such farming are directly procured from the farmer's fields and go straight away to the customer through the online platform. Rakesh is an agricultural graduate and attended entrepreneurial training at Shriram Gramin Sanshodhan Va Vikar Patishthan and currently employs two people and generates INR 2 lakh in annual turnover.

4.5.16 Big Farmer, Jammu and Kashmir: farm inputs for apple cultivation

Mr Mir Gohwar is a postgraduate in agricultural sciences and having worked in agri companies for 5 years, he had sufficient experience and motivation to start his own venture. After joining the training program of the Indian Society of Agri-Business Professionals, he thought of starting a farm input shop for apples as Srinagar is the hub for apple production. He caters to 5000 farmers across 100 villages by providing them with complete scientific knowledge on apple cultivation, pests, diseases, and their identification and control. He collects an annual turnover of INR 1 crore with a workforce of only two people.

4.5.17 Skyline Nutrition Pvt. Ltd, Karnal: provider of veterinary aids

This company was started by Dr Ajay Tomar who had been part of the training program at the Indian Society of Agri-Business Professionals, Karnal and also holds a doctorate in animal nutrition. The company supplies veterinary products in eight states of India and has clients abroad also. The company manufactures and markets veterinary products, medicines, nutritional supplements, and feed supplements using the latest technology and equipment. It employs 13 people and generates an annual turnover of INR 1.5 crore.

4.5.18 JK Agro, Karnal: bee-keeping and training center

This company was started by Mr Jai Kunvar Sheoran who is a retired officer from the agriculture department in Haryana. After getting training from the Indian Society of Agri-Business Professionals at Karnal, he ventured into the bee-keeping business across Haryana, Rajasthan, and Himachal Pradesh. Moreover, as a bee-keeping consultant, Jai Kunvar also runs an integrated bee-keeping development center sponsored by an Indo-Israel project. JK Agro supports 500 farmers across 50 villages with a workforce of five people and turnover of INR 50 lakh.

4.5.19 ShivShambhu Agro Search, Maharashtra: bamboo cultivation

Bamboo cultivation is quite underestimated in India despite its huge benefits, but Mr Nilesh Dattatray Nandre was able to tap into it. Bamboo is in high demand in the real estate sector, paper industry, and handicraft industry. Another advantage of growing bamboo is that it does not require any fertilizers or pesticides and its leaves act as a good feed for animals. Nilesh is a graduate in agriculture and gained entrepreneurial skills at Maharashtra Industrial and Technical Consultancy Organization (MITCON), Pune in order to establish this company. The company procures the bamboo produce from 350 farmers from nine villages and provides them with complete consultancy for cultivation. With a staff of eight, the company has been making a turnover of INR 25 lakh annually.

4.5.20 RK Patel Agriclinics and Agribusiness Center, Uttar Pradesh: floriculture with marigold

After observing the round-the-year demand for marigold, Mr Ramesh Kumar Patel thought of building a farm input and consultancy shop for the marigold crop. Being an agricultural graduate, he attended entrepreneurial training at Shree Maa Guru Gramodhoyag Sansthan, Varanasi. He began by selling seeds and pesticides and later concentrated all his efforts on the promotion of floriculture and started selling seedlings of marigold along with the required agri-inputs to facilitate 250 farmers from eight villages to earn a turnover of INR 30 lakh in a year.

4.6 Conclusion

Agripreneurship development in India has provided a platform to those individuals who have a quest for knowledge in agriculture and the risk-taking ability to convert it into a viable business. This kind of revolution was much needed in the agriculture sector as it has the potential to expedite national economic growth while simultaneously providing employment and income to a larger, vulnerable section of the society. The agriculture sector suffered with an array of limitations, some were inherent like seasonal dependence whereas some were man-made like the reluctance to utilize new methods and techniques. However, the Ministry of Agriculture and Farmers Welfare, Government of India has been mildly successful in eradicating these hindrances through effective ground-level initiatives which has ensured holistic agripreneurship development in the country within a timespan of just 10 years. The initiatives launched by the government comprehensively address the various issues concerning the agriculture sector: (i) the establishment of funds for financing infrastructure development in agriculture; (ii) launching widespread missions promoting and encouraging farmers to develop allied sectors of agriculture like horticulture, apiaries, dairy farming, bamboo cultivation, and mushroom plantations; (iii) providing digital portals for marketing and other assistances; and (iv) building an intensive network of training centers that can reach the remotest farmer. The AC & ABC Scheme has been the most successful in reaching the farmers, making them entrepreneurial by training them to build on the local resources. Maharashtra, Tamil Nadu, Haryana, and Uttar Pradesh have been the centers for agripreneurship development. Further, financial assistance by NABARD and other commercial banks have accelerated this development. Thus, the initiatives of the Indian government have been commendable as they have emphasized giving education, training, auxiliary services, and avenues for diversification of income for agricultural communities. The agriculture sector should have all these resources just as much as the industrial and service sectors in order to improve its productivity and profitability.

References

Adapa, S. and Yarram, S.R. (2022) Interplay of contextual dimensions – evidence from Agripreneurs and Agriventures in India. *International Journal of Business and Economics* 21, 177–192.

Bairwa, S.L., Lakra, K., Kushwaha, S., Meena, L.K., Kumar, P. *et al.* (2014, March) Agripreneurship development as a tool to upliftment of agriculture. *International Journal of Scientific and Research Publications* 4. Available at: https://www.ijsrp.org

Boye, M., Ghafoor, A., Badar, H. and Ali, A. (2022, December 23) An understanding of the Agripreneurial knowledge and motivation among gambian young Agripreneurs. *International Journal of Management Research and Emerging Sciences* 12(4), 36–57. DOI: 10.56536/ijmres.v12i4.286.

Chand, K. K. (2019) Agripreneurship: A tool for economic development of India in the new millennium. *International Journal on Recent Trends in Business and Tourism* 3(4), 19–25.

Dollinger (2003) *Entrepreneurship: Strategies and Resources*, 3rd ed. Prentice Hall, United States of America.

Garima, N., Dhingra, A., Centobelli, P. and Cerchione, R. (2023, April 24) Factors and activities considered by first generation agripreneurs for agri-business sustainable development: A study of haryana, India. *Sustainability* 15(9), 7109. DOI: 10.3390/su15097109.

MANAGE (National Institute of Agricultural Extension Management) (2018) *Enterprising Agripreneurs in Rural India*. National Institute of Agricultural Extension Management (MANAGE), Hyderabad, India.

Nwibo, S.U., Mbam, B.N. and Biam, C.K. (2016) Department of agricultural economics, management and extension, Ebonyi State University, Abakaliki, Nigeria, and department of agricultural economics, university of agriculture, Makurdi, Benue state, Nigeria. "Determinants of Agripreneurship among the Rural Households of Ishielu Local Government area of Ebonyi State". *Journal of Biology, Agriculture and Healthcare* 6, 3–10.

Otache, I. (2017) Agripreneurship development: A strategy for revamping Nigeria's economy from Recession. *African Journal of Economic and Management Studies* 8(4), 474–483. DOI: 10.1108/ajems-05-2017-0091.

Sharma, J.P. and Singh, R. (2006, December) Developing agripreneurs: Challenges for extension. *Journal of Community Mobilization and Sustainable Development* 1. Available at: https://www.IndianJournals.com

Sindhu, K., Gopi Krishna, T., Rambabu, P. and Reddy, G.R. (2016) Constraints faced by Agripreneurs in managing agro based enterprises and suggestions to overcome. *The Andhra Agricultural Journal* 63, 978–981.

Sreedhar, Ch.J. and Siva Sri Mani Kanta, T. (2022, October 25) Problems and prospectus of agripreneurs-an empirical study. *International Journal of Innovative Research in Engineering & Management* 210–215. DOI: 10.55524/ijirem.2022.9.5.29.

Uplaonkar, S. and Biradar, S. (2015, September 1) Development of agriculture in india through Agripreneurs. *International Journal of Applied Research* 1(9), 1063–1066. Available at: https://www.allresearchjournal.com

Usanga, U.J. (2012) Developing entrepreneurial mindset and skills. *Sustainable Agriculture: Challenges and Prospects* 1.

WTO (World Trade Organization) World Trade Report 2023 (2023) Re-globalization for a Secure, Inclusive and Sustainable Future. WTO.

Useful Websites

https://agriwelfare.gov.in
https://indianexpress.com/article/opinion/clumns/revealing-indias-actual-farmer-population-7550159/lite/
https://pib.gov.in
https://pib.gov.in/PressReleaseIframePage.aspx?PRID=2002012
https://pib.gov.in/PressReleaseIframePage.aspx?PRID=2032838
https://www.agriclinics.net/AboutScheme.aspx
https://www.fao.org
https://www.manage.gov.in/aboutUs/ourOrganization.asp
All websites last accessed October 2025

5 Harvesting Innovation: Women Agripreneurs Revolutionizing with Technology

Sumathi Annamalai[1]*, Deepika S.R.[1], Angappa Gunasekaran[2] and Annet H. de Lange[3]

[1]School of Business and Management, Christ University, Bangalore, India; [2]School of Business Administration, Penn State Harrisburg, Middletown, Pennsylvania, USA; [3]Open University, Faculty of Psychology, Heerlen, the Netherlands

5.1 Introduction

In today's scenario, the business world is inherently vulnerable, requiring entrepreneurs to develop and implement effective strategies to navigate uncertainties. Key elements such as diversification, sustainability, and technological advancement necessitates constant evolution of new business models for entrepreneurs to remain competitive and successful. The entrepreneurial landscape is significantly shaped by rapid technological advancements, with AI and digital platforms driving innovation and scalability (Giuggioli and Pellegrini, 2023). Sustainability and social impact have become central as entrepreneurs need to increasingly adopt business models prioritizing environmental and social responsibility (Rosário and Figueiredo, 2024). Along with these factors, access to resources, market conditions, and personal resilience significantly influence entrepreneurial experiences and shape how their businesses adapt and thrive amidst challenges.

Entrepreneurship has been a fundamental aspect of human history for centuries, serving as a cornerstone of economic development and innovation and has been recognized as a key driver of wealth creation (Loza de Siles, 2011), characterized by the ability to take calculated risks in the pursuit of new opportunities. Through innovation and strategic risk management, entrepreneurs have historically transformed industries and generated economic growth. However, it was not until the early 1960s that entrepreneurship began to emerge as a formally studied business phenomenon, gaining recognition in academic and professional circles (Bowen and Hisrich, 1986).

Furthermore, female entrepreneurship has been largely overlooked in academic literature (Buttner and Moore, 1997), with little research dedicated to understanding the unique challenges and contributions of women entrepreneurs. Much of the literature, even until the late 1980s and early 1990s, focused solely on male entrepreneurs, often failing to differentiate study participants by gender. When women were included, studies typically involved small samples of professional women or had other limitations that hindered a comprehensive

*Corresponding author: sumathi.a@christuniversity.in

 Agripreneurship, Digital Inclusion and Sustainability: Rural Innovation and Growth in the Global South (eds Dutta and Trivedi)
DOI: 10.1079/9781836992646.0005

understanding of women's entrepreneurship (Buttner and Rosen, 1988). Only recently has this oversight started to be addressed, which has resulted in an increasing body of work highlighting the significance and influence of female entrepreneurs in the global economy. Their accomplishments challenge societal norms and create opportunities for future generations, inspiring other women and accelerating broader changes in society that lead to gender equality. Simultaneously, systemic obstacles such as gender bias, restricted funding, and a dearth of mentorship opportunities present particular difficulties for women entrepreneurs (Isaga, 2019). The difficulties they face are frequently exacerbated by cultural norms and societal expectations, which can limit their capacity to manage work and personal obligations.

Combining the terms "agriculture" and "entrepreneurship," the term "agripreneurship" first appeared in the late 20th and early 21st centuries, and is still developing today (Otache, 2017). It developed in response to the agricultural industry's increasing demand for business savviness. With the increasing urgency of issues such as climate change and food security, the idea of the "agripreneur" emerged to describe the role of those who turn farming into a profitable and dynamic industry by bringing entrepreneurial principles to it. The phrase captures the transition from conventional farming to a more contemporary, market-driven strategy that prioritizes sustainability and financial viability.

In the words of Aleke *et al.* (2011, p. 70), "an agripreneur is a business owner who is self-employed and seeks to create wealth within the agricultural industry." According to Carr and Roulin (2016), "an agripreneur identifies viable business opportunities in agriculture, gathers resources, establishes and manages the resulting agricultural enterprise." In simple words, they are an individual who owns an agricultural business (agribusiness) (Bose, 2013; Nagalakshmi and Sudhakar, 2013). Agripreneurs are also called "agripreneurial farmers" or "agricultural entrepreneurs" (Otache, 2017).

Similar to the scenario of limited research studies among women entrepreneurs, women agripreneurs are also mostly ignored in academic literature, which often prioritizes the experiences of male agripreneurs and fails to adequately recognize the unique contributions and challenges faced by women in agriculture. This gap is driven by historical biases, underrepresentation of women in formal agricultural sectors, and a lack of gender-disaggregated data. As a result there is a lack of adequate literature focusing on women agripreneurs. This has led to a scenario where the current policies and existing research evidence do not fully support their specific needs, which continues to increase gender disparities in agriculture. Hence, the current scenario highlights the need for more gender-sensitive research studies.

With this context in mind, our goal is to learn more about the inspirational stories of women agripreneurs and the driving forces behind their success, passion, and motivation. This chapter looks at the creative methods they have so far employed, the sustainable methods they have found success with, and how those methods have influenced their business ventures. The chapter also discusses the primary obstacles faced by these women agripreneurs and the strategies they use to overcome them. For legislators, other stakeholders, and aspirational agripreneurs, this chapter provides insightful information. These observations will be helpful in guiding the stakeholders in utilizing technology to promote agricultural advancement. By means of this investigation, the chapter hopes to highlight the noteworthy influence of women agribusiness owners and their part in cultivating a future in agriculture that is both equitable and sustainable. In order to write this book chapter, the authors located these agripreneurs, asked them to participate in the study, and shared their experiences.

5.2 Methodology

Participants are an opportunity sample of agripreneurs who live in remote locations in Tamil Nadu and are active on social media platforms in doing their businesses. The participants were recruited by locating them through social media platforms and three inclusion criteria were considered: they must be agripreneurs, women, and operating from any village or town located

Box 5.1. Vignette of the women agripreneurs who participated in the interviews.

Participant 1: She is in her early 20s, started honeybee farming and honey business following her college days. She is a passionate agripreneur, hailing from Karupagoundanpalayam, Thiruchengodu, Namakkal district in Tamil Nadu.
Participant 2: She is in her mid-20s, a young mother, and operates from Salem district in Tamil Nadu. Her business venture markets seven varieties of millet biscuits with 100% natural ingredients.
Participant 3: She operates her business venture from Vanjipalayam, Iduvai, in Tiruppur district. She focuses on value-add products from gooseberries, sapota, and guava cultivation.
Participant 4: She operates from Kanchikovil village in Erode district. Her business venture focuses on collecting native seeds, producing more seeds, and marketing them. She is an engineer and IT employee turned agripreneur.
Participant 5: She operates from Konalam village, Tiruvallur district. Her business venture focuses on agritourism, centered around her farm, with 200 types of trees, rearing donkeys, camels, cattles, etc.
Participant 6: A project which started with making goat milk soap to avoid allergic effects on the skin for her family member, has transformed into a successful business venture today. She is from Gobichettipalayam, Erode district, Tamil Nadu.
Participant 7: She operates from Nochipatty village in Karur district, Tamil Nadu. She cultivates moringa drumstick on her farm and using that she makes around 32 value-added products like moringa leaf soup powder, moringa seed oil, moringa flower powder, honey, moringa soap, etc.
Participant 8: She operates from Anuparpalayam village, and does integrated farming. She exports turmeric powder with high curcumin content (mentioned on the pack itself), to several foreign countries and operates several agribusiness ventures.
Participant 9: She is a polymer chemistry scientist with a successful career in Europe, who returned to India to become an agripreneur. She operates near Dharapuram, a small town in Tamil Nadu and has more than 400 trees on her farm. Her focus is currently on value-added products and plans to venture into agritourism.

in Tamil Nadu. Box 5.1 provides the vignette of the participants' profiles. Informed consent was sought before they were interviewed. The interviews briefly covered their agripreneurial journey, triggers and drivers in becoming agripreneurs, the sustainable business practices they are adopting, the challenges they face, and the mitigation strategies they are using. Data was coded using QDA Miner Lite software. The study employed a reflexive thematic analysis framework, following Braun and Clarke's six-step process to develop themes.

5.3 Passion, Drive, and Motivation

The journey of women into agripreneurship, fueled by a passion for sustainability and innovation, is driven by a blend of personal experiences, economic necessity, and a deep-rooted connection to the land. Investigating past studies reveals an intriguing finding: a mix of extrinsic and intrinsic factors motivate women agripreneurs to start businesses. According to Surainita *et al.* (2022), intrinsic motivation encompasses psychological traits and aptitudes like a strong desire and aptitude to support economic growth and broaden creative opportunities.

In India, the rise of women agripreneurs has been fueled by government schemes and programs that offer extrinsic motivation (Kumar *et al.*, 2012). Extrinsic factors – namely government support and the acknowledgement of women's significance in economic development – cannot be undervalued, even while intrinsic factors like individual drive and aptitude play a big part. According to Kumar *et al.* (2012), there are push and pull factors that motivate women to launch their own businesses and provide financial support for their families. The broader economic and social development of the nation as well as the empowerment of women depend on these motivating factors (Kumar *et al.*, 2012; Shetty and Hans, 2019; Surainita *et al.*, 2022).

From the women agripreneurs involved in this study, similar motivational and drive factors emerged. Their ambitions for self-expression, financial security, ecological responsibility, and communal welfare serve as their driving forces. The agripreneurs are more empowered to investigate such creative endeavors because of their access to digital technologies and the networks of support they have gained from them. They resist social norms and pave the way for the development of a more sustainable agricultural future via their endurance and adaptability.

5.3.1 Identity and respect: a motivational driver

A major source of motivation for women agripreneurs is the need for a unique identity and recognition. By following their passion, they can become respected and recognized for their contributions while making a distinctive mark in a field that has historically been dominated by men. Living in a remote village and practicing integrated farming through natural processes, participant 8 recalls that she feels very happy and respected when elderly farmers ask for her advice, follow it, and put it into practice. Their search for self, propels them to innovate and embrace environmentally conscious practices, demonstrating their dedication to both their business and the environment. They also impart their knowledge to others who are open to experimenting and trying their techniques. Achieving great heights in agripreneurship brings personal fulfilment and elevates their status within their communities. Ultimately, this pursuit of a recognized and respected identity fuels their passion and determination to excel in sustainable agriculture.

5.3.2 Financial empowerment: a driving force

Financial independence and economic pressures are powerful motivators for women pursuing agripreneurship. Many women seek to escape financial dependency and achieve self-sufficiency by starting their own agricultural ventures. Participant 2, who has a business venture marketing millet biscuit as a value-added product, was a home maker and a parent focusing on giving nutritious food (millets) to her children. She says the transition to agripreneur occurred when she wanted to contribute to her family income and also become financially independent. Economic pressures, such as the need to supplement household income or address financial challenges, drive some women to explore innovative and profitable farming practices. Participant 7 recollects that after she lost some of her sister's family in a tragic accident, she had to embrace and take care of her sister's children, which necessitated her exploring earning opportunities. One way to establish reliable sources of income and take charge of one's financial destiny is through agribusiness. Furthermore, women who invest in sustainable practices that boost profitability and support environmental conservation do so frequently because they are pursuing economic stability. The desire for financial independence gives women the confidence to make choices that are consistent with their long-term objectives and values. They can fortify themselves against economic ups and downs and generate growth prospects for their communities by engaging in agripreneurship. In the end, the desire for financial independence is a strong incentive for women to pursue agripreneurship and attain both economic and personal empowerment.

5.3.3 Increasing social status via agripreneurship: what motivates them?

Many women agripreneurs are motivated to succeed in agribusiness not only for financial gain but also to set an example for others in their social circles and encourage them to pursue similar endeavors. By succeeding, they are able to defy gender norms and become respected businesswomen in an otherwise predominantly male industry. Furthermore, their standing among peers is enhanced when the community recognizes their efforts in environmental conservation and sustainable practices. Successful women agribusiness owners usually assume leadership positions in their communities, where they acquire respect and influence. Their families take great pride in their accomplishments, which improves their reputation as a whole. Their status is increased by social media and local press coverage, which spreads their influence beyond their close friends and neighbors. This appreciation and support gives them the drive and inspiration they need to overcome obstacles and continue to succeed in their business venture. Participant 8 related her proud moment when, from a remote village in the Erode district, her turmeric powder with a high curcumin level was exported to many foreign countries. The villagers praised and felt proud of her. In a similar example, Participant 7 recalls being referred to as "Murungai Arasi," or "Moringa Queen," a title she received for her business venture's successes.

5.3.4 Inspiring communities: social motivation for their agripreneurial efforts

Women agripreneurs cite a strong sense of purpose and fulfillment from contributing to society as one of their main sources of motivation. These women frequently farm sustainably, sustaining not only their livelihoods but also biodiversity and environmental conservation. According to Participant 2, through her business, she guarantees healthy organic food to her customers, that contributes to the well-being of the community itself. The dedication of these women to eco-friendly methods contributes to the fight against global warming, improving the environment for coming generations. In addition, a large number of female agripreneurs concentrate on resource sustainability, critical local water issues, and water conservation methods. They frequently engage in educational outreach, passing along knowledge of environmentally responsible practices and sustainable farming to others. According to Participant 8, she was appointed as the leader of a self-help group with the purpose of guiding and educating like-minded women in her village about beekeeping. Sharing knowledge like this encourages a sustainable culture and gives others the ability to make decisions that are less harmful to the environment. Women who work in agriculture are often involved in community-building initiatives, like setting up cooperative societies and farmers' markets, which promote social cohesion and boost local economies. Participant 4 recollects several seed collective exhibitions and meetings, organized by, or associated with, her and her team. Their entrepreneurial ventures create job opportunities, contributing to rural development and reducing poverty. Ultimately, their dedication to societal contribution not only enhances their personal fulfillment but also drives positive change in their communities and beyond. Participant 5 feels proud that when she started her agritourism project in a remote village she took the initiative of connecting with the minority communities living in that place, trained them and gave employment in her venture.

5.3.5 Digital technology: a catalyst for women's agripreneurial success

The ease of digital technological availability, and an interest and fascination in using them, is a significant motivation for women to become digital agripreneurs as it simplifies access to vital information, resources, and markets. Previous research studies also convey that knowledge and comfort in accessing these resources act as a motivational factor and gives confidence to women agripreneurs (Shetty and Hans, 2019). Out of nine agripreneurs interviewed, all of them are accessing digital technologies for their business ventures successfully. Participant 1, who is involved in honeybee farming, uses digital technology to connect with her associates and position honeybee boxes in various farms. She has developed a network group, and trains her associates to place the boxes and collect honey regularly, through the usage of digital technology. Similarly, Participant 7, an agripreneur, proudly shares how digital technologies have been instrumental in her journey from a tenth-grade student to a successful entrepreneur, enabling her to create and sell around 32 value-added products from drumstick leaves, seeds, and flowers (popularly called moringa leaf powder, oil, etc.). The ease of usage and availability has enabled most of the agripreneurs to depend on these technologies for better farm management, collection information, and efficient resource utilization, making sustainable practices more feasible and profitable. Furthermore, digital platforms enable direct sales to consumers, enhancing income opportunities and reducing dependency on others for marketing their produce. The integration of digital technology thus empowers women agripreneurs to innovate, optimize operations, and drive sustainable agricultural growth.

5.3.6 Productive engagement: leveraging their free time

Agripreneurship provides a productive and satisfying outlet for many women who are looking for meaningful ways to spend their free time. They can use their spare time to their advantage by starting agricultural businesses and fostering innovation and growth. This method helps people develop their skills and accomplish goals while also fostering personal growth. By dedicating their time and resources to agribusiness, women are building profitable, long-lasting businesses that reflect their values and areas of interest. This constructive use of leisure time ultimately promotes economic empowerment as well as personal fulfillment. Women are also given a sense of direction and purpose by this fruitful involvement in agripreneurship. Making the most of their time, it allows them to strike a balance between their financial opportunities and personal hobbies. Furthermore, because they can actually see the fruits of their labor and investment, this method gives them a sense of accomplishment and fulfillment.

5.3.7 Staying connected: the social benefits

Many women agribusiness owners discover that their jobs serve as a potent diversion from loneliness and a way to maintain their interest in both their work and the outside world. The majority of the female agribusiness owners these authors spoke with continue to operate their farms independently of towns or village settlements. This limits the interactions that one has with other people to some degree. Their business ventures keep them active and linked to both the physical and digital worlds. Due to the hands-on nature of farming and agribusiness, feelings of isolation can be fought by developing a daily routine that requires constant attention and involvement. Participant 9 is a 40-year-old single woman who moved back to India after having a prosperous scientific career in Europe. She claims that her project keeps her focused and occupied all day long. Additionally, using sustainable farming methods frequently puts these women in closer contact with the natural world, which can be a gratifying and healing experience. Agripreneurs' jobs usually involve interacting with a range of stakeholders, such as suppliers, customers, and members of the community, which helps to create a feeling of community and connection. Social interaction is essential for mental health because it fosters the development of friendships and support systems. Additionally, participating in local markets, agricultural fairs, and cooperative societies provides further opportunities for social engagement and professional growth. Through these activities, women agripreneurs not only find purpose and engagement but also contribute to the vitality of their communities, enhancing their own sense of self-worth and fulfillment.

5.4 Sustainable Practices Identified

Through interactions and interviews with women agripreneurs, this study learnt about a number of sustainable practices being implemented. Many of these techniques are successfully used by farming communities that prioritize environmentally friendly, sustainable farming methods, and they are documented in the body of existing literature. This section consolidates the key sustainable practices adopted by all the nine women agripreneurs interviewed.

The first focus is that their business venture itself is oriented to a sustainable business idea. For example, two agripreneurs focus on honeybee farming, embracing indigenous practices. Honeybee farming enhances pollination and yields by providing a consistent and reliable source of pollination for crops. Bees collect nectar and pollen from flowers, transferring pollen from one bloom to another, which is essential for the fertilization process. This increased pollination leads to higher fruit and seed production, as plants are more effectively fertilized. These agripreneurs set their honeybee boxes in various agricultural lands; either they seek permission or the farmers invite them to keep honeybee boxes on their land so that they get a higher yield of the crops they are cultivating. By introducing honeybee colonies to farms, farmers can significantly boost the quantity and quality of their harvests. Additionally, the presence of bees helps to support biodiversity, promoting the health of the entire ecosystem and improving overall crop resilience.

The second focus is on adopting sustainable practices whenever possible in their business venture. Participant 2 is manufacturing and marketing millet biscuits. Her strict regime is to procure only those millet crops that are produced using nature-based farming methods. She has established a network of farmers who supply her regularly and she visits these farms for regular inspections. Participant 3, who focuses on organic produce, namely gooseberries, guava, and sapota, rears cattle only for manuring her farm and no synthetic inputs are used on her farm. She extensively uses innovative methods as sustainable practice that supports the environment and keeps the uniqueness in the project. She uses a solar dryer and converts the fresh fruits into dry fruits and markets them. This technology enables her to save time as it is fast in adding value to the fresh produce, the hygienic approach protects the produce quality, and it makes the produce available for a longer period for the consumers. Another innovative practice adopted is allowing the consumers to directly visit the farm, to choose and pluck the fruits directly from the trees, thereby eliminating any kind of middlemen and wastages.

Overall, two approaches were identified among women agripreneurs connected to sustainable practices. One approach is that the business venture itself is sustainable, contributing to sustainability, ecological protection, and promoting biodiversity. Examples for these can be honeybee farming, earthworm farming, seed collectives' business ventures, nature-based agritourism projects, etc. Another approach involves sustainable practices, as much as possible, in otherwise normal business ventures. An example for this is the millet biscuit business venture, where only millets that are produced through nature-based farming methods are used.

5.5 Breaking Ground Against the Odds: Challenges and Mitigations of Women Agripreneurs

Women agripreneurs face a unique set of challenges in the agricultural sector, often stemming from deeply rooted gender biases and societal expectations (Bilgrami *et al.*, 2022). A few previous research studies point out that despite the growth in the number of female agripreneurs, their roles remain underutilized and they encounter several obstacles (Qadri *et al.*, 2022).

5.5.1 Ownership and financial support

Limited access to land ownership remains a significant barrier as cultural norms often favor men in inheritance and land distribution. Unequal land rights and limited access to productive resources restrict the ability of women to fully engage in mechanized agricultural tasks (Charitha *et al.*, 2023). Even when women do acquire land, they frequently encounter difficulties in accessing credit, financial services, and government support due to gender discrimination, lack of collateral, bureaucratic obstacles, and lack of awareness (Manimegalai and Sundaravaradarajan, 2022). For example, money for setting up solar dryers and manufacturing hubs for making value-added produce like cream, lotion, oil, capsules, etc. all need financial support. Among the agripreneurs this study interacted with, either they have ownership through their family settlements, or they get strong support from family members whenever collateral legal procedures arise. However, women agripreneurs do manage to acquire a lot of subsidies from government schemes. Previous research studies have indicated that women agripreneurs can access funding (Rani and Sundaram, 2024) from government initiatives and schemes. Overall, ownership and financial support are key challenges that need to be addressed proactively by women agripreneurs.

5.5.2 Access to education and training opportunities

Considering the remote locations in which many women reside and operate, they have limited access to education and training opportunities, which are crucial for adopting new farming techniques and business practices. Skill development in both farming techniques and business management is vital, and building connections for mutual support is important. Furthermore, women agripreneurs face challenges in accessing markets, where they often encounter gender-based biases and lower prices for their produce. For most of these requirements, they are expected to travel outside their comfort zone, explore new places, and connect with several stakeholders. Currently they mitigate these issues with the support of several digital platforms to gain information and knowledge, use appropriate digital platforms to virtually network and get guidance, and use such digital platforms to directly sell their produce. They use e-commerce websites, social media, and mobile apps to market produce and connect with customers. Online training provides access to webinars, tutorials, and courses to enhance their skills. Online communities and forums offer platforms for sharing experiences and advice. Today, most of the women agripreneurs extensively use digital technologies in operating their businesses end to end.

5.5.3 Work–life balance

Alexander (2020) in her research study indicated that women agripreneurs face difficulties in balancing their work and family. This problem emerged as one of the challenges among the women agripreneurs this study interacted with. Most of them conveyed that balancing family responsibilities with the demands of running a farm or agribusiness can be particularly taxing, leading to time constraints and added stress. Societal norms often require them to balance traditional roles with entrepreneurial dreams. The dual burden of doing business and taking care of household responsibilities constrains their entrepreneurial potential (Charitha *et al.*, 2023). Participant 4, who is involved in collecting, preserving, multiplying, and selling native seeds needs to travel regularly to discover new sources or new native varieties of seeds or bulbs possessed by indigenous farmers. Along with that, she also participates in seed collectives' exhibitions and meetings and also conducts awareness programs for various stakeholders. Family time, traveling to remote locations in search of native seeds, work–life balance while participating in various exhibitions inclusive of traveling and staying away, all these demand a lot of attention, and balancing family responsibilities with these demands is considered stressful. Some native seed possessors/owners prefer not to share the native seeds considered to have a superior edge, and time is needed to convince them, influence them, and build trust, which may affect the agripreneur's work–life balance.

5.5.4 Social expectations

Social expectations often limit women's participation in decision-making processes at the community and household levels, restricting their ability to influence agricultural policies and practices. Sometimes social stigmas like "you are too young and how can I trust you in this venture?" or at times "you are a woman, bring your husband or father along with you for discussions" occurs even today for a few women agripreneurs.

5.5.5 Upscaling business operations

Upscaling the business is a challenge faced by some women agripreneurs. Once they establish their business venture at a small-scale level, the growth trajectory is upscaling the business and to expand. Changing the business model from small to large needs financial support, technical support to establish larger factory set-ups, bigger machines and equipment to be customized, and then setting it up and making it operational. Participant 2 has established herself a name in the millet biscuits business venture. Every day, in her kitchen, she makes two varieties of millet biscuits for distribution. The demand for her millet biscuits is greater than the number she could make in her kitchen every day. She supplies in total six varieties of millet biscuits directly to her customers who book their orders directly and wait in an online queue to purchase their preferred variety. She understands that she needs to upscale her operations quickly and convert her kitchen to a factory set-up to utilize this business growth. A lack of knowledge, information, and support system in developing a factory, designing customized equipment, and installing and upscaling the whole business venture is the major challenge she is handling currently.

5.5.6 Family members' concerns, resistance, and support

Some women agripreneurs have only one operating space, their farm, that acts as their office, manufacturing space, living space, and family space. When they introduce something innovative in their business venture and in their workspace, it affects the other family members and resistance may come from within. For example, Participant 3 introduced a practice where the customers can come to the farm and pick the fruits they want, and then weigh and purchase them. This is part of a sustainable practice where one eliminates the storage cost, transportation cost, and other overheads, transiting the fruits from the farm to the marketplace to the customers. However, encouraging family members to accept this innovative practice was a struggle since reasons like intrusion into private space, strangers being allowed into the farm, and safety were voiced. Sometimes, family members are supportive and their presence plays a major role in mitigating safety challenges. Participant 1 says that she must travel to remote villages and farms during the night to set-up honeybee boxes. She is always accompanied by her father during such visits. Carrying, loading, and unloading many honeybee boxes also needs support and this is mitigated through her family members. She narrates that encountering dangers like stepping on a snake at nighttime in remote locations are part of the normal risks she takes.

5.5.7 Fluctuations in demand

Fluctuations in demand for produce create significant challenges for women agripreneurs, leading to unstable revenue streams. Seasonal variations, market volatility, and unpredictable consumer preferences often result in inconsistent sales, making it difficult to plan financially. This uncertainty can strain resources, as women agripreneurs may struggle to cover operating costs or to invest in future growth. Without stable revenue, securing loans or other forms of financial support becomes even more challenging, further limiting their ability to expand or sustain their businesses. For example, Participant 7 makes around 32 value-added products from moringa (drumstick tree) leaves, vegetables, and flowers, such as powder, cream, oil, soap, honey, etc. A few items sell fast whereas others have stagnant sales. Also, there are fluctuations in sales over the months. She is adopting a practice of maintaining very minimum stock and producing more stock after the order is confirmed. Securing financial support, upscaling the facility, and repayments are challenging in such scenarios. Participant 5, who focuses on agritourism also faces such fluctuations as her occupancy rates are irregular across different months and maintenance of the accommodation for agritourists during lean periods calls for additional resource allocation. Some women agripreneurs have realized that for a few value-added products, they have a limited market or limited awareness among potential customers. For example, jamun seed powder, moringa seed oil, etc. This leads to less upscaling possibilities for this product, or the need to promote them more aggressively to create awareness, or to explore new value-added products. To mitigate these risks, many women agripreneurs diversify their business ventures, explore new value-added products, or seek out alternative markets to create more consistent income streams.

5.5.8 Cluttered digital space

Digital platforms are widely used by many entrepreneurs today and they have become cluttered spaces with potentially several or many similar products and, often, information overload. Creating an identity for their brand is a real challenge that women agripreneurs say they are facing today. Gaining a niche or gaining attention demands brand-building expertise, which is considered to be a challenge by a few of the participants of this study.

5.5.9 Climate changes

Climate variations and water scarcity play a critical role in the quantity and quality of produce, presenting a significant challenge for women agripreneurs and affecting the value-added products they manufacture. Unpredictable weather patterns, such as droughts, floods, and unseasonal rains affect their agribusinesses leading to reduced yields and inferior product quality. Tamil Nadu is one of the states vulnerable to erratic rainfall patterns and frequent droughts. Women agripreneurs, who often rely on rain-fed agriculture, are disproportionately affected by these climate-induced risks, leading to reduced crop yields (Saravanakumar *et al.*, 2022) and income instability. This directly impacts their income, as lower quality produce often fetches lower prices in the market. Moreover, severe weather can occasionally harm their farming equipment and infrastructure, which raises the cost. Implementing the necessary changes is challenging due to limited resources and limited knowledge about climate adaptation. In order to better withstand the effects of climate change and protect their livelihoods, many women are embracing diversification in their business ventures as a coping mechanism. A multi-business venture has been built by Participant 8, an integrated farmer, to generate various revenue streams. These include exporting her value-added produce, running training programs, providing mentoring and consulting services, and creating and obtaining patents.

Overall, despite all these challenges, many women agripreneurs demonstrate remarkable resilience, leveraging community support networks and innovative approaches to overcome obstacles. By promoting gender equality in agriculture, providing targeted training, and improving access to resources, stakeholders can help unlock the full potential of women agripreneurs and foster more inclusive and sustainable agricultural practices. They also experience economic empowerment by contributing financially to their households and gaining independence. Access to finance remains a challenge but overcoming it can lead to significant growth. Agriculture-based entrepreneurship empowers women by generating their own income and improving their families' living standards. It enhances their skills in farming, marketing, and using digital tools, and enables them to emerge as leaders and role models in their communities. Their efforts contribute to the overall development of rural areas through sustainable practices.

5.6 Conclusion

Women agripreneurs develop business models that incorporate several crucial elements. They offer a strong value proposition by providing organic, high-quality produce that meets consumer demand for healthy and sustainable food. Their customer segments include urban consumers, health-conscious individuals, and niche markets interested in natural farming-based produce. They leverage digital platforms like WhatsApp, Facebook, and Instagram for marketing and direct sales, alongside traditional markets. Their revenue streams are generated through direct sales, subscriptions for regular delivery of produce, and agritourism. Key resources include access to land, water, seeds, and knowledge about sustainable practices, while key activities encompass farming, marketing, networking, and continuous learning. They form partnerships with local non-governmental organizations (NGOs), government bodies, and other agripreneurs for support and resources, and manage costs through efficient resource use, shared services, and minimal reliance on chemical inputs.

Several factors shape the entrepreneurial experience of these women agripreneurs. Access to resources, including land, water, and financial support, is crucial. Knowledge and training in sustainable farming techniques and digital literacy are essential, as is social support from family and the community. Market access, enabled by digital platforms, and supportive government schemes also play significant roles. The adoption of technology for farm management and marketing is another critical factor.

Overall, through the practices and factors embedded in their business model, the women agripreneurs in Tamil Nadu aim to enhance their market reach, increase profitability, promote sustainability, empower women, and foster innovation. By expanding their customer base beyond local markets and boosting income through direct sales and premium pricing for natural farming produce, these women are setting new benchmarks. Their commitment to environmentally friendly farming practices, combined with their use of digital technologies and innovative approaches, strengthens their economic and social status and inspires others in the field.

To conclude, rural entrepreneurship based on agriculture, with women involved, plays a pivotal role in driving rural development and empowering women in developing countries. By fostering entrepreneurship in rural areas, communities can create jobs, reduce poverty, and improve local economies, leading to more sustainable and self-reliant societies. Women, who often form the backbone of rural economies, benefit immensely from entrepreneurial opportunities, as these initiatives allow them to gain financial independence, contribute to household income, and enhance their social status. Digital technologies have enabled this growth tremendously. However, the journey of women agripreneurs in rural areas is fraught with challenges, including limited access to resources, education, and markets, as well as deep-rooted gender biases. Despite these hurdles, women entrepreneurs have shown remarkable resilience, often turning to innovative and sustainable business practices that not only boost their own livelihoods but also contribute to the broader goal of rural development.

Supporting women as agripreneurs can have a ripple effect, as they tend to reinvest their earnings into their families and communities, thereby enhancing education, health, and overall well-being. To amplify these benefits, it is essential for governments, NGOs, and the private sector to provide targeted support, including access to finance, training, and mentorship, tailored to the unique needs of rural women. The anticipated growth in women's share of entrepreneurship, if supported by effective policies, could lead to a more gender-inclusive business ecosystem in India (Murthy, 2010). By addressing these challenges and creating an enabling environment for women agripreneurs, developing countries can unlock the full potential of rural areas and drive inclusive growth. Ultimately, the intersection of rural agribusiness ventures, rural development, and women's development holds the key to transforming the economic and social landscape of developing countries. Empowering women agripreneurs in rural areas is not just an economic imperative; it is a pathway to achieving gender equality, reducing poverty, and fostering sustainable development. As more women rise to become successful agripreneurs, they will inspire others, creating a cycle of growth and empowerment that benefits entire communities and future generations.

References

Aleke, B., Ojiako, U. and Wainwright, D.W. (2011) ICT adoption in developing countries: Perspectives from small-scale agribusinesses. *Journal of Enterprise Information Management* 24(1), 68–84.

Alexander, E.M. (2020) Obstacles Encountered and Overcome by Female Agricultural Entrepreneurs in Niche Markets. Master's thesis, Purdue University.

Bilgrami, F., Purohit, H. and Tanwar, V. (2022) A theoretical assessment of problems faced by Women Entrepreneurs (in micro industry): A study of Indian scenario. *International Journal for Global Academic & Scientific Research* 1(4), 30–41.

Bose, V. (2013) Empowering youth for sustainable agribusiness: A study of Kerala. *Journal of Rural and Industrial Development* 1(2), 15.

Bowen, D.D. and Hisrich, R.D. (1986) The female entrepreneur: A career development perspective. *Academy of Management Review* 11(2), 393–407.

Buttner, E. and Rosen, B. (1988) The influence of entrepreneur's gender and type of business on decisions to provide venture capital. In: *Proceedings*, Southern Management Association.

Buttner, E.H. and Moore, D.P. (1997) *Women entrepreneurs: Moving beyond the glass ceiling*. Sage, Beverly Hills, CA.

Carr, S. and Roulin, A. (2016) An exploration of Agripreneurship, scope, actors and prospects. *Agenda* 2063, 16.

Charitha, S., Kumari, V. and Shukla, Pragati. (2023) Documentation of women leaders as farmers/entrepreneurs, women-led agri start-ups and fpos. In: *MANAGE Center for Gender In Agriculture, Nutritional Security and Urban Agriculture. National Institute of Agricultural Extension Management (MANAGE)*. Hyderabad, Telangana, India.

Giuggioli, G. and Pellegrini, M.M. (2023) Artificial intelligence as an enabler for entrepreneurs: A systematic literature review and an agenda for future research. *International Journal of Entrepreneurial Behavior & Research* 29(4), 816–837.

Isaga, Nsubili. (2019) Start-up motives and challenges facing female entrepreneurs in Tanzania. *International Journal of Gender and Entrepreneurship* 11(2), 102–119.

Kumar, R., Singh, S. and Singh, N. (2012) Successful Women entrepreneurs in India: Attributes based analysis. *Asian Journal of Managerial Science* 1(1), 57–62.

Loza de Siles, E. (2011) Female entrepreneurship theory: A multidisciplinary review of resources. *Journal of Women's Entrepreneurship and Education, Institute of Economic Sciences (Belgrade, Serbia)* (1–2), 26–64.

Manimegalai, M. and Sundaravaradarajan, K.R. (2022) Problems faced by Women agri-entrepreneur in Tamil Nadu: An economic analysis. *European Online Journal of Natural and Social Sciences* 11(4), 1290–1298.

Murthy, P.S.R. (2010) Challenges faced by Women entrepreneurs in globalized era. *Available at SSRN 1650583* 1–10.

Nagalakshmi, T. and Sudhakar, A. (2013) Agripreneurs: A case study of Dharmapuri farmers. *International Journal of Science and Research* 2(8), 208–214.

Otache, I. (2017) Agripreneurship development: A strategy for revamping Nigeria's economy from recession. *African Journal of Economic and Management Studies* 8(4), 474–483.

Qadri, F.A., Shadab, S. and Khan, A. (2022) Role and prospects of women entrepreneurs in the agriculture sector of India. In: *Driving Factors for Venture Creation and Success in Agricultural Entrepreneurship*. IGI Global, pp. 250–260.

Rani, V.S. and Sundaram, N. (2024) Opportunities and challenges faced by women entrepreneurs for sustainable development in India. *Revista de Gestão Social e Ambiental* 18(3), 1–20.

Rosário, A.T. and Figueiredo, J. (2024) Sustainable entrepreneurship and corporate social responsibility: Analysing the state of research. *Sustainable Environment* 10(1), 1–18. DOI: 10.1080/27658511.2024.2324572.

Saravanakumar, V., Lohano, H.D. and Balasubramanian, R. (2022) A district-level analysis for measuring the effects of climate change on production of rice: Evidence from Southern India. *Theoretical and Applied Climatology* 150(3), 941–953.

Shetty, S. and Hans, V.B. (2019) Women entrepreneurs in India: Approaches to education and empowerment. *International Journal of Research and Analytical Reviews* 1–10. DOI: 10.2139/ssrn.3346540.

Surainita, M.S., Fatah, F.A. and Azam, N.M. (2022) Motivational factors of women entrepreneurs in agribusiness entrepreneurial development in Malaysia. *Journal of Agricultural Studies* 10(4), 113–124.

6 Financial Resilience Strategies of Farmer Producer Organizations in Rural India

Veena Suresh[1*], Sreejith S.S.[1], Vivek S.[2] and Erika Valerio[3]
[1]Department of Management Studies, National Institute of Technology Calicut, Kozhikode, India; [2]Department of Social Work, Amrita Vishwa Vidyapeetham, Amritapuri, India; [3]School of Environmental and Rural Science, University of New England, Armidale, Australia

6.1 Introduction

In India, agriculture is a growing sector with a lot of promise and challenges. To increase agricultural productivity and to give rural marginal farmers more authority, the Indian government has launched a number of initiatives. Farmer producer organizations (FPOs) are being formed and promoted in India as one such programme. FPOs are collective organizations of primary producers in small and marginal farms across India. FPOs help farmers to respond well to the ever-changing agri-market landscape. FPOs have the power to enhance members' socio-economic status (Gautam and Mallaiah, 2024) and improve rural livelihoods, incomes, and agricultural production (Gorai *et al.*, 2022). Through FPOs, farmers can pool their resources and production and negotiate better prices for agricultural inputs and outputs. Bulk purchasing of inputs reduces cost. Farmers can access group loans via FPO membership status. Collective transportation and marketing can lower costs and provides maximum profit to the farmers directly by controlling the interference of third-party people in the agri-markets. There have been studies that report economic benefits offered by FPOs to the farmers through collective action. However, it is not clear whether these economic benefits were for a shorter period or were prolonged, ensuring financial security to the farmers, as there are studies that quote the financial challenges of FPOs as well. Nevertheless, there have been cases of FPOs that survived financial setbacks and their strategy of survival has been less explored in the academic literature. Therefore, the objective of the study on which this chapter is based is to explore the community-driven strategies employed by moderately functioning farmers' groups to achieve financial resilience. The study adopted a qualitative case study investigation to explore the financial strategies adopted by rural farmers (FPO members) to ensure the financial resilience of the FPOs. The study findings provide insights into the complex interaction of existing financial barriers, challenges, and community-driven financial strategies used by rural farmers in their collectives. It is helpful for the FPO-promoting agencies to enhance the functioning

*Corresponding author: veena_p220035ms@nitc.ac.in

 Agripreneurship, Digital Inclusion and Sustainability: Rural Innovation and Growth in the Global South (eds Dutta and Trivedi)
DOI: 10.1079/9781836992646.0006

of moderately functioning FPOs in their locality. It also provides confidence to the rural farmers that there is scope for moderately functioning FPOs to become financially resilient by creating indigenous strategies. Moreover, understanding the strategies employed by the sampled FPOs to manage financial resources, lower risks and takes advantage of opportunities that can teach stakeholders in the rural agriculture sector valuable lessons for developing sustainable agribusiness strategies. This will help achieve some of the sustainable development goals (SDGs), specifically SDG1, no poverty, and SDG2, zero hunger.

6.2 Literature Review

6.2.1 History and origin of farmers' producer organizations in India

In India, the concept and evolution of FPOs began in the early 2000s based on the proposal made by the Alagh Committee in 1999. India's National Agricultural Policy (NAP) of 2000 included the development of FPOs as a significant strategy to reduce the need for mediators to access markets. Large-scale steps to institutionalize and support FPOs were taken in 2003. The evolution of FPOs has four phases; the first phase began in 2003 and continued until 2010 when farmers started registering under FPOs. The second phase (2011–2013) got policy attention at the national level. The government appointed the National Bank for Agriculture and Rural Development (NABARD) and the Small Farmers Agribusiness Consortium (SFAC) as promoting institutions for FPOs. These agencies started building the capacity of FPO members and encouraging them to achieve the expected outcomes from FPOs. A widespread registration of FPOs was seen in the third phase from 2014 to 2017 based on the FPO 2013 policy guidelines published by the government. To enhance the efficiency of FPOs throughout India, NABARD suggested the creation of a Producers Organisation Development and Upliftment Corpus Fund in 2014. To facilitate the establishment and expansion of FPOs, the Ministry of Agriculture and Farmers Welfare of the Indian government launched the 'Scheme for Promotion of FPOs' in 2016. Agriculture extension organisztions conducted various training programs like state agriculture universities, state departments of agriculture, and Krish Vikas Kendras to improve the efficiency of FPOs.

In the fourth phase (2017–2020), FPOs were developed as a farming enterprises, and most state governments supported the growth of FPOs. A report on doubling farmer's incomes was published in 2019 to promote the benefits of FPOs for farmers. In the current development phase (from 2020), the Indian government plans to promote 10,000 FPOs throughout the country to increase agricultural output (Ministry of Agriculture and Farmers Welfare, 2023). The potential benefits of FPOs lined the way to the announcement of the Central Sector Scheme of 'Formation and Promotion of 10,000 FPOs' authorized and launched by the government of India with a total budgeted investment of INR 6865 crore (Ministry of Agriculture and Farmers Welfare, 2021). The statistical evidence published by the Ministry of Agriculture and Farmers Welfare shows the number of FPOs registered under NABARD and SFAC as being 3904 and 898, respectively, until 2022 (Ministry of Agriculture and Farmers Welfare, 2021). Recent data from the Tata Cornell Institute FPO platform for India show a boom in the number of registered FPOs in 2022 (Tata-Cornell Institute for Agriculture and Nutrition, 2024). In 2022, the government also announced a new program called the Credit Guarantee Fund, which offers grant assistance to FPOs that are promoted. However, the financial struggles of FPOS are a pertaining issue in agribusiness.

6.2.2 Bibliometric literature review analysis of FPO studies

A bibliometric literature review analysis of FPO studies for a 5 year period (2019–2023) was conducted to understand the significance and need for study objectives. The key authors of FPO studies are Mukherjee Anirban, Premlata M. Singh and Shubha Kumari from the Indian Agricultural Research Institute (ICAR) and Shambu Prasad from the Institute of Rural Management Anand (IRMA). Interestingly,

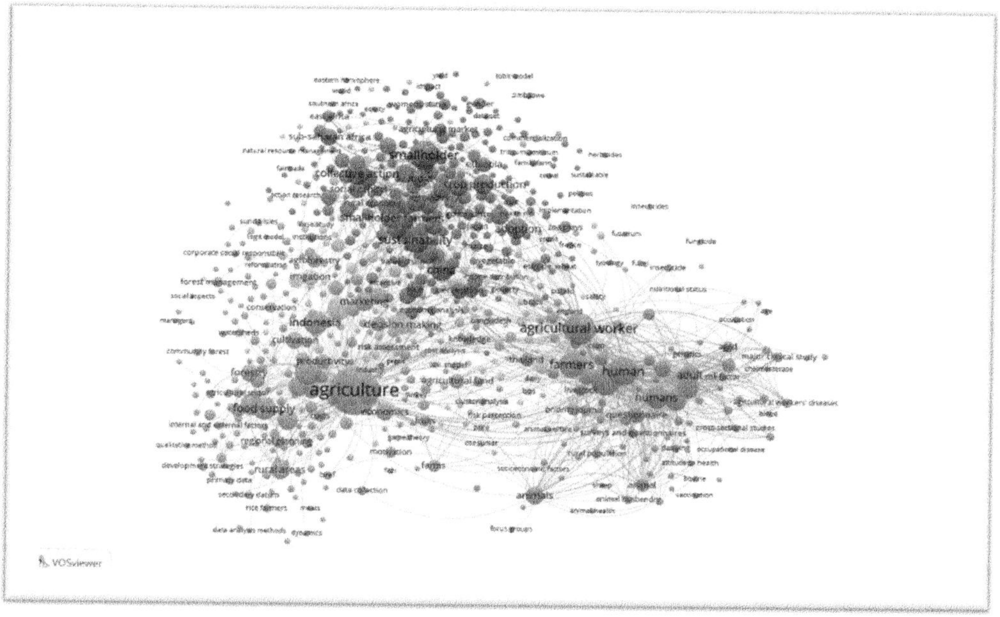

Fig. 6.1. Network visualization of FPO studies (2019–2023) generated from VOS viewer software.

these institutions topped the number of publications related to FPOs in the bibliometric analysis. Apart from this, the Tata Cornell Institute and Azim Premji University are involved in recent research projects in uplifting the FPOs.

The clusters of topics researched and the trending areas within FPO studies were analyzed and produced using VOS viewer bibliometric data analysis software's network and overlay visualization techniques, shown in Figs 6.1 and 6.2, respectively. The network visualization indicates agriculture production, market supply chain, smallholder farmers, agriculture workers, agri-markets, crop production, sustainability, collective action, and food supply as significant topic clusters. The overlay visualization shows current trending topics such as food supply, marketing, smallholder farmers, sustainability, rural areas, and perception of members. The sustainability aspect is shown as an upcoming area of research among the FPO studies; however, the financial or economic aspect of FPOs is not seen as a topic cluster, which indicates the presence of a smaller number of studies in the economic area.

Further, when the sustainability cluster in network visualization in Fig. 6.3 is examined more closely, it can be seen that sustainability is linked to a few areas like crop production, agriculture, supply chain management, smallholders, collective action, and agriculture workers. The holistic understanding of FPO sustainability based on the financial dimensions of SDGs is barely seen in the academic literature.

Some studies elaborate the potential of FPOs in achieving SDGs. It is believed that if farming becomes profitable, it will positively impact migration and yield optimistic results for SDG1. A study in 2021 briefly described the SDGs connected to agriculture that can be touched by FPCs, based on the SDGs and the role of FPCs in promoting the same (Mourya and Mehta, 2021).

The FPOs promise to realize the SDGs by promoting sustainable agriculture, improving food security, empowering farmers, promoting gender equality, promoting responsible water usage and sustainable land management practices, fostering partnerships and collaborations, and actively participating in collective action towards achieving the SDGs.

As FPOs play a crucial role in realizing the SDGs by committing to various actions and initiatives, it is vital to explore the financial

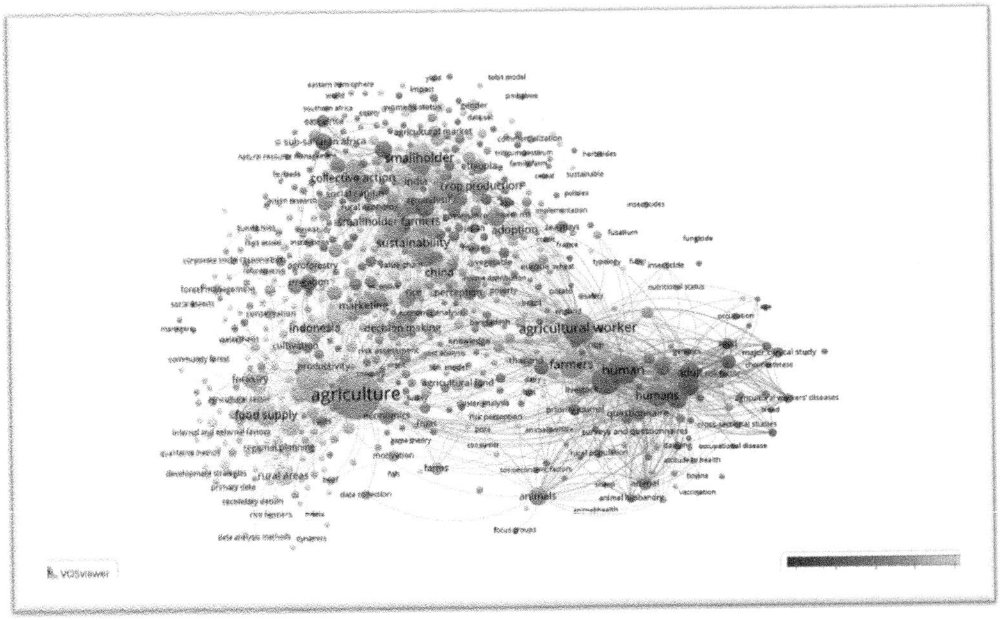

Fig. 6.2. Overlay visualization of FPO studies (2019–2023) generated from VOS viewer software.

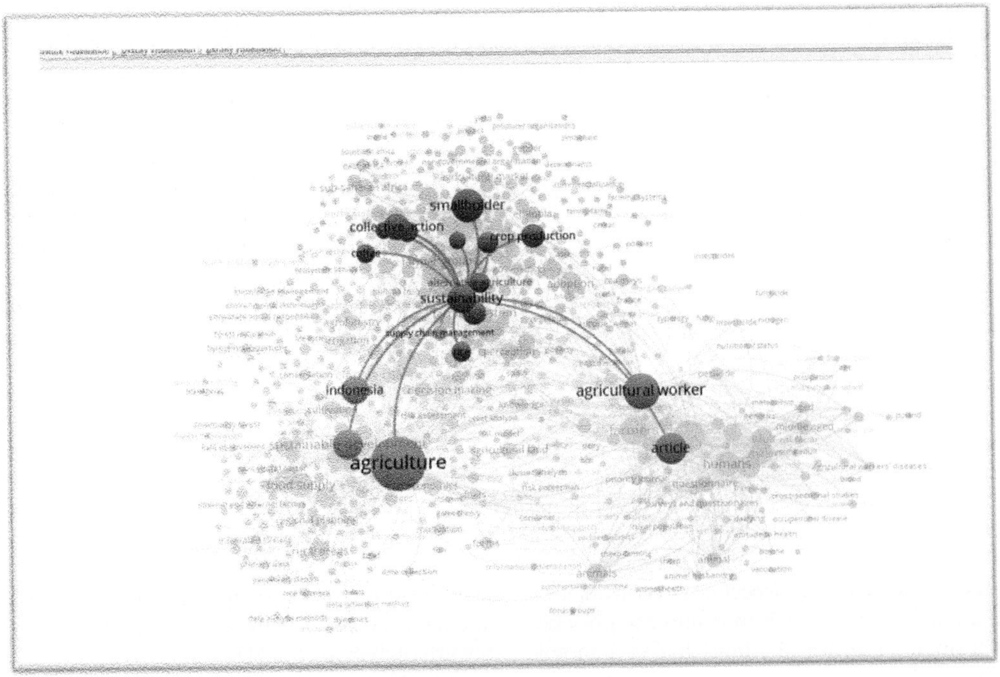

Fig. 6.3. Network visualization of the sustainability area of FPO studies (2019–2023) generated from VOS viewer software.

resilience of FPOs to sustain their daily economic activities.

6.2.3 Financial barriers and challenges for farmers' collectives

Financial constraints of farmers are a global concern faced by farmers in developing countries around the world (Lindsjö *et al.*, 2021). India's farmer collectives deal with several financial challenges that impair their livelihoods and productivity (Kujur and Gauraha, 2019). These limitations include restricted access to finance and credit, a lack of infrastructure and market knowledge, and trouble negotiating rewarding prices for their goods. Moreover, the working capital needs of farmers' collectives cannot be satisfied by the limited share capital obtained from membership fees (Nithya and Vaishnavi, 2022). Consequently, these collectives are forced to depend on other funding sources, which might not always give their requirements top priority or enough money. Worse are the high borrowing charges associated with the rural finance services that farmers' collectives can access.

The ability of farmers' collectives to grow and flourish is further hampered by a lack of entrepreneurship skills and inadequate infrastructure (Kumari and Bisht, 2023). Farmers' collectives frequently have a lack of business management expertise and financial literacy, which makes it difficult for them to efficiently manage their finances and make the most use of their resources (Mourya and Mehta, 2021). The financial difficulties that Indian farmers' collectives face are complex and call for all-encompassing solutions that take into account market intelligence, loan availability, financial literacy, infrastructural development, and business management expertise (Trivedi *et al.*, 2023).

A study that analyzed the secondary data from the Ministry of Corporate Affairs of India reported that the existing FPOs are in financial distress, without sufficient external funding (Dhineshwari *et al.*, 2021). Research by Sharma (2022) and Prasad *et al.* (2017) highlighted common issues that moderately functioning FPOs deal with, such as restricted financial availability, poor governance frameworks, deprived infrastructure, and insufficient technical capability. Some case studies showed the average financial performance of producer organizations in the dairy and farm sector (Kumar *et al.*, 2021; Wadkar *et al.*, 2021). Kakati and Roy's (2019) study on the financial sustainability of FPOs found only an average financial health.

Furthermore, the financial difficulties faced by farmer collectives are also a result of sociocultural limitations and unfavorable climate changes. Ullah *et al.* (2020) quoted an interesting association between the characteristics of farmers and the nature of credit sources they depend on for farming. It is noteworthy that farmers with greater assets, more farming experience, and better awareness of credit sources depend on banks for credits; older farmers with larger farmland and higher farm income were more likely to depend on informal credit sources, indicating the necessity of having wider credit options to cadre the needs and characteristics of groups of farmers.

Ranjan's (2014) study observed that farmers may use many approaches to enhance their financial resilience, including risk mitigation and economic growth, based on their individual risk preferences. Another study highlighted that the phenomenon of financial inclusion – encompassing the provision of bank accounts, mobile money services, and credit – has a significant role in enhancing resilience within agricultural households (Bekele, 2023).

A study on understanding the impact of financial inclusion on farmers' resilience by Bekele (2023) suggested that awareness of financial services, and facilities to access these services at community levels and reduced cost to farmers' households, helps in building the resilience of farmers to climate shocks. Kumari and Garg (2023) claimed that long-term credit facilities are helpful for farmers to ensure sustainable agriculture rather than short and medium-level credit services. Chaiya *et al.* (2023) stated that financial credit constraints in farming limit the farmers in adopting innovative technologies, which negatively impacts agriculture productivity. Kakati and Roy's (2019) research highlighted the significance of enhancing smallholder farmers' financial resilience as a means of reducing the effects of external stressors such as market volatility and climate unpredictability. According to the case study of the Maharashtra Farmers Producer Company Ltd, the production and efficiency of farmers' collectives can be increased by investing in contemporary machinery and technologies when their finances are secure (Wadkar *et al.*, 2021).

These financial barriers and challenges in FPOs paved the way for the proposition that there are hidden resources and capabilities for farmers to handle their own financial difficulties by themselves and to enhance the financial resilience of the collective. However, a robust theoretical framework is necessary to corroborate whether the set of resources and capabilities of the farmers' collectives are benefiting the farmers' collectives or not. The researchers of this study examined the strategic choices and resource management strategies made by farmers to become financially resilient using the resource-based view (RBV) theory.

6.2.4 Resource-based view theory

Resource-based view theory posits that an organization's performance and competitive advantage can be elevated from its unique resources and capabilities (Roxas and Chadee, 2011). RBV highlights valuable, rare, inimitable, and organized (VRIO) resources as crucial for the competitive performance of firms or organizations (Adnan *et al.*, 2018). It has been applied to various settings, including small and medium enterprises (Jaafar and Abdul-Aziz, 2005), export firms (Roxas and Chadee, 2011), and family business groups (Manikutty, 2000).

A systematic literature review of the use of RBV in rural properties states human and organizational resources as the most strategic resources for improving the performance of a farm (Carvalho *et al.*, 2014). RBV theory was applied to showcase the potential of urban agriculture as a competitive resource (Fuzinatto and Santos Junior, 2020) Negreti-Campos *et al.* (2022) RBV analysis proposed the involvement of family members in the collective action of farmers' cooperatives as an internal resource that enhances the performance of rural-based organizations. Gibson *et al.* (2021) stated community as a strategic resource for the sustainability of firms. A recent study by Malik and Kajale (2024) on empowering small and marginal farmers in FPOs underlines the scope of RBV theory as a possible solution for resource management in FPOs. However, research on RBV theory in farmers' collectives is limited. There are not enough studies that investigate the capacity of FPO members to endure and bounce back from financial setbacks while preserving the core operations of FPOs. Therefore, the researchers for this study aimed to explore the financial resilience strategies adopted by the sampled FPOs using the RBV approach to set an example for similar FPOs in the country to assess their capabilities and strategic resources to confidently face financial as well as other sustainability challenges.

6.3 Methodology

The qualitative case study method adopted in this study enabled the researchers to get an in-depth understanding of the FPO's institutional support system, governance, and financial management. Special focus was given to the financial performance of FPOs and the sustainability challenges related to financing. The complex interaction of existing financial barriers, challenges, and community-driven financial strategies used by rural farmers can only be captured by a qualitative in-depth investigation of the financial strategies adopted by farmers to mitigate the financial challenges.

6.3.1 Sampling

The researchers purposively selected FPO representatives from the states of Kerala and Tamil Nadu in South India because of their regional and linguistic familiarity to conduct an in-depth field study. Previously established rapport with sampled FPO farmers enabled the researchers to understand the complex interaction of internal and external resources and strategies that help the farmers become resilient to financial shocks and uncertainties. FPOs were chosen based on the year of establishment (completed minimum of 3 years) because the financial operations and resources of a collective or a group can only be

Table 6.1. Sample scorecard used for purposive sampling for the study.

Criteria	1	2	3	4	5	Total score
	Low				High	
Years of operation						
Number of active members						
Diversity of crops/products						
Financial health						
Market linkages						
Governance structure						
Technology adoption						
Environmental practices						
Access to government support schemes						
Extent of support from the promoting agency						
Total score of FPO						

observed once it reaches the performing stage of group development. Secondly, FPOs involved in the agricultural production of crops, vegetables, and plantations were included in the study to ensure the homogeneity of groups sampled for the study. FPOs engaging in dairy, horticulture, and fish farming were excluded. A scorecard was used for choosing FPOs for the study. FPOs are scored based on the criteria shown in Table 6.1 from highest to lowest, and the FPOs that scored lowest were chosen for the study. This scorecard helped to screen moderately functioning FPOs that were most relevant to the research objectives.

The criteria were as follow.

1. *Years of operation.* FPOs that have been operational face a variety of financial challenges compared with those of a newly established FPO. Therefore, the researchers chose FPOs who had completed at least 3 years of functioning and a maximum of 8 years of experience. The range is 3–8 years of working experience in which 3 years is rated as low and 8 as high.
2. *Number of active members.* Impacts organizational dynamics and resource availability.
3. *Diversity of crops/products.* Affects market resilience and risk management.
4. *Financial health.* Key indicator of sustainability. Assessed briefly through revenue and profit generated by the FPO.
5. *Market linkages.* Crucial for deciding the profit gained by FPOs.
6. *Governance structure.* Influences decision making and member engagement. Includes adherence to FPO policies, proper organization structure, and documentation of FPO activities.
7. *Technology adoption.* Indicates adaptability and efficiency. Includes the usage of the agri online portal and adoption of new technological tools/machines in agri production and procurement.
8. *Environmental practices.* Relevant to long-term sustainability. Includes community-based environmental practices related to the preservation and protection of natural resources and the extent of use of pesticide and waste management practices.
9. *Access to government support schemes.* Can impact financial and operational sustainability. Extent of financial support received from government bodies and access to other support services related to agriculture like subsidy for pesticides, buying machinery, agri loans, etc.
10. *Extent of support from the promoting agency.* Such agencies help FPOs to work effectively, however there were cases of FPOs becoming ideal once the support is withdrawn.

The steps in using the scorecard were:

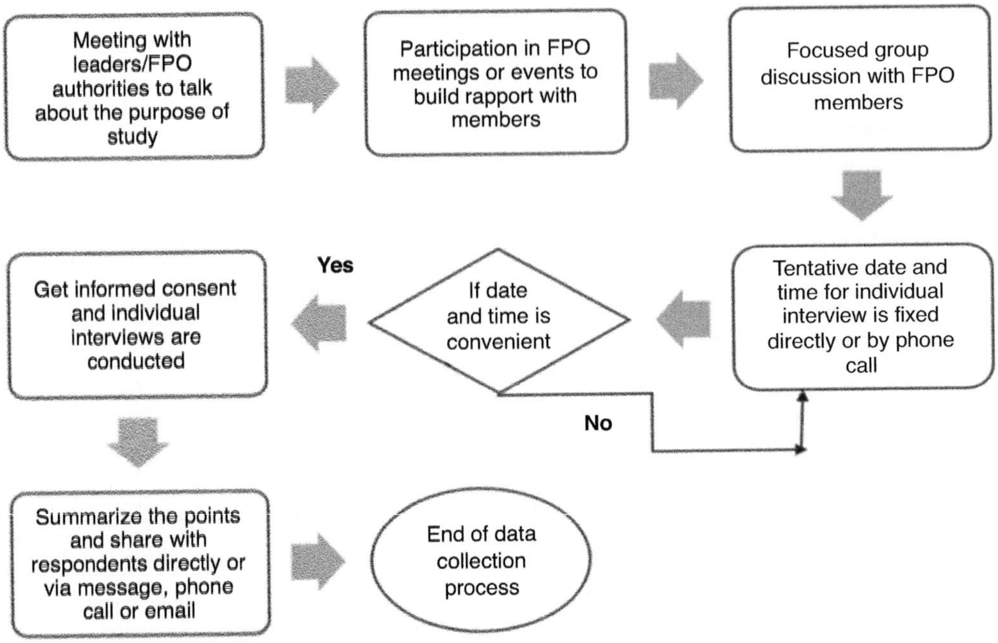

Fig. 6.4. Flow chart of the data collection process of the study.

1. For each FPO, score each criterion from 0 to 5 (0 being lowest, 5 being highest).
2. Sum all scores to get a total score for each FPO.
3. Select the FPOs with the lowest total scores for the study.

6.3.2 Data collection

The researcher conducted a focused group discussion and semi-structured interview with a selected representative sample of key stakeholders of FPOs, including FPO members, the board of directors, and promoting institution project coordinators, to collect evidence about their perception of the financial performance of the FPOs. Around 19 members were interviewed to attain the saturation point in the qualitative study. A combination of information from field visits, focused group discussions, and field interviews were used to examine the aim of the study. The data collection process is shown in Fig. 6.4.

The details of the FPOs chosen for the study are shown in Table 6.2 and the name of the FPO is coded to ensure anonymity of the organization studied. The respondents were concerned about displaying the name of their FPOs because they worried whether the name and branding built by their FPO would be lost if they exposed their financial struggles to the outside community. Therefore, the researchers explained to the respondents that those details would be confidential and took the data with the informed consent of the participants.

The socio-demographic details of the 19 interview respondents selected for the study are shown in Table 6.3. About 40% of the participants were female. Ninety-five percent of respondents were in the age group above 45 years of age, which might be an indication of the ageng population in the farm sector. About 90% of respondents had less than or had barely completed their secondary education, which should be taken into consideration when assessing the challenges faced by farmers in comprehending FPO policy guidelines and institutional settings. More than half of the respondents were of poor economic status with their primary income depending on farming, and all the respondents were married, which triples their burden of dependency and the significance of having a stable and consistent income to look after their families. The vulnerabilities in age, education, income status, and dependency of family members reinforced the significance of understanding the financial resilience of FPOs since FPOs sometimes struggle to gain profits for their

Table 6.2. Details of sampled FPOs.

	FPO code F1	F2	F3	F4	F5
FPO location	Palakkad, Kerala	Calicut, Kerala	Nenmara, Kerala	Nagerkovil, Tamil Nadu	Vellukuri, Tamil Nadu
Year of registration	2016	2019	2018	2017	2017
Total number of members	150	45	250	100	120
Main crop of the FPO	Rice and vegetables	Banana and vegetables	Rice and vegetables	Rice and banana	Rice and banana

Table 6.3. Socio-demographic details of the interview respondents.

Categories	Frequency	Percentage
Male	11	58
Female	8	42
Age of participants (years):		
35 to 44	1	5
45 to 54	8	42
55 to 64	8	42
Above or equal to 65	2	11
Education status:		
Completed primary education 4th grade pass	4	21
Completed upper primary education 7th grade pass	4	21
Completed secondary education 10th grade pass	9	47
Completed graduation degree	2	11
Economic status:		
Antyodaya Anna Yojana beneficiaries	6	32
Below poverty line	5	26
Above poverty line	4	21
Non-priority	4	21
Primary earner's income-generating activity:		
Full-time farmer	10	53
Non-agricultural work	9	47
Marital status:		
Married	19	100
Mean number of members in the family	4	NA

members when the majority of their members are already suffering from socio-economic vulnerabilities.

6.3.3 Data analysis and measurement

NVivo data analysis software was used to code and analyze the interview transcripts of the study. Content analysis was applied to generate codes and themes for the study. The frequency of reference of the codes was considered to identify significant themes that emerged from the coding and sorting of data transcripts.

6.3.4 Ethical considerations during the study

Given the vulnerabilities of the population under study, there were several ethical considerations to be taken into account when the financial resilience measures of farmers' collectives were examined. The researchers ensured that participants were adequately informed about the nature, goal, and possible risks and benefits of the study in order to obtain their informed consent. Taking into account the possible power dynamics inside the collectives, this is especially crucial.

By anonymizing data and making sure that sensitive information is kept private, the researchers safeguarded participants' privacy. They needed to respect and acknowledge the customs, values, and cultural beliefs of the farmers' communities. This entailed being aware of linguistic obstacles and making sure that the dialog was suitable for the participants. They also needed to make certain that the research procedure was just and equal, and that the community and participants shared equally in the advantages of the study. The community was involved in the study process at every stage, from the planning stage to the distribution of results. This guaranteed the research's applicability, cultural fit, and community-benefiting nature.

6.4 Results and Discussion

The study investigated the existing resources and capabilities available for moderately functioning FPOs using RBV theory. The key resources and whether the farmers have the capabilities to benefit from those resources are discussed here.

6.4.1 Economic resources

The Indian government offers equity grant schemes for FPOs that help eligible farmers to access grants equal to their shareholder members' equity contribution in their FPOs. This credit is a VRIO resource offered by the government specifically to FPOs in India. The allocated fund can be utilized for asset creation, as working capital, and for enhancing operational capacity. Grants help FPOs enhance their net worth, leveraging their eligibility to apply for group loans from financial institutions. A matching-equity grant of INR 2000 per farmer member was announced for the central sector plan of 2019, with a maximum threshold of INR 15 lakh per FPO, to resolve the shortage in capital amount for the formation and promotion of FPOs (Government of India, 2023). However, FPOs are committed to increasing their membership to a maximum of 300 to get the full benefit of the equity grant. The project managers of promoting agencies report that:

> "It is very difficult to mobilize 2000 rupees as the farmers are from poor socioeconomic backgrounds. They think that this money is too high for them to contribute and will be gone and not returned to them in any way."

This indicates the poor capability of farmers to acess the equity grants offered to FPOs. There is the possibility of getting agriculture loans for FPOs from financial institutions, depending on the amount of cultivation they do. However, existing FPO members reported that the banks are reluctant to lend money to the FPOs due to a lack of sufficient collateral. Project managers shared that information unevenness, a lack of procedural documents for applying for loans, and prior experiences with poor repayment of loans are the critical reasons why these institutions choose not to lend to the FPOs. The lack of confidence from financial institutions is also an indication of the poor capability of farmers to repay the loans. Similarly, other studies (Sawairam, 2016; Chauhan *et al.*, 2021) suggested that FPO members need to build effective financial management to win over the confidence of members and potential investors.

Existing research indicates that there is a positive correlation between financial literacy and behavior, which in turn improves financial resilience through the enlargement of asset ownership and income diversification (Agatha *et al.*, 2023). An experimental study from rural Rwanda claimed that financial literacy training has the potential to enhance the financial knowledge, savings, and borrowing behavior of farmers (Sayinzoga *et al.*, 2016).

However, these researchers were curious to know how the FPO members survived financial incapabilities amid these financial barriers and challenges. It was found that the diversification of income from agricultural and non-agricultural areas is the resilient strategy adopted by FPOs.

6.4.2 Diversification of income sources by FPOs to become resilient to financial shocks

Farmers started diversifying their income streams beyond traditional agricultural activities, such as agro-processing, value-added services, and non-farm enterprises. Some members become part-time farmers and they mobilized alternative sources of income. Moreover, several FPOs forged partnerships with government agencies, financial institutions, and private sector stakeholders to access technical expertise, funding opportunities, and market linkages.

All the moderately functioning FPOs sampled for the study had a variety of crops to cultivate on a rotation basis to ensure agricultural engagement and income. When asked about how the farmers managed to find income during the off season, the majority of them shared that they go for non-agricultural labor work. Respondent 12 shared his idea of how their FPO (F2) managed to find income for FPO members in all seasons:

> Initially our FPO collected and sold bananas to outside markets and we used to get good prices through collective bargaining. However, during the time of the pandemic, our market was closed due to travel restrictions and we thought of converting the bananas to chips and powder so that storage time can be prolonged and income will be more. This strategy eventually became successful and now we earn income from both means.

The story shared by F4 FPO from Tamil Nadu is different. They were worried that they could not depend on the FPO for consistent income. But they maintained membership in order to access group loans. The members use this loan amount for meeting their expenses related to hospital visits, marriages of sons or daughters, educational fees of their children, and buying household equipment like refrigerators, televisions, or washing machines. The repayment of the loan will be done once they receive profit from selling the rice they cultivate in the market. FPOs in Tamil Nadu started mobilizing more members to join FPOs by offering the opportunity to access group agri-based loans from the FPOs. People started joining the FPO when they heard about this opportunity and the FPOs were able to sustain the activities with the active involvement of members. People are more interested in accessing a loan from the FPOs than in making profits from their products. Therefore, FPOs that struggle to increase their membership base can adopt this strategy of highlighting the scope of getting agricultural loans from them, which may significantly influence the active participation of members in the group; this then impacts the human resources available in the FPOs.

When some FPO members diversified their income by engaging in non-agricultural work during the agricultural off season, another set of members engaged in value addition of their crops for better pricing. Additionally, when some members could not find consistent income from FPOs, they continued their membership in the FPO because they found that their extra household expenses could be managed with the help of FPO group loans.

Similar trends of farm diversification have been observed using a resource-based view in rural farm organizations in European countries and it was confirmed that diversified farms were successful in possessing resources needed for their firm's success (Rantamäki-Lahtinen, 2009).

6.4.3 Human resources

The second resource identified in FPOs is the human resource. The FPO scheme is an effective program to mobilize farmers in a locality under a registered organization. The collective action of farmers under the FPOs is unique because only primary producers in a locality are eligible to be registered as FPO members, which purposefully restricts the involvement of non-producers in an area to be part of the decision-making process of the collective. Farmers in FPOs use the services of their family members in agricultural activities since FPOs are regional and it is often more convenient for family members to engage in farm activities. Respondent 2 in F5 remarked:

> I am the official member of FPO but when I go for non-agricultural jobs in town, my wife and children help in watering the plantations. My children are studying at college and they use their free time to engage in FPO activities like documenting monetary transactions of FPO members, report writing and sharing the report to FPO promoting agency. All the computer-related activities in FPO are done by my son since we do not have computer skills.

This indicates that the insufficient managerial and technical skills of farmers may be compensated for by the involvement of their educated children. This cannot be replicated in all FPOs. Some FPOs lack family support and do not have the technical expertise to fulfil managerial works. Consequently, FPOs in Tamil Nadu were good at their technical skills due to the effective monitoring and handholding of group leaders within the groups. A female FPO member from Tamil Nadu commented about how they have learned about banking transactions:

> I was unaware of how to start a bank account and was afraid to visit the bank but our group leader gave the right guidance and accompanied us to the bank and gradually we women got the confidence to do all bank transactions by ourselves. The involvement in FPO helped us to learn these social skills.

A blend of skilled and unskilled human resources with formal help (support of FPO-promoting agencies) and informal help (support of family members) is helping farmers perform well in the institutional framework of FPO policies and regulations in India. In the same vein, labor management in farm organizations was researched by Mugera and Bitsch (2005) based on RBV theory and they found that human resource management qualifies as a source of competitive advantage for farm organizations.

6.4.4 Organizational resources in FPOs

Organizational resources in FPOs include the share capital contributed by members, the crops, pesticides, infrastructure and warehousing facilities, bookkeeping documents, and members' passbooks. FPOs have registers to document the financial transactions of the FPO every week. However, these registers are not in the local language of farmers, which means the farmers are unable to fulfill the documentation process efficiently. Some FPOs find it difficult to scale up their operations due to infrastructure and financial shortages. On the other side, there are FPOs, especially those with good leadership and business competence, who have been able to scale up their operations and extend into various agricultural value chains, which has led to improved financial outcomes (Bikkina *et al.*, 2018; Gauraha *et al.*, 2021; Kumari *et al.*, 2021). Respondent 2 said about the organizational governance of their FPO:

> There is a need to set up robust governance frameworks and financial literacy training classes for FPO members to achieve financial resilience. More farmers can be attracted to FPOs if we can ensure the economic well-being of farmers by adopting creative strategies for the financial growth of FPOs by themselves.

These challenges from the field corroborate recent findings from a systematic literature study on the financial constraints of agriculture finance in underdeveloped and developing countries (Khan *et al.*, 2024). This literature study proposed an integrated framework of financial constraints that are broadly classified as demand-side constraints, supply side constraints, and infrastructural constraints.

Thus, it is significant to note how FPOs mitigate this organizational resource shortage. The farmers unknowingly adopted institutional bricolage as a resilient strategy to perform well in the resource-constrained environment of FPOs.

6.4.5 Institutional bricolage as a resilient strategy for human resource development in FPOs

Institutional bricolage was one approach adopted by all these FPOs to sustain their activities. In organizational settings, institutional bricolage is defined as the innovative rearranging and combining of pre-existing social, cultural, and institutional elements to create new practices, artefacts, and identities (De Koning and Cleaver, 2012, p. 281). It is informed by local practices, critical viewpoints, the need for resourcefulness and risk management, and a tendency to blend formal and informal, modern and traditional aspects (Sakketa, 2018). Similarly, FPO members rearranged the organizational functions of FPOs when they found the prescribed institutional guidelines set by the government were time consuming to achieve. Respondent 6 remarked:

> FPO policy guidelines are sometimes difficult for us to follow. Therefore, what we do is we frame our management rules to fulfil the institutional requirements of FPO policy guidelines, for example when the FPO registers were in a non-native language, we made a register in our local language so that everybody can understand and fill in the details easily.

Farmers produced FPO registers in their language and templates to document the activities of their FPOs. When markets were down, they found opportunities in local markets instead of going to government marketing platforms. They started running microfinance credit facilities by mobilizing small amounts of money from FPO group members to fund agricultural or personal expenses.

Weak FPO governance structures were revised by the members themselves according to the social and cultural norms of their locality. FPOs in Palakkad initially started forming groups by involving only full-time farmers but later realized that the membership base could be scaled up without such restrictions. They started welcoming part-time farmers as well. The part-time farmers were, for example, retired bank officers or people employed in non-agricultural work, who helped the development of FPO in many ways. For instance, one retired bank employee was skillful in preparing all the bank-related documents that needed to be submitted to the bank to avail financial aid effectively. Some part-time members were heavy vehicle drivers who could mobilize the transportation of agricultural produce at nominal rates. A small revision in the membership guidelines of FPOs to include part-time farmers also allowed a tremendous growth for the FPOs. This can be linked to the approach of institutional bricolage in farmers' collectives where the existing governance systems of FPO are reorganized or revised to accommodate new developments for the organization.

The conceptual framework of financial sustainability challenges and strategies adopted by farmers' communities to mitigate the challenges is depicted in Fig. 6.5.

The community-driven strategies helped farmers to become financially resilient. Financial resilience is critical to the success and long-term viability of farmers' collectives. It helps businesses handle unanticipated expenses, fluctuations in commodity prices, and financial hardships. Furthermore, financial resilience allows farmers' collectives to improve infrastructure, invest in state-of-the-art agricultural technologies, and increase overall productivity (Bizikova *et al.*, 2020). Financially resilient farmers' collectives are better equipped to withstand shocks and recover, ensuring the continued existence of their enterprises and means of subsistence. Furthermore, the collective financial resilience of farmers may contribute to greater food security and a decline in poverty.

6.4.6 Policy implications and recommendations

There are two main recommendations.

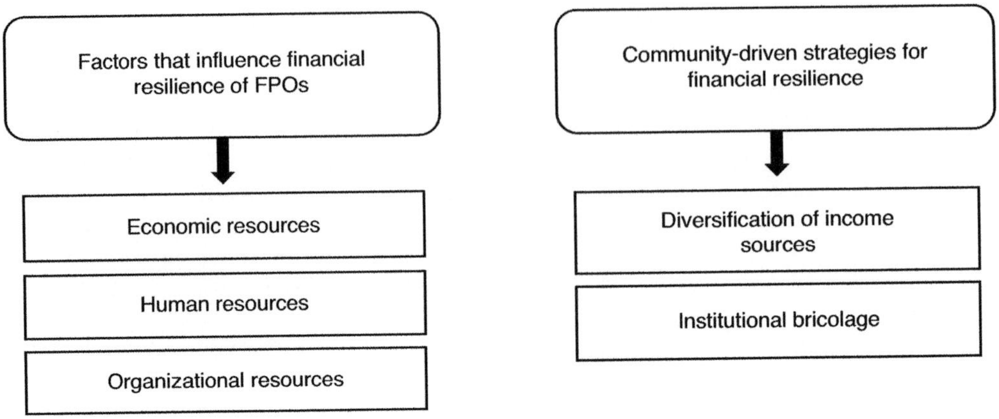

Fig. 6.5. Conceptual framework on financial resilience and sustainability of FPOs.

1. *Strengthening financial infrastructure*. Policy makers should focus on improving access to credit, enhancing governance structures, and providing technical assistance to FPOs to strengthen their financial resilience.
2. *Promoting collaborative partnerships*. Initiatives that facilitate partnerships between FPOs, government agencies, and private sector stakeholders can enhance resource mobilization, knowledge sharing, and market access for sustainable agricultural development.

Overall, the study underscores the critical role of financial resilience in enabling FPOs to navigate sustainability challenges and to advance agricultural development objectives. By understanding and promoting effective financial strategies, stakeholders can support the growth and impact of FPOs in promoting sustainable agriculture and improving livelihoods in rural communities.

6.4.7 Limitations of the study

In the study discussed in this chapter, only a few key stakeholders from five sampled FPOs were purposively selected for the interviews based on the prior knowledge and familiarity of the researchers with the respondents. However, some respondents expressed their concerns about sharing the financial challenges of their FPO. Although the researchers explained the ethical considerations of keeping anonymity and confidentiality of data shared, some of the respondents hesitated to share their actual opinions openly with the researcher. Some respondents found it difficult to explain the challenges on record (with the voice recorder switched on) but were comfortable in sharing the same when not being recorded. The findings of this study are only indicative and cannot be generalized to the entire Indian context since the sustainability challenges of FPOs are multifaced and unique to the context factors.

6.5 Conclusion

This chapter sheds light on the intricate relationship between financial resilience and sustainability outcomes within moderately functioning FPOs in South India. Through an in-depth analysis of financial challenges, strategies, and their impacts, several key insights have emerged. When the financial sustainability of farmers' collectives is impacted by these many challenges and factors, it is imperative to understand the community-driven strategies adopted by the farmers to resist or mitigate the negative impacts and sustainability challenges. By diversifying income sources and institutional bricolage approaches, FPOs can mitigate financial risks, strengthen market linkages, and improve livelihood opportunities for their members.

Furthermore, the chapter demonstrates the positive association between financial resilience and the adoption of sustainable agricultural practices. Financially resilient FPOs are better positioned to invest in environmentally friendly farming techniques, such as organic farming and water conservation, contributing to the conservation of natural resources and the promotion of agroecological sustainability. The research provided insight into the paths towards achieving sustainability in the face of inherent restrictions and uncertainties by investigating the financial mechanisms employed by FPOs that perform moderately.

The implications of this study extend beyond academic discourse to inform policy and practice in the realm of agricultural development. Policy makers, practitioners, and FPO leaders can leverage the insights gained from this research to design targeted interventions that strengthen the financial resilience of FPOs, promote sustainable agriculture, and enhance the socio-economic well-being of rural communities. In essence, this study underscores the imperative of integrating financial resilience into the broader agenda of agricultural sustainability. By empowering FPOs with the financial tools, resources, and support they need, transformative change can be catalysed towards a more resilient, inclusive, and sustainable agricultural sector in South India and beyond.

In addition to contributing to the academic discourse on the sustainability of agribusiness organizations, the research findings offer valuable insights for policy makers, practitioners, and leaders of FPOs who seek to enhance the financial resilience and efficacy of agribusiness organizations. The world struggles with the requirement of modifying agricultural systems for a sustainable future, but the experiences of moderately operating rural farmers' collectives offer hope and inspiration in the battle for rural livelihood resilience and agricultural sustainability.

References

Adnan, M., Abdulhamid, T. and Sohail, B. (2018) Predicting firm performance through resource based framework. *European Journal of Business and Management* 10(1), 31–40. Available at: www.raoso ft.com

Agatha, M.K., Nurfadilah, R.L.A. and Pandin, M.Y.R. (2023) The effect of financial literacy and financial behavior on financial resilience in MSMEs (Study of MSME actors in Ngagel Village, Surabaya). *Finance: International Journal of Management Finance* 1(2), 66–75. DOI: 10.62017/finance.v1i2.25.

Bekele, W.D. (2023) Enhancing farmers' resilience to climate change-induced impacts through financial inclusion in sidama region, Southern Ethiopia. *Climate and Development* 16(2), 149–160. DOI: 10.1080/17565529.2023.2203675.

Bikkina, N., Turaga, R.M.R. and Bhamoriya, V. (2018) Farmer producer organizations as farmer collectives: A case study from India. *Development Policy Review* 36(6), 669–687. DOI: 10.1111/dpr.12274.

Bizikova, L., Nkonya, E., Minah, M., Hanisch, M., Turaga, R.M.R. *et al*. (2020) A scoping review of the contributions of farmers' organizations to smallholder agriculture. *Nature Food* 1(10), 620–630. DOI: 10.1038/s43016-020-00164-x.

Carvalho, D.M.D., Prévot, F. and Machado, J.A.D. (2014) The use of the resource based view theory based in rural properties: A systematic review of the literature. *Revista de Administração (São Paulo)* 49(3), 506–518.

Chaiya, C., Sikandar, S., Pinthong, P., Saqib, S.E. and Ali, N. (2023) The impact of formal agricultural credit on farm productivity and its utilization in Khyber Pakhtunkhwa, Pakistan. *Sustainability (Switzerland)* 15(2). DOI: 10.3390/su15021217.

Chauhan, J.K., Adhikary, A. and Pradhan, K. (2021) Identification of constraints associated with farmers' producer organisations (FPOs). *International Journal of Current Microbiology and Applied Sciences* 10(1), 1859–1864. DOI: 10.20546/ijcmas.2021.1001.217.

De Koning, J. and Cleaver, F. (2012) Institutional bricolage in community forestry—setting out an agenda for future research. In: *Forest-People Interfaces: Understanding Community Forestry and Biocultural Diversity*. pp. 1–317. DOI: 10.3920/978-90-8686-749-3.

Dhineshwari, S., Selvam, S. and Prabakaran, K. (2021) Performance analysis of the farmer producer companies in Western Tamil Nadu, India using Altman's Z-score. *Madras Agricultural Journal -TNAU 108*. DOI: 10.29321/MAJ.10.000534.

Fuzinatto, N.M. and Santos Junior, S.Í.L.V.I.O. (2020) Urban farming as competitive resource in food services: An evaluation through the resource-based view theory. *Turismo: Visão e Ação* 22(1), 2–23. Available at: http://www.scielo.br/scielo.php?script=sci_arttext&pid=S1983-71512020000100002&lang=pt

Gauraha, A. K., Sahu, Y., Jaiswal, R., Sahu, T.K. and Chandrakar, M.R. (2021) Production, marketing and processing of black rice: A case study of ojashvi farmer producer organization, Kurud district Dhamtari. ~ *653* ~ *The Pharma Innovation Journal* 10(8), 653–658. Available at: http://www.thepharmajournal.com

Gautam, S. and Mallaiah, L.C. (2024) Enhancing farmer's income and farmer producer organizations' (fpos) in india. *Saudi Journal of Economics and Finance* 8, 91–101.

Gibson, C.B., Gibson, S.C. and Webster, Q. (2021) Expanding our resources: Including community in the resource-based view of the firm. *Journal of Management* 47(7), 1878–1898. DOI: 10.1177/0149206320987289.

Gorai, S.K., Monika Wason, R.N.P., Rao, D.U.M. and Paul, R.K. (2022) Factors contributing to the stability of the farmer producer organisations: A study in West Bengal. *Indian Journal of Extension Education* 58(2), 91–96. DOI: 10.48165/ijee.2022.58218.

Government of India (2023) Constitution of farmer producer organizations. *Ministry of Agriculture & Farmers Welfare* 2013. Available at: https://pib.gov.in/PressReleaseIframePage.aspx?PRID=1942484#:~:text=In

Jaafar, M. and Abdul-Aziz, A.-R. (2005) Resource-based view and critical success factors: A study on small and medium sized contracting enterprises (SMCEs) in Malaysia. *International Journal of Construction Management* 5(2), 61–77. DOI: 10.1080/15623599.2005.10773075.

Kakati, S. and Roy, A. (2019) Financial sustainability: A study on the current status of farmer producer companies in Northeast India. *The IUP Journal of Management Research* 18(2), 23–42.

Khan, F.U., Nouman, M., Negrut, L., Abban, J., Cismas, L.M, *et al.* (2024) Constraints to agricultural finance in underdeveloped and developing countries: A systematic literature review. *International Journal of Agricultural Sustainability* 22(1). DOI: 10.1080/14735903.2024.2329388.

Kujur, P. and Gauraha, A.K. (2019) Business performance of farmer producer organizations (FPOs) in Chhattisgarh plains. *Journal of Pharmacognosy and Phytochemistry* 8(4), 3476–3480.

Kumari, S., Bharti, N. and Tripathy, K.K. (2021) Strengthening agriculture value chain through collectives: Comparative case analysis. *International Journal of Rural Management* 17(1_suppl), 40S–68S. DOI: 10.1177/0973005221991438.

Kumar, S., Sankhala, G., Kar, P. and Sharma, Ph.R. (2021) An appraisal of financial sustainability of dairy-based farmer producer companies in India. *Indian Journal of Extension Education* 57(4), 115–119. DOI: 10.48165/ijee.2021.57425.

Kumari, A. and Garg, V. (2023) Impact of credit on sustainable agricultural development in India. *Journal of Sustainable Finance and Investment* 13(1), 560–571. DOI: 10.1080/20430795.2021.1964811.

Kumari, N. and Bisht, K. (2023) Identification of core competencies for managing farmer producer identification of core competencies for managing farmer producer organizations. *Indian Journal of Extension Education* 59, 108–112. DOI: 10.48165/IJEE.2023.59320.

Lindsjö, K., Mulwafu, W., Djurfeldt, A.A. and Joshua, M.K. (2021) Generational dynamics of agricultural intensification in Malawi: Challenges for the youth and elderly smallholder farmers. *International Journal of Agricultural Sustainability* 19(5–6), 423–436. DOI: 10.1080/14735903.2020.1721237.

Malik, S. and Kajale, D. (2024) Empowering small and marginal farmers: Unveiling the potential and addressing obstacles of farmer producer organizations in India. *Research on World Agricultural Economy* 5(1), 32–47. DOI: 10.36956/rwae.v5i1.994.

Manikutty, S. (2000) Family business groups in india: A resource-based view of the emerging trends. *Family Business Review* 13(4), 279–292. DOI: 10.1111/j.1741-6248.2000.00299.x.

Ministry of Agriculture and Farmers Welfare (2021) Central Sector Scheme "Formation and Promotion of 10,000 new Farmer Producer Organizations (FPOs)" of Rs. 6865 crore, Press Information Bureau, Government of India, 09 FEB 2021 6:06PM by PIB Delhi. Available at: (accessed 2 June 2022).

Ministry of Agriculture and Farmers Welfare (2023) *Ministry of Agriculture and Farmers Welfare*. Constitution of Farmer Producer Organizations. Available at: (accessed 12 October 2023).

Mourya, M. and Mehta, M. (2021) Farmer producer company: India's magic bullet to realise select SDGs? *International Journal of Rural Management* 17(1_suppl), 115S–147S. DOI: 10.1177/0973005221991660.

Mugera, A.W. and Bitsch, V. (2005) Managing labor on dairy farms: A resource-based perspective with evidence from case studies. *International Food and Agribusiness Management Review* 8(3), 18.

Negreti-Campos, A.S., Dall Evedove, A.C.L. and Smith, A.E.B. (2022) Resource-based view: An analysis of cooperatives involving family farmers. *Revista Ibero-Americana de Estratégia* 21(1), 1–38.

Nithya, S. and Vaishnavi, P. (2022) Challenges faced by farmer producer organisations (FPOs) - a review. *Journal of Agricultural Extension Management* XXIII, 131–140.

Prasad, R., Shivay, Y.S. and Kumar, D. (2017, January) Current status, challenges, and opportunities in rice production. *Rice Production Worldwide* January, 1–32. DOI: 10.1007/978-3-319-47516-5_1/COVER.

Ranjan, R. (2014) Multi-dimensional resilience in water-scarce agriculture. *Journal of Natural Resources Policy Research* 6(2–3), 151–172. DOI: 10.1080/19390459.2014.898872.

Rantamäki-Lahtinen, L. (2009) The success of the diversified farm - resource-based view. *Agricultural and Food Science* 18(1), 1–10. DOI: 10.23986/afsci.5969.

Roxas, H.B. and Chadee, D. (2011) A resource-based view of small export firms' social capital in a Southeast Asian country. *Asian Academy of Management Journal* 16(2), 1–28.

Sakketa, T.G. (2018) Institutional bricolage as a new perspective to analyse institutions of communal irrigation: Implications towards meeting the water needs of the poor communities. *World Development Perspectives* 9, 1–11. DOI: 10.1016/j.wdp.2017.11.003.

Sawairam, P. (2016) Training need assesment of members of farmer producer organisation and producer companies in India with reference to Pune district. *We'Ken- International Journal of Basic and Applied Sciences* 1(3). DOI: 10.21904/weken/2016/v1/i3/103461.

Sayinzoga, A., Bulte, E.H. and Lensink, R. (2016) Financial literacy and financial behaviour: Experimental evidence from rural Rwanda. *Economic Journal* 126(594), 1571–1599. DOI: 10.1111/ecoj.12217.

Sharma, D.R. (2022) Farm producers organizations a collectiveness approach, study & review of scheme. *International Journal For Multidisciplinary Research* 4(4), 517–531. DOI: 10.36948/ijfmr.2022.v04i04.059.

Tata-Cornell Institute for Agriculture and Nutrition (2024) Identifying the number of fpos promoted in India. *TCI*. Available at: https://tci.cornell.edu/?publications=identifying-the-number-of-fpos-promoted-in-india

Trivedi, P.K., Manjur, A. and Satpal. (2023) Farmer producer organisations in North India: Potentials and challenges. *International Journal of Rural Management* 19(3), 379–398. DOI: 10.1177/09730052221107730.

Ullah, A., Mahmood, N., Zeb, A. and Kächele, H. (2020) Factors determining farmers' access to and sources of credit: Evidence from the rain-fed zone of Pakistan. *Agriculture (Switzerland)* 10(12), 1–13. DOI: 10.3390/agriculture10120586.

Wadkar, S., Ingale, P. and Thakur, B. (2021) Financial performance evaluation: A case study of MAHA-Farmers Producers Company Ltd. *Journal of Co-Operative Management* 56(3), 14–26.

7 Empowering Indigenous and Rural Women through Community-Based Sustainable Tourism Development: Building Resilient Communities

Smitha Amarail Mohanan[1*], Anuradha Sathiyaseelan[2], Suja John[1] and Shuaibu Chiroma Hassan[3]

[1]Department of Tourism Management, Christ University, Bangalore, India; [2]Department of Psychology, Christ University, Bangalore, India; [3]Department of Home Economics, Isa Kaita College of Education, Dutsinm, Nigeria

7.1 Introduction

Community involvement is crucial for achieving sustainable development goals (SDGs) (United Nations, 2019), focusing on empowering individuals, inclusivity, and gender equality. Rural communities often face challenges in implementing sustainable development initiatives, but women play a significant role in achieving economic empowerment. The Self-Employed Women's Association in India has made significant progress in elevating women's economic status (Mehra, 1997). However, there is a lack of noticeable improvement in rural women's status in India. To create more opportunities for women in the tourism sector, researchers should focus on identifying various dimensions of women's empowerment and addressing insufficient levels of involvement in tourism literature (Movono and Dahles, 2017; Elshaer *et al.*, 2021).

Community-based ecotourism is gaining recognition as an effective form of sustainable development that aims to enhance local livelihoods, preserve the environment, and promote cultural heritage. Gender inequality was recognized by the United Nations Development Programme (UNDP) as a significant factor leading to a decline in human development. As a response, "gender equality" was designated as a SDG within the United Nation's 2030 agenda (Bastian *et al.*, 2019). The study this chapter is based on investigates women's empowerment and involvement in Uravu Organization's ecotourism activities, focusing on local livelihoods and cultural preservation.

Uravu, a non-profit organization in Kerala, focuses on fostering rural empowerment through sustainable solutions. Established in 1996, it collaborates with individuals, governments, and businesses to drive sustainable development initiatives. Uravu emphasizes the bamboo sector and promotes sustainable development goals, including handicrafts for rural women and skill development training. The organization also conducts workshops and skill development programs.

*Corresponding author: smitha.am@res.christuniversity.in

 Agripreneurship, Digital Inclusion and Sustainability: Rural Innovation and Growth in the Global South (eds Dutta and Trivedi)
DOI: 10.1079/9781836992646.0007

The research reveals that community-based ecotourism (CBET) is the most effective way to empower rural women, promoting balanced household responsibilities and environmentally friendly practices. The Uravu project, despite its limited presence in India, aligns with CBET theories and highlights the connection between women's empowerment and sustainable development through CBET activities.

The objectives of the study were twofold.

1. To understand the influence of Uravu's CBET activities on women's empowerment.
2. To discover the outcomes of women's empowerment achieved through the practice of CBET activities by Uravu.

7.2 Community-Based Sustainable Tourism Development

Tourism development represents a comprehensive approach to environmental preservation. Sustainable tourism aims to provide viable income and employment opportunities for local communities while minimizing environmental impact and respecting traditional values (Bhuyan, 2014). This shows that sustainable tourism development is the same as community-based ecotourism, theoretically and practically. The World Tourism Organization defines sustainable tourism as:

"Sustainable tourism development meets the needs of present tourists and host regions while protecting and enhancing opportunities for the future. It is envisaged as leading to management of all resources in such a way that economic, social, and aesthetic needs can be fulfilled while maintaining cultural integrity, essential ecological process, biological diversity, and life support system" (Cernat and Gourdon, 2011).

Here, we can see that community-based ecotourism and sustainable tourism development protect the community and biodiversity of the destination.

Tourism development in Bandarban, Bangladesh is expected to benefit both ethnic and non-ethnic local people by generating employment, alleviating poverty, and maintaining an ecological balance. To ensure sustainability and protect the unique culture and environment, community-based tourism (CBT) is suggested as a recent approach to connect tourism actors with the local community and the poor (Hasan and Islam, 2015). CBT integrates numerous activities and facilities with the involvement of the local community. Individuals engaged in the CBT program can be homestay families, tour guides, artisans, performers, community leaders, local restaurant operators, farmers, and others fulfilling significant roles (Hasan and Islam, 2015).

7.3 Community-Based Ecotourism

Community-based ecotourism acts as a social transformation tool for local communities, especially in developing countries. The rural area of Sabha, Malaysia has enhanced the livelihood of local communities in terms of economic, sociocultural, and environmental aspects, which has led to social transformation (Kunjuraman *et al.*, 2022). CBET is a distinct approach to tourism that grants the local community authority over its development and administration, ensuring that the majority of the benefits are reinvested in the community. For more than three decades, CBT has been advocated as a means of development, satisfying the social, environmental, and economic requirements of local communities by providing tourism products (Goodwin and Santilli, 2009).

CBET is a promising economic development strategy, especially in rural areas, requiring key prerequisites like economic advancement, environmental protection, cultural preservation, effective marketing, and political support (Kiss, 2004; Akbar *et al.*, 2022). The effectiveness of CBE initiatives hinges on well-executed participatory management schemes that establish collaborative connections between essential stakeholders, such as protected area administrative personnel and local communities. Additionally, these initiatives work to nurture community empowerment. Therefore, to enhance community empowerment, it is advisable to involve women and motivate them in activities related to CBET.

In rural areas where the primary means of sustenance rely on natural resources like forests, there exists a potential danger of excessively exploiting these resources. Most of the population engages in occupations like farming, logging, mining, hunting, and gathering; the continuous pursuit of their economic endeavors might ultimately lead to the exhaustion of these valuable natural assets. Nonetheless, CBET offers a more enduring approach to overseeing these natural resources, as it encourages conservation through active engagement of the local communities (Mensah, 2017). However, there is a need for further exploration into the impact on livelihoods resulting from the implementation of CBET projects. This is particularly relevant for women, as their empowerment has a broader community influence than men.

CBET is a promising initiative for generating income for local communities and promoting conservation despite its limitations, such as limited geographic coverage, limited community participation, and limited connections between biodiversity enhancements and commercial viability in the tourism sector.

7.4 Women's Participation in Community-Based Ecotourism

The concepts of CBET and women's empowerment converge significantly within the domain of sustainable development and conservation endeavors. The involvement of women in community development initiatives has become indispensable, contributing to the advancement of inclusive development and the decentralization of development planning (Nutsugbodo *et al.*, 2020). CBET is increasingly acknowledged as a sustainable development approach to fostering local livelihoods, environmental preservation, and cultural heritage. Nevertheless, similar to other development projects, the extent of participation and benefits are influenced by contextual factors and social structures, including gender. A Vietnamese case study contributes valuable insights to the growing knowledge base on ecotourism as a means of sustainable development (Tran and Walter, 2014). It illuminates explicitly the challenges of integrating a gender perspective into CBET and emphasizes the potential positive outcomes for local women. Furthermore, the study demonstrates the effectiveness of employing gender analysis tools commonly used in gender and development projects to address the unique context of CBET.

The green belt movement in the Malang Regency, Indonesia initially involved just men but has been transformed by Ramadhoan and his team, who have incorporated gender inclusivity into the model (Ramadhoan *et al.*, 2023). This initiative empowers local women to tackle environmental challenges and preserve biodiversity, contributing to global efforts to adapt to climate change and mitigate disasters.

7.4.1 Women's empowerment

Women across various societies and in all aspects of life often find themselves in an unequal position and with a lower status. Therefore, it becomes crucial to enable them by ensuring equal opportunities. The concept of empowerment, which encompasses dimensions such as psychological, social, political, and economic aspects, finds its origins in Scheyvens' pivotal study in 1999. This study was the pioneering effort to apply insights from the psychology and development literature's understanding of empowerment to the context of ecotourism, building upon the groundwork laid by Friedmann (1992) and Perkins and Zimmerman (1995). Empowerment, a multifaceted social process, is vital in granting individuals control over their lives.

Moreover, it can be described as a transformative process that cultivates personal and collective power, enabling individuals to address significant issues in their lives, communities, and society. From the latter half of the 20th century, scholars in universities, as well as national and international platforms, have increasingly recognized the significance of women's empowerment. However, it was not until the declaration of the "women's decade" in 1975 that this concept began to be firmly integrated into governmental policies and programs (Mandal, 2013).

Women exhibit a higher tendency to perceive themselves as empowered compared with males. The SDGs prioritize the empowerment of women as a critical area of concentration and as a yardstick for societal change (Miedema *et al.*, 2018). Although sustainable tourism hinges on achieving gender equality and empowerment, a majority of the existing literature has predominantly explored these ideas using qualitative approaches (Boley *et al.*, 2016). CBET is an alternative form of sustainable tourism; the research on CBET and women's empowerment has yet to be explored in more detail. Skill development programs and training will help to improve self-confidence and enhance the participation and involvement of women in different community-based activities.

While women's empowerment in the tourism sector is in its early stages, it has already left a significant impact as a cornerstone of community involvement. The active involvement of women in tourism, especially in developing countries, has notably enhanced economic conditions in financially challenged areas. This has led to the promotion of sustainable resource utilization and an increase in planning authority (Mohanty and Chandran, 2018). Women must participate in all project stages, including planning, development, implementation, and overall engagement. That will improve the organization's productivity and increase women's empowerment. While India is making strides in its economic advancement, there is a pressing need to simultaneously focus on social and human development and promote women's empowerment. The participation and empowerment of women plays a crucial role in reinforcing women's rights, granting them autonomy over their lives, and exerting a positive influence on society. Economic empowerment of women is a key driver for achieving sustainable development. The empowerment of women and the establishment of gender equality significantly amplify developmental endeavors. Aligned with SDG5, India has committed to ensuring that, by 2030, women and girls contribute as equal partners to the nation's growth and progress. On a global scale, it has been recognized that achieving gender parity in economic participation could result in a remarkable US$11 trillion addition to the world's annual GDP if the contribution of women matched that of men (Tiwari and Malati, 2023).

Training and skill development also catalyze rural women's empowerment (Sindhi, 2012). Women's empowerment through education and training, self-confidence, and decision-making skills are enhanced by involving marginalized rural women in cultural heritage tourism activities. For example, participation in embroidery tourism by rural Hui women in China brings economic advantages and fosters social, psychological, educational, and political empowerment. Employee training is identified as a pivotal factor in this process (Su *et al.*, 2023). In this case study, the role of employee training holds great importance.

Kerala's government is implementing a gender and development initiative focusing on women's empowerment and involvement in governance decision making and economic growth, empowering women with influence over their life choices. The government is coordinating efforts and endorses the initiative, acting as a catalyst for transformative changes. The success of Kudumbashree could be applied to designing community development programs for women and marginalized communities (Venugopalan *et al.*, 2021). (Kudumbashree is a program initiated by the State Poverty Eradication Mission (SPEM) of the government of Kerala, with the aim of eradicating poverty and empowering women.)

Empowerment is an evolving process based on resources, agency, and achievements (Kabeer, 1999; Sharaunga *et al.*, 2019). The study this chapter discusses examines the impact of CBET activities in the Uravu organization on women's empowerment. It examines how these activities contribute to female employees' economic, political, social, and psychological empowerment within Uravu. Assessing women's agency and resource access is crucial for measuring their empowerment.

7.5 Uravu Organization

Uravu is the largest bamboo nursery in South India, offering over 50 species to support eco-restoration programs and bamboo-based production units. Committed to promoting bamboo for its environmental and economic benefits, Uravu views bamboo as a sustainable crop that guarantees income with minimal risk. The nursery not only supplies saplings but also manages plantations, aligning with the central government's original initiative to double farmers' income by 2022. Uravu has set industry benchmarks with over 500 bamboo products, training women artisans to create a wide range of items. The organization conducts training programs and workshops in bamboo art, crafts, furniture making, and construction, partnering with agencies like the UNDP and National Bank for Agriculture and Rural Development (NABARD) (Uravu, 2019).

Customized workshops are offered to design students, school groups, and artisans, contributing to skill development in the bamboo sector. Uravu collaborates with Vythiri Resort to promote responsible tourism in Wayanad, Kerala emphasizing eco-friendly construction, community involvement, and the preservation of local culture and heritage. Bamboo interiors and sustainable architecture are key activities, with Uravu using specific bamboo species known for their strength and esthetic value. With 25 years of expertise, Uravu meticulously processes and cures bamboo poles for construction, ensuring only the best materials are used. The focus is on innovative, sustainable designs that blend traditional and modern techniques, harmonizing with the natural environment (Uravu (Uravu Indigenous Science and Technology Study Centre), (2019)).

The primary focus of this organization is the empowerment of indigenous women through sustainable, community-based activities. A key feature of the organization is its commitment to a zero-waste policy, which underscores its environmental stewardship. No harmful by-products are produced during the manufacturing process, and all residual bamboo waste is converted into fertilizer. This approach not only minimizes environmental impact but also contributes to soil enrichment, making the end product beneficial for plant growth. Furthermore, bamboo products produced by the organization serve as natural air purifiers by sequestering carbon dioxide and releasing oxygen, thus enhancing air quality while promoting sustainability.

7.6 Methodology

In line with the exploratory nature of this study, qualitative research methods have been adopted to form the foundation of this chapter. Braun and Clarke's (2006) thematic analysis was followed to analyze the data. The chosen approach for this research is a case study, as it allows for a comprehensive evaluation of the situation by applying inductive reasoning, aiming to address a specific problem. The selection of Uravu Organization as a case study is based on its close ties to CBET activities in Kerala, India. This positioning makes it ideal for conducting an enlightening investigation into the empowerment of rural women within the organization. Employing a case study approach allows for an in-depth exploration of a specific issue, enabling a comprehensive understanding of the situation through inductive reasoning.

Primary and secondary data sources were used. Primary data included individual semi-structured interviews with female employees. The participants included 12 individuals aged 18 years and older who are either married, single, or widowed. Interviews took place between October and November 2021. Thematic analysis was applied according to Braun and Clarke's six steps (Braun and Clarke, 2006). First, consent was obtained for collecting the data. After transcribing the data, consent was again taken from the respondents before analyzing the data. The transcript was read multiple times to familiarize the researchers with the content, and codes were identified, classified into the final codes, and grouped into themes.

Secondary data sources used included published research articles, books, and websites.

A total of 12 individuals were interviewed, with the main criteria for recruiting participants being their willingness to take part. Throughout the process, the researchers placed a strong emphasis on respecting the rights of the participants. They were given the freedom to exit the interview or decline to answer any questions they found uncomfortable. On average, each interview lasted approximately 1 hour. The interviews were audio-recorded after obtaining consent from the participants. Interviews were conducted in the Malayalam language, later translated into English. Respondents were asked to identify their age group, marital status, educational attainment, and work experience within Uravu.

Table 7.1 reveals that the majority of the working women fall within the age groups

Table 7.1. Respondents' age, marital status, education, and length of experience at Uravu.

Respondent	Age (years)	Marital status	Education	Experience (years)
R1	46–55	Divorcee	10th grade	23
R2	36–45	Married	10th grade	17
R3	26–35	Married	Plus two (includes 11th and 12th grades)	5
R4	46–55	Married	10th grade	17
R5	46–55	Married	Pre-degree	14
R6	46–55	Married	Bcom	13
R7	46–55	Widow	Degree incomplete	15
R8	46–55	Husband left	7th standard grade	15
R9	36–45	Married	10th grade	10
R10	36–45	Widow	10th grade	15
R11	36–45	Married	9th grade	15
R12	36–45	Married	10th grade	13

of 36–45 and 46–55 years. Among these respondents, there is one divorcee, two widows, seven married women, and one employee whose husband has left her. The educational attainment of most employees in these age groups is at or below the 10th standard, with only three having pursued education beyond this level. Furthermore, a significant portion of these employees have more than 15 years of work experience within the organization.

7.7 Results and Analysis

The respondents were asked open-ended questions, and their responses, originally recorded in Malayalam, were transcribed into the Malayalam language. These transcriptions were then reviewed and verified by the respondents. Once consent was obtained, the transcriptions were translated into English. Subsequently, the data were analyzed to identify global themes, organizing themes, and basic themes (Table 7.1).

Table 7.2 lists the key themes highlighted in the transcriptions. The respondents were asked a series of questions designed to elicit detailed insights into their experiences with Uravu. These included inquiries such as, "How did you first learn about Uravu?" and "Have you worked with the organization before?" Additionally, they were asked, "How did Uravu assist you during the Covid-19 pandemic?" and "Is your current income sufficient for living expenses, and are you able to save from your salary?" Questions regarding their workplace environment were also posed, including, "How do you feel about the working atmosphere here?" Furthermore, respondents were asked to reflect on their personal experiences with adversity: "Was there a moment in your life when you felt depressed or devastated by life events? If so, how did you manage?" Lastly, the respondents were queried about their participation in decision-making processes within Uravu; specifically, "Does Uravu consider your opinion in decision making processes?" Many respondents confirmed that Uravu conducts meetings that encourage their participation, and they expressed their feelings of empowerment – economically, socially, politically, and psychologically – as a result of their involvement with the organization.

7.7.1 Women's empowerment

The elimination of poverty relies heavily on women's empowerment, which has been highlighted by various international and national organizations. They emphasize the importance of women's participation in society, decision making, education, and health (Sahu and Singh, 2012). The literature on women's empowerment does not provide a singular definition; instead, it encompasses a range of conceptualizations,

Table 7.2. Global theme, organizing themes, and basic themes

Global theme	Organizing themes	Basic themes
1. Women's empowerment	1.1 Working atmosphere	1.1.1 Cooperative colleagues 1.1.2 Less work pressure 1.1.3 Incorporating opinions in decision making 1.1.4 Happy to come to work regularly 1.1.5 Self-confidence
	1.2 Financial independence and security	1.2.1 Financial independence 1.2.2 Self-decision making in purchasing items 1.2.3 Salary savings 1.2.4 Insufficient salary
	1.3 Employee training and community cohesion	1.3.1 Employee training 1.3.2 Improved communication skills
	1.4 Awareness of social and political issues	1.4.1 Lack of representation for women's voices in their society 1.4.2 Non-responsive to political issues 1.4.3 Less political involvement

which can be viewed as a process or an outcome, a goal, or a method to achieve a goal, or as a capability (Sharaunga *et al.*, 2019). Women form the majority of the community population. If women are empowered, the community is also empowered. Most of the sustainable development goals (SDGs) can be attained by women, and women's empowerment can facilitate three of the eight SDG goals. For example, a study in Africa showed that land ownership by women reduced the gender gap (Njoh and Ananga, 2016). This chapter identifies the influence of Uravu's CBET on the empowerment of women and evaluates the outcomes of women's empowerment achieved through this practice. The organizing and basic themes are now examined in more detail.

7.7.2 Working atmosphere

Positive workplace conditions play a crucial role in influencing the well-being of employees, impacting both men and women (Ma and MacMillan, 1999). A study conducted in Korea suggests that job satisfaction is shaped by factors such as the physical environment, negative social interactions, occupational status, and the quality of management (Lee and Park, 2021). Interestingly, the level of employee job satisfaction at Uravu does not predominantly hinge on financial remuneration, as evidenced by employees choosing to remain with the organization despite receiving offers from other sectors.

A pressure-free working atmosphere leads to improved productivity and keeps workers happy. The code "Happy to come to work regularly" states that participants are experiencing psychological well-being and mental healing during the time of daily work. "Cooperative colleagues" is another factor that helps them share their problems and is mentioned as being part of relaxation for them. "Less work pressure" in Uravu helps them to work more peacefully. One of the respondents said that Uravu asks them when they receive an order and only commits to it when they are assured that the employees are able to finish the work on time. "Incorporating opinions in decision-making," with the organization taking their opinions into decision making, is another factor that 90% of the respondents said made them feel part of the institution. A good working atmosphere and supportive colleagues are important factors for the psychological well-being of working women.

Uravu fosters a positive work environment, promoting long-term employee retention through cooperative colleagues, reduced work pressures, and decision-making opportunities, enhancing self-confidence and loyalty.

7.7.2.1 Cooperative colleagues

Social interdependence theory suggests that cooperative activities with colleagues fosters mutual support, enhancing workplace productivity. Research shows that support from colleagues positively impacts teachers' performance, as also found with other organizations and employees (Wolgast and Fischer, 2017).

Carless's (2004) research indicated that employees' perception of their work environment directly influences their job satisfaction and intrinsic motivation, highlighting the importance of environmental factors. Also, individuals form attachments to organizational practices and characteristics of their work environment (Schneider, 1983). Employers' cooperation and understanding, as well as also organizational support, enable the psychological empowerment of rural women. Women experience psychological empowerment when they develop a belief in their capacity to exert influence over their own circumstances. This newfound confidence enables them to initiate personal and societal transformations, as mentioned by Stromquist (1995a).

Even during the challenging period of lockdown due to Covid-19, Uravu demonstrated their support for their employees. Recognizing the financial difficulties faced by the employees who were unable to work, the organization provided them with financial assistance.

Respondent 12 stated, "I have the privilege of working alongside supportive colleagues with whom I can share my concerns and challenges. Even during the lockdown period, Uravu has continued to provide me with valuable support. The work environment is incredibly friendly, to the extent that I can momentarily forget about the problems I face at home during working hours."

7.7.2.2 Less work pressure

Less work pressure leads to women being better able to maintain their work–life balance. A review paper on the work–life balance of women employees found that there are various challenges and issues faced by female employees (Sudha and Karthikeyan, 2014). Working from home can contribute positively to job satisfaction levels by alleviating the pressure associated with traditional office environments (Tabassum and Sarker, 2021).

Work stress has been identified as a factor resulting in poor performance (Sudha and Karthikeyan, 2014). According to a study conducted by Shiva in 2013, women in high-ranking positions face the pressure of returning home on time, managing household chores, and taking care of family matters. This additional burden often results in increased stress levels, leading to various health problems. The primary source of stress stems from communication with their superiors (Shiva, 2013).

Uravu demonstrates a high level of cooperation by seeking employees' opinions before accepting orders, ensuring that they can meet the product's deadline.

Respondent 8 mentioned, "Uravu does not impose pressure on us, as they consult with us before committing to work orders, ensuring that we can complete the projects on time. This approach is beneficial for us and leads to a stress-free environment."

7.7.2.3 Incorporating opinions in decision making

According to a study conducted in Nigeria by Agwu and Olele (2014), there is a strong link between employees' active participation in decision-making processes and their heightened commitment to the organization. This increased commitment ultimately translates into improved organizational productivity.

Respondent 8 mentioned, "Uravu does not impose pressure on us, as they consult with us before committing to work orders, ensuring that we can complete the projects on time. This approach is beneficial for us and leads to a stress-free environment."

Respondent 1 mentioned, "Yes, they conduct meetings and encourages us to make new bamboo products using our own ideas."

Taking into account the ideas and opinions of employees is an integral part of incorporating their decisions, which in turn fosters a greater sense of attachment to the organization.

7.7.2.4 Happy to come to work regularly

Employee satisfaction and happiness are crucial factors to monitor regularly in various industries, especially in healthcare. In this sector, it is particularly important for healthcare marketers to collaborate closely with human resource departments to understand and influence the employees' work environment and to uphold high levels of job satisfaction (Atkins *et al.*, 1996). Collaboration between employees and employers can enhance job satisfaction and happiness, contributing to overall organizational success and fostering a conducive work environment.

Respondent 8 mentioned, "I consistently make it a point to come to Uravu without fail due to the pleasant atmosphere it offers, regardless of any challenges I may face."

Respondent 3 said, "I am mainly coming here to forget the problems in my house. There are numerous issues at home, but during my time here in Uravu, from the moment I arrive until the moment I depart, I experience genuine happiness and contentment."

7.7.2.5 Self-confidence

Self-confidence plays a pivotal role in women's empowerment (Soharwardi and Ahmad, 2020). Research shows men have higher self-confidence than women, impacting their entrepreneurial aspirations and beyond, which has been studied in various research contexts (Bengtsson *et al.*, 2005). In order to foster psychological empowerment among rural women, both cooperation and understanding from co-workers and organizational support are essential. Stromquist's (1995b) research indicates that psychological empowerment in women occurs when they develop confidence in their ability to influence their circumstances, leading to personal and societal transformations.

Some single mothers have demonstrated their ability to overcome challenging circumstances and provide support for their families; respondent R1, R7, and R8 are single mothers.

Respondent 8 mentioned, "I faced many problems I got married at a young age (18), but unfortunately, my husband left me. I have been living with my mother ever since. However, since I joined Uravu, I have undergone tremendous personal growth. My self-confidence has significantly increased over the past 20 years of working here."

Respondent 1 said, "I got married when I was 18 years old, but unfortunately, my marriage ended in divorce when I was just 19. Despite this setback, I took up the responsibility of caring for my family. Working at Uravu has significantly boosted my self-assurance."

Respondent 7 mentioned, "I have been employed at this place for a span of 15 years. Unfortunately, my husband had an accident that resulted in paralysis, which led me to seek employment at Uravu, conveniently located near my home."

The women expressed that they have experienced a notable increase in confidence and an enhanced sense of empowerment. They feel more self-assured in their decision-making abilities and believe they have grown personally and professionally.

Respondent 11 said, "I have the confidence that I can face the same better than before."

7.7.3 Financial independence and security

The empowerment of women and economic development are deeply interconnected. Economic development inherently leads to the empowerment of women, and in turn, empowering women leads to changes in decision-making processes that directly affect development (Duflo, 2012). Therefore, economic empowerment holds significant importance as a key aspect of empowering women. Nazier and Ramadan (2018) highlighted that women's empowerment is significantly influenced by their economic resources, employment status, financial independence, self-confidence, self-esteem, and access to education, thus supporting families and improving their overall well-being. A survey of Uravu employees revealed that financial independence and security significantly contribute to women's economic empowerment. This financial freedom allows them to meet their needs, invest in their children's education, and boosts self-confidence.

The final codes, which included financial independence, self-decision making in purchasing items, salary, savings, and insufficient salary, were collectively categorized under the theme of financial independence and security. These both can be combined into a common theme as economic empowerment. In addition, economic empowerment exhibited a secondary yet more pronounced positive effect on employees when compared with psychological empowerment. Furthermore, the study identified an interrelationship between psychological empowerment and economic empowerment influencing working women in Uravu.

7.7.3.1 Financial independence

Uravu women are gaining economic empowerment, and improving their psychological and social well-being. Their work equips them with confidence to face challenges and achieve financial independence, although there are difficulties in assessing its impact. Government reports often aim to portray an image of gender equality in terms of financial independence, making it difficult to gain a clear and accurate understanding of the actual situation concerning women's financial autonomy (Meurs and Ponthieux, 2015). A study in Mauritius found that sociological factors, lack of opportunities, domestic violence, and employer choice did not significantly impact women's financial independence, while the legal system significantly influenced it (Rughoobur-Seetah *et al.*, 2022). In Uravu, 90% of employees enjoy financial independence by spending their salary according to their own requirements.

Respondent 5 mentioned, "There is a significant contrast between my life before and after starting to work. Previously, I relied on my husband for everything, but now I am delighted that I can independently support my children's education and fulfil my own needs through my personal growth and learning."

7.7.3.2 Self-decision making in purchasing items

A stereotypical belief is that women are not good in risk financial decision making compared with men. In one study, the authors studied gender and household savings and investment decisions and the process through which couples, both married and cohabitating, make joint financial choices (Bernasek and Bajtelsmit, 2002). They observed that women are more likely to influence these decisions when they contribute a larger share of income to the household. However, as women's average earnings are lower than men's, this suggests that, on average, women are less involved in making household-level decisions.

The stereotype that women are risk averse in financial decision making has led to discrimination and hinders their success in the financial and labor markets (Schubert *et al.*, 1999). However, risky decisions depend on context and may be more biased than factual. Women's purchasing decisions reflect their empowerment, with 80% at Uravu having the freedom to spend their salary, primarily for household purchases and children's education.

Respondent 6 mentioned, "Because of Uravu salary, I am able to repay loans and support my family, particularly when it comes to my children's education."

Respondent 12 said, "Yes, I'll make a purchase for myself first, and afterward, I'll contribute to the household."

7.7.3.3 Salary savings

The fact that women's average earnings are lower than men's implies that, on average, women tend to be less engaged in household-level savings decisions (Bernasek and Bajtelsmit, 2002). At Uravu, it was observed that despite their limited salaries, more than 80% of employees are able to accumulate modest savings. This gradual progress instils confidence in the economy and contributes to their economic empowerment.

Responded 2 mentioned, "Yes, we make a modest deposit in the post office and also invest in chits."

Respondent 3 said, "We allocate a small sum for depositing in both the post office and chits. I will provide a portion of the amount to my husband for household expenses, while the remaining portion will be saved in the Kudumbasree saving scheme."

Respondent 6 mentioned, "Because of Uravu salary, I am able to repay loans and support my family, particularly when it comes to my children's education."

7.7.3.4 Insufficient salary

In a study researchers found that self-management training programs significantly improved salary negotiation skills among participants, reducing gender differences, and suggesting that mastering these skills could potentially empower women (Stevens *et al.*, 1993). Uravu's employees remain loyal despite potentially better offers due to its positive working environment, supportive colleagues, and culture, which alleviate external work pressures.

Respondent 1 stated "The salary I receive is inadequate for me to support our household since I am a single mother residing with my own mother. My mother is engaged in thozhilurappu, and both of us need to work in order to cover our family's expenses, as I have a daughter to take care of." (Thozhilurappu is the Mahatma Gandhi National Rural Employment Guarantee Scheme, which gives a legal guarantee of 100 days of wage employment in a financial year to adult members of a rural household.)

7.7.4 Employee training and community cohesion

The themes of employee training and community cohesion can be grouped into a major theme, social empowerment, which refers to the strengthening force that enhances social connections and improves the position of women within social systems (Mandel, 2013). Social empowerment involves developing autonomy, self-confidence, and collective action to transform marginalized individuals and reduce the perpetuation of poverty. Engaging impoverished individuals in local associations and intercommunity cooperation enhances skills, knowledge, self-perception, and social and economic empowerment (GSDRC (Governance, Social Development, Conflict and Humanitarian Knowledge Services), 2014). Scheyvens (2000) research delves into the concept of social empowerment by examining the level of community cohesion.

The theme of social empowerment emerged from the final codes of employee training, lack of knowledge of social issues, non-participation in social and political issues, and improved communication skills.

7.7.4.1 Employee training

Training has emerged as a pivotal factor in the ever-changing and competitive market landscape. It is the human capital that sets apart a great organization from a good one. Notably, organizations that invest in effective training and development for their workforce tend to yield both short- and long-term advantages (Nda and Fard, 2013). Proficient trainers equipped with fundamental work-readiness skills can substantially enhance the retention of women from historically under-represented groups entering formal employment for the first time (Ranganathan, 2018).

A study conducted by Bartel in 1994 on the impact of formal training programs on labor productivity in the manufacturing sector found that businesses operating below their 1983 levels experienced significant improvements between 1983 and 1986 following training programs, indicating the importance of employee training. Hence, employee training not only enhances organizational productivity but also boosts the confidence level of newcomers, as demonstrated in the experience of Uravu employees (Bartel, 1994).

Respondent 1 mentioned, "I heard that Uravu was giving training and I joined. I received training for six months. I haven't worked anywhere before Uravu."

Respondent 3 said, "I came here my own interest. I got training in making bamboo pen. I don't need to ask to anyone for my own needs. I feel so satisfied. I got training, I am mainly working on pen."

7.7.4.2 Improved communication skills

Enhancing communication skills empowers women to establish organizations, manage businesses, and develop skills in non-business areas. This empowers rural women, enabling them to become globally competent individuals, despite cultural differences in communication principles (e.g. Sharifian, 2010). Indigenous communities possess unique interpersonal communication traits, like a preference for minimal eye contact, indirect questioning, and gratuitous concurrence. However, these traits have sometimes been misinterpreted as defiance or with suspicion by criminal justice personnel (Eades, 2013). By improving effective communication skills, such misunderstandings can be overcome, fostering greater confidence in interacting with people from outside the community, particularly in the tourism sector.

Respondent 4 mentioned, "The experience is very good. Every day people come here while communicating with them, I feel so happy. Moreover, the working atmosphere in Uravu is very friendly."

Respondent 1 said, "Previously, I struggled with shyness and discomfort when it came to socializing with others. However, now I find myself interacting with everyone effortlessly, without any difficulties."

7.7.5 Awareness of social and political issues

Scheyvens's (1999) study evaluated community control over ecotourism initiatives, emphasizing the importance of women's political influence in fostering development and advancement (Sundström *et al.*, 2017). One of the participants in the Uravu study emphasized that they lack the means to voice their opinions regarding political matters outside their organization. They expressed concern about the rural setting, which tends to undervalue women's opinions and political involvement. The participant mentioned that Uravu, however, values their opinions and suggestions, incorporating them if they prove effective.

The themes of being unaware and non-responsive to political issues fall within the dimension of political empowerment. During the analysis it was identified that Uravu values the opinions of its employees, but there is a noticeable lack of representation for women's voices. Additionally, there is a trend of limited participation in social and political issues among the employees. When asked whether they had witnessed or participated in any advocacy events where they voiced their own rights and those of their community, almost all of the respondents provided a similar response, stating that they had not. When asked why they had not spoken for their community, they explained that their voices were not taken into consideration in decision-making processes, and they were not accustomed to engaging in political participation in their society.

7.7.5.1 Lack of representation for women's voices in their society

The World Bank, 2012 report highlighted the importance of women's voices, agency, and participation in achieving gender equality, focusing on four key aspects: access to resources, freedom of movement, protection from violence, and family decision-making power (Markham, 2013). However, despite these efforts, women around the world still face unequal opportunities for success, often living in societies that undervalue them compared with men and where their rights may be violated with little recourse (Sakho *et al.*, 2020). Regrettably, their voices often go unheard in society.

Looking at the brighter side, progress can be observed in significant domains, notably in increased economic participation and global leadership. However, it is essential to acknowledge the experiences shared by Uravu employees from certain societies, where their voices seem to hold little importance. Some mentioned having a voice within their families but not in the broader society. Unfortunately, this lack of societal voice is one of the factors hindering the empowerment of rural women.

In our interview, respondent 8 stated "No one, consider here our opinion ma'am especially in our place."

Despite progress in some areas, gender inequality still persists in various aspects of life across EU countries, encompassing the labor market, education, health, and political empowerment (Wieczorek-Szymańska, 2023).

7.7.5.2 Non-responsive to political issues

Political empowerment positively impacts ethnic and racial minorities' attitudes and participation, while women's representation in politics increases interest, participation, and political efficacy. However, women often face limited opportunities and lack encouragement (High-Pippert and Comer, 1998). Women, constituting over half India's population, have minimal political involvement due to psychological, social, and physical barriers (Singh, 2016). Notably, in the Uravu Organization, only a small minority show interest in politics, with a staggering 90% expressing disinterest and non-responsiveness to political matters.

Respondent 3 mentioned, "I have not yet intervened in any political matters."

Respondent 4 stated, "I have not witnessed any issues; I don't dare to raise voice if I witness such situations."

7.7.5.3 Less political involvement

Political participation serves as a crucial pathway to empower women, and their active involvement in decision-making processes holds the key to achieving genuine women's empowerment. By having a say in decision making, women gain the ability to influence matters that directly impact their lives and the broader community (Rao, 2023). Moreover, the impact of political marginalization on women's political empowerment in Balochistan, Pakistan, highlights the need for equal representation and active women's participation in politics (Kakar *et al.*, 2023).

The low involvement in politics of female employees at Uravu is attributed to factors such as lack of interest, support, resources, and opportunities.

Respondent 5 stated, "When it comes to matters within the house, I am willing to respond if someone disagrees with a statement by saying 'this is not like this.' However, outside of the house, I have not had many opportunities to do so yet."

Respondent 7 mentioned, "No, I didn't participate, as rural women, we face constraints that limit our involvement in political and social matters."

7.8 Conclusion

Community involvement is essential for achieving the SDGs (United Nations, 2019), which emphasize empowering individuals, fostering inclusivity, and promoting gender equality. Despite the challenges rural communities encounter in implementing sustainable development initiatives, women play a pivotal role in advancing economic empowerment. The Self-Employed Women's Association in India has made considerable strides in enhancing women's economic standing (Mehra, 1997). Nevertheless, there remains a noticeable lack of improvement in the status of rural women in India. This gap can be addressed by adopting the model of the Uravu Organization, which empowers women economically, psychologically, politically, and socially.

The findings of this study indicate a positive and significant level of empowerment among women employees, with 90% experiencing psychological and economic empowerment. However, a noteworthy observation is that these employees exhibit limited awareness of social and political issues, as well as a low level of engagement in related activities. This particular aspect requires focused attention from researchers, especially in rural, tribal, and Dalit women's cases, as they lack awareness of the issues affecting their communities. The reduced involvement in social and political spheres suggests limited accessibility and awareness in these areas. Consequently, the lack of political and social empowerment negatively impacts overall women's empowerment.

The majority of women employees who have been with Uravu since its inception continue to remain loyal. At the same time new employees, despite receiving training, tend to leave their jobs due to lower wages. They seek employment opportunities in other areas, as they have noted an increase in living expenses following the Covid-19 pandemic. However, there seems to be a lack of significant improvement in their social and political empowerment. To address this issue, active engagement in social and political activities is essential to enhance their empowerment in these spheres. Rural women are influenced and empowered through the community-based activities of Uravu, but while psychological and economic empowerment are evident, political and social empowerment are comparatively lower.

The economic dynamics of households are shifting, with women increasingly contributing their own earnings to cover essential domestic expenses such as food and children's education costs (Bryan and Mekonnen, 2023). Consequently, the empowerment of women stands to foster both community and national development. In this context, it becomes crucial for developing nations to prioritize women's empowerment, given its potential to enhance the country's economic progress.

As women gain economic empowerment, a cascade of positive effects unfolds, encompassing psychological, social, and political dimensions. Among these, psychological empowerment holds particular significance, as it bolsters women's self-assurance to participate in the workforce. The equilibrium between professional and personal life, along with the quality of the working environment, plays a pivotal role in influencing female employees. Creating favorable conditions both at work and within the family sphere can pave the way for women to attain psychological empowerment.

The scope of this study had certain limitations. It exclusively examined a single organization and confined its focus solely on women employees. A more comprehensive approach could involve researching male employees as well, allowing for a comparative analysis of empowerment between genders. Moreover, there is a need for increased attention to research on CBET and women's involvement. Such studies can yield benefits for community development, the tourism sector, and environmental conservation.

Uravu is primarily dedicated to empowering indigenous women through sustainable, community-based initiatives. A significant aspect of its mission is its adherence to a zero-waste policy, highlighting its commitment to environmental stewardship. The manufacturing process generates no harmful by-products, and any left-over bamboo waste is transformed into fertilizer. This method not only reduces environmental impact but also enriches the soil, as this final product is advantageous for plant growth. Additionally, the bamboo products created by the organization act as natural air purifiers, capturing carbon dioxide and releasing oxygen, thereby improving air quality and advancing sustainability.

Indigenous and rural women can be empowered through community-based sustainable tourism development, thus building resilient communities.

References

Agwu, M.O. and Olele, H.E. (2014) Perception survey of employees' participation in decision making and organizational productivity in Julius Berger Nigeria PLC, Bonny Island. *British Journal of Economics, Management & Trade* 4(4), 620–637.

Akbar, I., Abdreyeva, S., Artemyev, A., Tumazhanova, M. and Orynbasarova, G. (2022) Research on design and management of community-based ecotourism model in Aksu-Zhabagly Nature Reserve of Kazakhstan. *Journal of Geoscience and Environment Protection* 10(4), 33–48. DOI: 10.4236/gep.2022.104003.

Atkins, P.M., Marshall, B.S. and Javalgi, R.G. (1996) Happy employees lead to loyal patients. *Marketing Health Services* 16(4), 14.

Bartel, A.P. (1994) Productivity gains from the implementation of employee training programs. *Industrial Relations: A Journal of Economy and Society* 33(4), 411–425.

Bastian, B., Metcalfe, B. and Zali, M. (2019) Gender inequality: Entrepreneurship development in the MENA region. *Sustainability* 11(22), 6472. DOI: 10.3390/su11226472.

Bengtsson, C., Persson, M. and Willenhag, P. (2005) Gender and overconfidence. *Economic Letters* 86(2), 199–203.

Bernasek, A. and Bajtelsmit, V.L. (2002) Predictors of women's involvement in household financial decision-making. *Financial Counseling and Planning* 13(2), 39–47.

Bhuyan, Anil. (2014) Local entrepreneurship as an approach towards sustainable tourism development in the north-eastern region of India. In: Manhas, P.S., Gupta, D.R. and Gupta, A. (eds) *Strategic Development Policies and Impact Studies of Sustainable Rural and Community-Based Tourism*. Primus Books, New Delhi, pp. 3–22.

Boley, B.B., Ayscue, E., Maruyama, N. and Woosnam, K.M. (2016) Gender and empowerment: assessing discrepancies using the resident empowerment through tourism scale. *Journal of Sustainable Tourism* 25(1), 113–129. DOI: 10.1080/09669582.2016.1177065.

Braun, V. and Clarke, V. (2006) Using thematic analysis in psychology. *Qualitative Research in Psychology* 3(2), 77–101.

Bryan, E. and Mekonnen, D. (2023) Does small-scale irrigation provide a pathway to women's empowerment? Lessons from northern Ghana. *Journal of Rural Studies* 97, 474–484.

Carless, S.A. (2004) Does psychological empowerment mediate the relationship between psychological climate and job satisfaction? *Journal of Business and Psychology* 18, 4. DOI: 10.1023/B:JOBU.0000028444.77080.c5.

Cernat, L. and Gourdon, J. (2011) Is the concept of sustainable tourism sustainable? *Developing the Sustainable Tourism Benchmarking Tool* (200708).

Duflo, E. (2012) Women empowerment and economic development. *Journal of Economic Literature* 50(December), 1051–1079. DOI: 10.1257/jel.50.4.1051.

Eades, D. (2013) *Aboriginal Ways of Using English*. Aboriginal Studies Press.

Elshaer, I., Moustafa, M., Sobaih, A.E., Aliedan, M. and Azazz, A.M. (2021) The impact of women's empowerment on sustainable tourism development: Mediating role of tourism involvement. *Tourism Management Perspectives* 38, 100815.

Friedmann, John. (1992) *Empowerment: The Politics of Alternative Development*. John Wiley & Sons, New York.

Goodwin, H. and Santilli, R. (2009) Community-based tourism: a success? (Occasional Paper no.11, pp. 1–37) In: International Centre for Responsible Tourism.

GSDRC (Governance, Social Development, Conflict and Humanitarian Knowledge Services) (2014) Social and economic empowerment. Available at: https://gsdrc.org/topic-guides/voice-empowerment-and-accountability/supplements/social-and-economic-empowerment

Hasan, S.R. and Islam, M.S. (2015) Exploring bandarban through trekking trails: A new horizon for community-based tourism development. *European Scientific Journal* 11(5).

High-Pippert, A. and Comer, J. (1998) Female empowerment: the influence of women representing women. *Women & Politics* (4), 53–66.

Kabeer, N. (1999) Resources, agency, achievements: Reflections on the measurement of women's empowerment. *Development and Change* 30(3), 435–464.

Kakar, M.M., Rahim, M. and Ali, S. (2023) Balochistan women in politics: An examination of political participation and its effects on women empowerment in Balochistan, Pakistan. *Pakistan Journal of Humanities and Social Sciences* 11(2), 1051–1059. DOI: 10.1111/1467-7660.00125.

Kiss, A. (2004) Is community-based ecotourism a good use of biodiversity conservation funds. *Trends in Ecology & Evolution* 19(5), 232–237. DOI: 10.1016/j.tree.2004.03.010.

Kunjuraman, V., Hussin, R. and Aziz, R.C. (2022) Community-based ecotourism as a social transformation tool for rural community: A victory or a quagmire. *Journal of Outdoor Recreation and Tourism* 39, 100524. DOI: 10.1016/j.jort.2022.100524.

Lee, C. and Park, S. (2021) Changing factors of employee satisfaction with working conditions: An analysis of the korean working conditions survey. *Safety and Health at Work* 12(3), 324–338.

Ma, X. and MacMillan, R.B. (1999) Influences of workplace conditions on teachers' job satisfaction. *The Journal of Educational Research* 93(1), 39–47.

Mandal, K.C. (2013, May) Concept and types of women empowerment. *International Forum of Teaching & Studies* 9(2), 17–30.

Markham, Sarah. (2013) Women as agents of change: Having voice in society and influencing policy. *World Bank's Women's Voice, Agency, and Participation Research Series* 5.

Mehra, R. (1997) Women, empowerment, and economic development. *Annals of the American Academy of Political and Social Science* 554(1), 136–149.

Mensah, Isaac. (2017) Benefits and challenges of community-based ecotourism in park-fringe communities: The case of mesomagor of Kakum National Park, Ghana. *Tourism Review International* 21(1), 81–98.

Meurs, D. and Ponthieux, S. (2015) Gender inequality. In: Atkinson, A.B. and Bourguignon, F. (eds) *Handbook of Income Distribution*, Vol. 2. Elsevier, North Holland, pp. 981–1146.

Miedema, S.S., Haardörfer, R., Girard, A.W. and Yount, K.M. (2018) Women's empowerment in East Africa: Development of a cross-country comparable measure. *World Development* 110, 453–464.

Mohanty, P. and Chandran, A. (2018) Poverty alleviation and women empowerment through tourism development—an explorative study of model ventures. *Atna Journal of Tourism Studies* 13(1), 59–74.

Movono, A. and Dahles, H. (2017) Female empowerment and tourism: A focus on businesses in a Fijian village. *Asia Pacific Journal of Tourism Research* 22(6), 681–692.

Nazier, H. and Ramadan, R. (2018) What empowers Egyptian women: Resources versus social constraints. *Review of Economics and Political Science* 3, 153–175. DOI: 10.1108/REPS-10-2018-015.

Nda, M.M. and Fard, R.Y. (2013) The impact of employee training and development on employee productivity. *Global Journal of Commerce and Management Perspective* 2(6), 91–93.

Njoh, A.J. and Ananga, E. (2016) The development hypothesis of women empowerment in the millennium development goals tested in the context women's access to land in Africa. *Social Indicators Research* 128(1), 89–104.

Nutsugbodo, R. Y., Adjei Mensah, C., Amenumey, E. K., & Ansah Owusu, C. (2020). Women's participation in ecotourism development within the Kakum conservation area, Ghana: Implications for community planning. *Tourism planning & development*, 17(6), 693–710.

Perkins, D.D. and Zimmerman, M.A. (1995) Empowerment theory, research, and application. *American Journal of Community Psychology* 23, 569–579.

Ramadhoan, R.I., Hadi, K., Dewayani, E.K.U. and Argubi, A.H. (2023) Developing a model of women's empowerment for environmental adaptation and disaster mitigation in malang regency. *Sospol* 9(2), 160–175.

Ranganathan, A. (2018) Train them to retain them: work readiness and the retention of first-time women workers in India. *Administrative Science Quarterly* 63(4), 879–909.

Rao, V.D. (2023) A perspective study on Panchayati Raj institutions (with reference to women empowerment). *Galaxy International Interdisciplinary Research Journal* 11(6), 258–263.

Rughoobur-Seetah, S., Hosanoo, Z. and Soupramanien, L.D.B. (2022) Financial independence of women – The impact of social factors on women empowerment in small island developing states (SIDS). *International Journal of Organizational Analysis* 31(6), 2383–2408. DOI: 10.1108/ijoa-10-2021-2980.

Sahu, L. and Singh, S.K. (2012) A qualitative study on the role of self-help groups in women empowerment in rural Pondicherry, India. *National Journal of Community Medicine* 3(3), 473–479.

Sakho, S., Caruso, G. and Acuña, M. (2020) Progress without women is not possible. *World Bank Blogs*. Available at: https://blogs.worldbank.org/latinamerica/progress-without-women-not-possible

Scheyvens, R. (1999) Ecotourism and the empowerment of local communities. *Tourism Management* 20(2), 245–249.

Scheyvens, R. (2000) Promoting women's empowerment through involvement in ecotourism: experiences from the third world. *Journal of Sustainable Tourism* 8(3), 232–249.

Schneider, B. (1983) Interactional psychology and organizational behavior. In: Cummings, L.L. and Staw, B.M. (eds) *Research in Organizational Behavior*, Vol. Vol 5. JAI Press, Greenwich, CT, pp. 1–31.

Schubert, R., Brown, M., Gysler, M. and Brachinger, H.W. (1999) Financial decision-making: are women really more risk averse. *American Economic Review* 89(2), 381–385.

Sharaunga, S., Mudhara, M. and Bogale, A. (2019) Conceptualisation and measurement of women's empowerment revisited. *Journal of Human Development and Capabilities* 20(1), 1–25.

Sharifian, F. (2010) Glocalization of english in world englishes: An emerging variety among persian speakers of english. In: Omoniyi, T. and Saxena, M. (eds) *Contending with Globalization in World Englishes*. Multilingual Matters, Cleveland, UK, pp. 137–158.

Shiva, G. (2013) A study on work family balance and challenges faced by working women. *IOSR Journal of Business and Management* 14(5), 1–4.

Sindhi, S. (2012) Prospects and challenges in empowerment of tribal women. *IOSR Journal of Humanities and Social Science* 6(1), 46–54.

Singh, J. (2016) *Defending India*. Springer.

Soharwardi, M.A. and Ahmad, T.I. (2020) Dimensions and determinants of women empowerment in developing countries. *International Journal of Sustainable Development and Planning* 15(6), 957–964.

Stevens, C.K., Bavetta, A.G. and Gist, M.E. (1993) Gender differences in the acquisition of salary negotiation skills: The role of goals, self-efficacy, and perceived control. *Journal of Applied Psychology* 78(5), 723.

Stromquist, N.P. (1995b) Romancing the state: gender and power in education. *Comparative Education Review* 39(4), 423–454.

Stromquist, N.P. (1995a) Women's literacy and empowerment in Latin America. In: Torres, C.A. (ed.) *Education and Social Change in Latin America*. James Nicholas Publishers, Albert Park, Australia, pp. 47–61.

Su, M.M., Wall, G., Ma, J., Notarianni, M. and Wang, S. (2023) Empowerment of women through cultural tourism: perspectives of Hui minority embroiderers in Ningxia, China. *Journal of Sustainable Tourism* 31(2), 307–328.

Sudha, J. and Karthikeyan, P. (2014) Work life balance of women employees: A literature review. *International Journal of Management Research and Reviews* 4(8), 797.

Sundström, A., Paxton, P., Wang, Y.T. and Lindberg, S.I. (2017) Women's political empowerment: a new global index, 1900–2012. *World Development* 94, 321–335. DOI: 10.1016/j.worlddev.2017.01.016.

Tabassum, F. and Sarker, R.R. (2021) Impact of work from home on employee satisfaction during COVID-19: an empirical study on Bangladesh. *Shanlax International Journal of Management* 9(2), 43–55.

Tiwari, P. and Malati, N. (2023) Role of training in women empowerment: An empirical analysis. *Journal of Technical Education and Training* 15(1), 234–245.

Tran, L. and Walter, P. (2014) Ecotourism, gender and development in northern Vietnam. *Annals of Tourism Research* 44, 116–130. DOI: 10.1016/j.annals.2013.09.005.

United Nations (2019) Increased community-based engagement seen as critical to build climate action and achieve the Sustainable Development Goals. United Nations Sustainable Development. Available at: https://www.un.org/sustainabledevelopment/blog/2019/07/increased-community-based-engagement-seen-as-critical-to-build-climate-action-and-achieve-the-sustainable-development-goals

Uravu (Uravu Indigenous Science and Technology Study Centre) (2019) Livelihood support programme. Available at: https://www.uravu.in/home/livelihood-support-programme (accessed 29 August 2024).

Venugopalan, M., Bastian, B.L. and Viswanathan, P.K. (2021) The role of multi-actor engagement for women's empowerment and entrepreneurship in Kerala, India. *Administrative Sciences* 11(1), 31. DOI: 10.3390/admsci11010031.

Wieczorek-Szymańska, A. (2023) Gender inequalities in the light of EU gender mainstreaming activities. *Humanities and Social Sciences* 30(2), 147–160.

Wolgast, A. and Fischer, N. (2017) You are not alone: Colleague support and goal-oriented cooperation as resources to reduce teachers' stress. *Social Psychology of Education* 20, 97–114.

World Bank (2012) *World Development Report 2012: Gender Equality and Development*. World Bank, Washington, DC. DOI: 10.1596/978-0-8213-8810-5.

Appendix 7.1: Questionnaire

Dear Respondent, I am working on a PhD dissertation titled "Tribal women empowerment through community based sustainable tourism development" in CHRIST (Deemed to be University), Bangalore. I am in need of your valuable cooperation in answering this questionnaire. I shall not collect your names, and the collected information will be represented as the group study. Thus, all information provided will be confidential and will be used for research purpose only.

Section A: socio-demographic profile

1. Age
   - 18–25
   - 26–35
   - 36–45
   - 50–55
   - Above 60
2. Religion
   - Hindu
   - Christian
   - Muslim
   - SC/ST
   - Other (specify): ____________________

3. Place of stay: ____________________

4. Level of education
   - Primary
   - Upper primary
   - High school
   - Higher secondary
   - Undergraduate
   - Postgraduate
   - Illiterate
   - Can read
   - Can write

5. Marital status
   - Married
   - Single
   - Divorced
   - Widow
6. Family composition

No.	Relationship	Age	Occupation	Health	
1.					
2.					
3.					
4.					
5.					

6.					
7.					
8.					

7. Employed
   - Full time
   - Part time
   - Multiple job sites: ______
8. Nature of work: ______
9. Languages known: ______
10. Annual income:
    - Less than 25,000
    - 25,000–50,000
    - 50,000–100,000
    - Above 100,000

Section B

Economic empowerment

1. How do you know about this Uravu? Have you worked before? Is the amount sufficient for a living?
2. Are you able to save from this salary?
3. Was there any moment in your life where you longed to have a job? Or to find a source of income?
4. Is there any change in your food style, or living style after started earning for yourself?

Psychological empowerment

1. How far have you changed in terms of your level of confidence, esteem, and approach to life after coming to Uravu?
2. How do you feel about the working atmosphere here?
3. Was there any moment in your life where you felt depressed or devastated by life events? If yes, how did you manage? And how could you have faced it now?

Social empowerment

1. What were your feelings and thoughts when you met people outside your community? Were you able to interact with them easily? And how do you feel now?
2. How far has the social environment (unity, conflict management, gender discrimination, etc.) of your community has changed after the Uravu project?

Political empowerment

1. Can you share any experience in your life, where you were asked [an] opinion or suggestion regarding a matter in your community, family, and in Uravu?
2. [Did] Uravu consider your opinion in [the] decision-making process?
3. Have you witnessed or be[en] part of any advocacy event where you voiced for your [rights] and the rights of your community?
4. What do you think should be done if some outsiders interfere into your community or any government official denies your rights provided by the Government?

8 Exploring the Impact of Eco-Friendly Initiatives on Rural Tourism and Hospitality: A Comprehensive Review of Current Practices and Potential Areas for Innovation

Madhu Kumari[1], Sandeep Guleria[1], Suneel Kumar[2*], Nisha Devi[3] and Marco Valeri[4]

[1]Department of Tourism and Hospitality Management, Chandigarh University, Mohali, India; [2]Department of Commerce, Shaheed Bhagat Singh College, University of Delhi, Delhi, India; [3]Department of Commerce, Shri Ram College of Commerce, University of Delhi, Delhi, India; [4]Faculty of Economics, Niccolo Cusano University, Rome, Italy

8.1 Introduction

The achievement of sustainability remains a big problem for the rural tourist and hospitality sectors. There are many issues that are common, including excessive trash production, community misuse as discussed by Ghoddousi *et al.* (2018), and environmental degradation. According to the United Nations Environment Program (UNEP), 14% of the world's garbage is caused by the tourism sector and, if present consumption trends continue, the sector's energy use is expected to increase by 154%, greenhouse gas emissions by 131%, and water consumption by 152% by 2050. Research on the sustainability difficulties in rural tourism (RT) has increased significantly as a result of these environmental issues. The literature also emphasizes how RT has a great deal of potential to support rural communities helping make them sustainable through the efficient use of natural and cultural resources. Conservation of natural resources and increased rural income are two benefits of RT, as discussed by Lee and colleagues in Lee *et al.* (2021) and Mikhaylova *et al.* (2022). The important impacts of RT on rural communities are through processes such as creating jobs and supporting sustainable rural development. There is continued interest in policy and practice, for example the 2019 annual report by the UK's Ministry of the Environment, Food and Rural Affairs summarizes these results: RT activities supported 14% of jobs in rural areas and generated an estimated US$16.2 billion in 2019 in economic value for the UK through a new type of rural travel experience. The area began to receive official assistance from Asian nations such as China to promote tourism development. Nonetheless, there are important

*Corresponding author: suneel.kumar@sbs.du.ac.in

DOI: 10.1079/9781836992646.0008

limitations to this increase. Each rural area has a specific "carrying capacity" for tourism; going over this limit might make the social, cultural, and environmental aspects of the area less sustainable. Therefore, investigating the function of sustainability in RT is essential at this time. Numerous assessments have provided ample evidence of the increased interest in sustainable tourism. A wide range of subjects are covered in recent literature reviews, such as the evaluation of sustainability in tourist destinations, energy consumption in the tourism sector, and the use of big data and related approaches. Because RT is so important and has such a big impact on rural areas, it is imperative that the literature now in existence be compiled and synthesized in order to guide future research directions. Despite its importance, no thorough analysis has yet concentrated on the function of sustainability in RT. According to recent research, this analysis is essential to determining the main forces behind sustainability in rural tourism as well as its advantages and stakeholders.

Today's tourism and leisure offerings are diverse due to a wide range of factors. Individual passions, a fascination with traveling the world, cultural and recreational opportunities, the accessibility of tourist attractions and amenities, technological developments, and the attraction of particular locations whether characterized by their distinctive landscapes or cultural significance, are a few of these. Rural tourism has grown significantly in popularity over the past few decades, at a time of sustained expansion fueled by changes in rural economies and landscapes. Moreover, the transformation of the rural economy and landscape is closely related to the growth of RT. When this type of tourism is guided by the principles of sustainable development it is particularly satisfying. These guidelines are necessary to protect the natural environment in which tourism operates and also to promote the type of travel that has the least harmful impact on the environment and society. Many studies highlight the growing importance of RT around the world as a strategy that promotes the sustainable development of rural communities (Sharpley and Roberts, 2004). Protection of natural and cultural resources is a top priority of RT and tourists need to use these resources appropriately, and special emphasis is placed on the active participation of local communities (Mvula, 2001). Many experts emphasize the importance of local communities playing a key role in the development of sustainable RT (Falak *et al.*, 2014; Franzoni, 2015; Lee and Hsieh, 2016). Documents from the European Union highlight how important sustainability is to the competitiveness of travel destinations, particularly rural areas. In villages and communes, RT can have a significant, positive impact on sustainable regional development. It is renowned for having less of an impact on the environment than more conventional tourism strategies (Saarinen and Lenao, 2014; Serra Cantallops *et al.*, 2015; Korani and Shafiei, 2021).

8.2 Literature Review

Rural tourism and hospitality is the process of using the unique features of rural areas to attract tourists. This definition of rurality includes the customs, landscapes, culture, heritage, and other aspects that give a rural place its unique appeal to visitors. To offer a range of "authentic" experiences that tourists seek, such as warm hospitality, convenient transportation, scenic hiking trails, and other desired amenities, RT enterprises turn unique rural features into appealing, marketable experiences. Accordingly, sustainable RT is characterized as a type of tourism that attends to the requirements of visitors, host communities, and other stakeholders while balancing the needs of the present and future in terms of social, environmental, and economic concerns. The tourism industry, regional businesses, host communities, and the environment are some of these stakeholders. According to Ayhan *et al.* (2020) the origins of RT can be traced back to the 17th century, and the movement became formally recognized in the 1970s. Immersing visitors in the peaceful countryside is its main goal in providing a getaway from the hectic pace of metropolitan life. The creation of a wide range of RT enterprises aimed at satisfying and amusing tourists is the result of this need. Under the influence of encouraging laws and government initiatives, RT has changed dramatically over time, especially in Europe. Since RT has two effects on rural regions, it is still a controversial matter particularly with its expansion. It can lead to environmental and cultural degradation as well as sustainable development on an economic, social, and environmental level. This paradox emphasizes the significance of adopting a holistic approach when addressing sustainability. This chapter defines sustainability in terms of rural regions' social and economic dimensions as well as their environmental characteristics, drawing on the triple bottom line idea.

8.2.1 Function of rural tourism

Rural tourism is a type of travel centered on experiences related to outdoor recreation, farming, rural life and culture, fishing, and sightseeing. This approach, which aims to promote progress in rural communities while maintaining traditions, has been used as an effective development technique in both developed and developing countries. Promoting RT promotes a continuous relationship between people and the natural world. This results in collective change and better experiences. Hegarty and Przezbórska (2005) describe these phenomena as consisting of a wide range of activities defined in terms of physical behavior and time frames. These include agriculture, forestry, fishing, hunting, agritourism, and conservation of cultural heritage and rural life. RT as a key sector has increased its importance in enhancing rural development and became an important part of regional rural economies (Polukhina *et al.*, 2021). The Covid-19 pandemic also increased the demand for transient recreational activities for visitors from cities (Slater *et al.*, 2020). Marques *et al.* (2022) pointed out that the Covid-19 epidemic caused tourists to change their travel options from distant locations to nearby rural experiences, which represented a new opportunity for rural development. This change highlights the fact that RT has become a competitive alternative for economic revitalization, proving that rural development is no longer based solely on agriculture. Ironically, although urbanization has exacerbated problems in rural areas, it has also encouraged the development of RT as a way of strengthening the local economy (Lewis and Delisle, 2004). Urban residents can find a number of interesting attractions in RT, such as agricultural festivals, crafts, historical monuments, nature conservation, nostalgic memories, regional cuisine, and leisure and family opportunities (Getz, 2008; Christou, 2020). RT is a contemporary step towards achieving a balance between ecosystem protection and human activity from an environmental perspective. This strategy encourages farmers to use organic farming methods, green chemicals, and value-added goods such as land ethics in order to preserve the rural landscape, protect natural regions, and develop green consumerism, as discussed by Higham and Ritchie (2001). The promise of RT to garner additional investment from policy makers and support from people is highlighted by its profound and important contributions to community development (Marques, 2006).

8.2.2 Promoting sustainability in rural tourism

Sustainability in the tourist industry is typically associated with development issues and is characterized as "new tourism" (Hampton, 2005), the central feature of which is its minimal environmental effect. In rural regions, sustainable tourism is frequently limited to small-scale enterprises, local cuisine and crafts, transportation, and farms that welcome visitors. The need for nature conservation often accompanies the expansion of RT, which emphasizes a commitment to sustainable practices. Its absence may have an impact not only on the offered tourism product but also on the future standard of living in the local community (Berry and Ladkin, 1997). Sustainable tourism, which encompasses tourism in rural areas as well, places an emphasis on the preservation of natural and cultural resources while also emphasizing visitor responsibility and the necessity for local community involvement (Huete-Alcocer and Valero-Tévar, 2021). Environmental sustainability has emerged as a prominent characteristic of modern RT, and pro-environmental conduct is acknowledged as a means of attaining this objective (Jamal and Higham, 2021). Such development goes beyond only maintaining economic expansion. In rural areas, environmental, economic, cultural, and social concerns are all taken into consideration under the notion of sustainable development. Erokhin *et al.* (2014) emphasized that every element plays a crucial role, with RT representing a fresh facet of sustainable development in the travel and tourism industry (Palacios-Florencio *et al.*, 2021).

Gilbert and Tung (1990) emphasized that farmers play an important role in RT since they provide tourists with lodging and establish the circumstances for numerous leisure activities within a farm setting. It is evident that RT has several facets due to the variety of products provided for visitors as well as the farming operations. Bramwell and Lane (2012) identified various types of tourism, including ecotourism, sports tourism, health tourism, nature tourism, and other active pursuits such as hiking, mountain biking, horseback riding, and fishing. They also mentioned heritage-related tourism, including folk and ethnic tourism, cultural and traditional tourism, and educational tourism. Pechlaner *et al.* (2015) further explored the link between tourism and the rural environment, a connection emphasized by Milman *et al.* (1999). A few people have noted that RT is year-round rather than just seasonal, and the limited size of this industry has been highlighted in the research on the topic thus far. Since the lockdown, when this market's superb organization was discovered, this kind of tourism has been especially valued. Roberts *et al.* (2017) defined RT as a small-scale, well-managed sector with high educational value. Given this, it is critical to develop proper policies and management techniques to boost it. The European Union emphasizes the need to develop rural areas. The Cork Declarations 1.0 and 2.0 outline key policy objectives for the future of rural areas, such as addressing rural migration, combating poverty, promoting employment, and meeting demands for quality, safety, health, and personal development in order to improve the overall quality of rural environments. Even more, Declaration 2.0 promotes entrepreneurship, investment, and innovation as means of advancing prosperity in rural communities. It highlights the necessity of making investments in their vitality, which would entail doing away with digital exclusion.

The preservation of the environment is a key requirement, and it should be reflected in the growth of the local economy, ecotourism, and, more generally, leisure activities in rural regions. These ideas have an indirect as well as direct impact on the growth of RT. Within this framework, the development of the rural communal economy is considered to be significantly aided by sustainable RT. The promotion of climate action is key, as is the expansion of knowledge and innovation, especially with regard to the production of sustainable energy from renewable sources (Widawski *et al.*, 2023). By increasing rural residents' earnings, giving farmers job options, and encouraging the commercialization of regional and agricultural products, it improves living circumstances (KC and Parajuli, 2014). Moreover, increasing productivity in rural regions is seen to be possible through the use of sustainable RT and can thrive through non-agricultural activities (Mutlu Danaci and Eser, 2013; ; Korani and Shafiei, 2021). RT, in particular, provides opportunities for achieving socio-cultural, economic, and environmental sustainability in regions (Erokhin, 2014; Yun, 2014; Ivona, Antonietta, 2021). This component might be crucial in establishing global food security. Globally, small farmers' ability to survive in the real world is impacted by both trade liberalization and climate change in rural agricultural areas (Wright and Middendorf, 2010) Thus, it is even more crucial to adhere to the tenet of creating sustainable RT, which includes social fairness and financial viability (Mitchell and Hall, 2005).

8.2.3 Principles of sustainable development in rural tourism

Three aspects of sustainability are present in tourism: economic, social, and ecological. The first is about managing income and developing institutions and commercial endeavors; the second is about working together, being innovative, and being creative; and the third is about being aware of the environment, managing environmental resources in a way that promotes rational use, and managing tourism in the area (Utami *et al.*, 2023). According to Andrei *et al.* (2012), in order to meet present demands without jeopardizing the ability of future generations to use the same resources, sustainable rural activities must be in harmony with the ecological, social, economic, and environmental domains. Sustainable development – which echoes the Brundtland Report's 1987 definition of development that satisfies current demands while allowing future generations to meet their own – requires balanced progress in all four sectors. Blancas *et al.* (2011), Trukhachev (2015), and Apin and Hassan (2023) all support this view point. As discussed by He (2012), sustainable development has three qualities. First and foremost, in order to prevent damaging future generations, consideration should be given to both present and future development. Second, the advancement of one facet must not impede or stifle the advancement of other aspects. Third, it is crucial to protect natural resources and the broader natural environment while promoting economic growth. Ensuring that the economy, society, and natural resources coexist harmoniously is essential. According to this perspective, sustainable tourism development is becoming more recognized as a useful and politically acceptable instrument for the growth of the travel industry (Sharpley, 2003). This aligns with a perspective that views tourism as the primary means of eradicating poverty and using it as a tool to improve local populations' quality of life (Blackstock, 2005; Croes and Vanegas, 2008). This also applies to RT, which has goals that are essentially in line with those of sustainable development, such as ensuring sustainable consumption and production and eradicating poverty, as well as encouraging inclusive and sustainable economic growth and full and productive employment and work.

8.2.4 Enhancing sustainability in rural tourism for local communities

The natural beauty of rural areas, together with the typically preserved cultural traditions of the region and their cultural heritage history, make them desirable destinations for tourism development. As discussed by Štetić (2012), it follows that sound socio-economic growth may benefit greatly from sustainable tourism. However, the development of RT cannot be sustained solely by nature or traditional folk culture. The attitude of the local community towards the region's tourism industry plays a crucial role in any prospective success in this field (Bimonte and Punzo, 2016). Success is impossible to achieve without the active participation of villagers and acceptance of the region's products and diverse tourism forms (Andriotis and Vaughan, 2003). The social, economic, and environmental elements of sustainable RT can be greatly enhanced by community involvement in decision making concerning tourism products, information exchange, and the expansion of knowledge about the tourism business and its operations (Zou *et al.*, 2014). Consequently, it would seem that RT plays a significant role in the growth of rural communities (Anderson, 2015). Research highlights the critical role that rural dwellers play as a community that is conscious of and convinced of the need to preserve natural resources (Lee *et al.*, 2021; Mikhaylova *et al.*, 2022). Many researchers have noted that, in addition to stressing the value of protecting the environment as an essential resource for present and future generations, a sustainable approach also calls for the equitable distribution of economic benefits within the community (Chin *et al.*, 2017; Ghoddousi *et al.*, 2018). The local population can also have a big influence on visitors' plans to return to the area and their use of the rural environment (Lo *et al.*, 2019). Therefore, it is crucial to foster their innovation, creativity, managerial, social, and environmental awareness (Ataei *et al.*, 2020).

It is believed that the participatory approach (Tosun, 2006) makes it possible to apply the principles of sustainable tourism by giving locals more chances to benefit from tourism in larger and more sustainable ways. This bottom-up belief (Yan and Morpeth, 2015) then drives the need to protect nature and resources. As a result, people are becoming more conscious of the fact that being involved in the decision-making process serves their own interests because all preparation activities have an impact on communal life. In terms of tourism development, participation translates into a chance to voice aspirations, concerns, and hopes (Timothy and Tosun, 2003). The purpose of this kind of involvement is to both safeguard locals and the region's resources from the effects of tourism development and to get access to the benefits of RT development (Timothy, 1999). Since locals are the primary stakeholders in RT locations, community involvement is also crucial. A primary focus should be on training stakeholders, since a lack of knowledge and involvement could prevent RT from being implemented in a sustainable way. The interdependent relationship between rural community development and RT has been acknowledged. Going one step further, Du and Su (2011) contend that rural people should be involved in and promote tourism development.

Rural community members actively participate in RT, regardless of their level of involvement. They are more than just spectators; their participation promotes a sustainable rural environment, supports responsible use of cultural and natural resources, and improves farmers' economic conditions and living standards, hence increasing employment chances. All of this is possible as long as the rural population takes part in the RT initiatives created specifically for the area.

8.2.5 Techniques for developing rural tourism while protecting the environment

The idea that "green water and green mountains are gold and silver" must be incorporated when creating rural tourist projects, with an emphasis on "ecological benefits". With this strategy, resource integration is addressed through coordinated planning and design, and RT is matched with regional businesses. Developing sustainable tourism in a way that respects decolonization norms, maintains tourism resources, and strikes a balance between growth and the protection of the environment and production systems is essential. The competitiveness, attractiveness, and influence of any tourism company are strongly influenced by the state of its infrastructure. RT is particularly dependent on its robust natural features, so infrastructure design is critical to avoid disrupting its sustainable growth. Infrastructure modernization with an emphasis on the "three modernizations" – ecological, green, and sustainable – must be given the highest priority in the regions to guarantee the successful development of RT. All of this includes agribusiness transformation, rural industrial growth, and infrastructure investment and design that work together. Emphasizing the unique brands of RT such as "one city, one product" or "one village, one product" can solve the problems of "homogenization" and increase the attractiveness of infrastructure. Furthermore, it is necessary to have a comprehensive, methodical, and coordinated approach to the development of tourism infrastructure, which includes catering, accommodation, transport, guides, shopping, and attractions. This includes the integration of "circular economy" principles to promote environmental protection, green tourism, and low-carbon consumption.

It is important to carry out environmental impact assessments (EIAs) and environmental audits (EAs) in regions designated for the development of RT, as the resources for RT are not completely "smoke-free" and are not infinite. It is recommended that developers and operators undertake a classification assessment to ensure they have the necessary skills and credentials to create and manage tourism businesses in an environmentally responsible manner. These assessments help balance economic growth with social and ethical issues. It is important to prioritize the creation and marketing of environmentally friendly products, as well as the use of devices and equipment that do not emit any pollutants. The establishment of public education initiatives such as "public education halls" is necessary to promote environmental stewardship and to increase tourists' understanding of environmental protection. In addition, environmentally irresponsible tourism should have financial or employment consequences. To monitor the effects of tourism on the environment, "ecological monitoring stations" should be built in known RT locations. To guarantee that RT grows within the ecological capacity and meets the requirements of local sustainability, appropriate environmental protection measures can be implemented by anticipating the environmental impacts and risks associated with tourism development and evaluating "eco-economic fitness."

8.3 Materials and Methodology

A systematic literature review (SLR) approach can be used to examine the literature on RT and sustainability and used for educational record problems, gaps, obstacles, and directions for

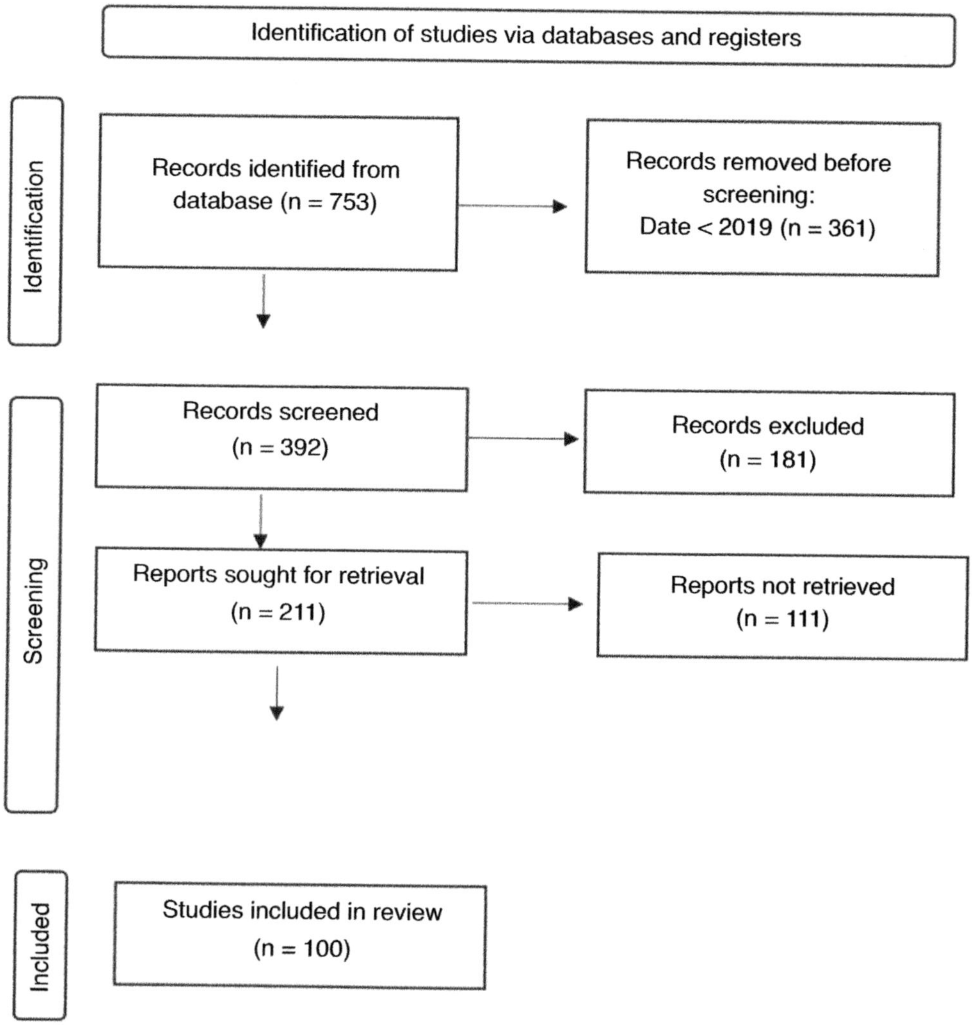

Fig. 8.1. Flowchart of the screening process.

future education. The SLR method has several advantages over traditional review techniques such as narrative review. These benefits include reduced bias and commissions, and improved audit quality, organization, and reproducibility. The issue of sustainable tourism is an area where SLR approaches are very common. The flowchart created in the authors' study discussed in this chapter also adheres to the PRISMA (preferred reporting items for systematic reviews and meta-analyses) guidelines (Fig. 8.1), which were slightly modified for clarity and completeness. A summary table was created with bibliographic information for the review looking at 100 research studies and using Microsoft Excel (Table 8.1). Two steps were used to guarantee thorough coverage of pertinent literature.

8.3.1 Finding the right keywords

A methodical three-step procedure was followed in the identification of keywords. First, the Google Scholar database was searched using the phrases "rural tourism" and "sustainability."

Table 8.1. The number and sources of research papers.

Online publishers' databases	No. of research papers found
Science Direct (sciencedirect.com)	35
Web of Science (webofscience.com)	16
Google Scholar (googlescholar.com)	17
Research Gate (researchgate.com)	21
ABDC (abdc.edu.au)	11
Total	100

Other keywords including "sustainable rural tourism," "village tourism," "socially responsible," and "green practices" were found and added from the top 100 results, as determined by Google Scholar's default algorithm. Science Direct, Web of Science, Research Gate, and the Australian Business Deans Council (ABDC) journal rating recommendations stated that these revised keywords should be utilized in the second part of the search process, which involved searching top tourism journals. A panel of experts from academia, practice, and policy validated the final set of keywords: sustain*, rural tourism, village tourism, sustainable rural tourism, green practices, Albergo Diffuso, environment-friendly, eco-friendly, and responsible.

8.3.2 Methods for identifying and screening literature

The final set of keywords yielded 753 results from Web of Science and Scopus; these databases are useful for conducting systematic literature reviews in the tourist sector. Reviews, conference proceedings, non-peer-reviewed publications, and novels were not considered during the search phase. The titles and abstracts of the publications were then evaluated for relevancy, yielding 211 results. Further refining was conducted by studying the entire texts of these papers, which resulted in the final selection of 100 research studies for review.

8.4 Discussion

This study makes a theoretical addition by investigating how RT experiences influence tourists' green consumption after their visit. It emphasizes that unique RT experiences, a strong connection to nature, and increased environmental knowledge are all important aspects impacting tourists' green consumption. While prior studies suggested links between these constructs, they had not been rigorously tested until this study. This concept was tested primarily in the setting of RT. The findings provide important insights into the elements that influence green consumption and shed light on the mechanisms by which RT experiences foster ecologically friendly behaviors. The findings of this study are consistent with earlier research and have both theoretical and managerial implications.

8.4.1 Theoretical implications

This work adds significantly to the body of recent literature on tourism in a number of ways. First, by adding the viewpoint of RT experiences, it improves comprehension of tourists' green consumption. Although prior research has predominantly examined the link between RT and environmental sustainability from pre-visit or on-site perspectives, with a focus on the direct impact of nature experiences on green consumption, there has been limited exploration into how RT experiences influence green consumption after the visit. This study addresses this gap by applying S-O-R theory to explore how RT experiences affect visitors' green consumption in the post-visit context. RT provides new insights and theoretical contributions to understanding the mechanisms that promote pro-environmental behavior over time. Travelers can make great memories that last while interacting with the environment, and the bond with nature in the future should be strengthened by these reminders. A stronger sense of connection with nature and greater environmental awareness can have a positive impact on green tourists' RT consumption. For this reason, travelers who have good memories about their experiences while traveling in the countryside may be more aware of environmental issues and this results in a higher intention to consume green even after the visit.

The scale used in this investigation covered a range of activities, many of which extend beyond nature. Four aspects of the RT experience used in this study were recreation, escape, education, and beauty, including the impact on green consumption. The results of this study are also consistent with previous research, with esthetic implications for promoting more inclusive behavior, providing environmentally friendly and enjoyable holidays, and clarifying the importance of education.

Overall, the study shows that positive memorable rural experiences create a strong relationship between green consumption and RT. It promotes a connection with nature and increases environmental awareness. This strengthens the findings of sources that show a relationship between positive experiences in nature and actions that promote environmental well-being. The study recommends that tourism managers use memorable RT experiences to promote green consumption intentions among tourists. However, the limited impact of memorable experiences on environmental awareness may be due to their timing since many tourists spend less than a day in rural areas.

8.4.2 Practical implications

The study emphasizes how crucial unique RT experiences are to strengthening visitors' bonds with the natural world, which raises awareness of environmental issues and promotes environmentally friendly behavior. It is vital for tourist marketers to create and promote these kinds of unforgettable experiences. These experiences – which include hedonism, refreshments, local customs, cultural significance, learning, novelty, and engagement – have the potential to greatly increase visitors' environmental consciousness and pro-environmental behavior. Urbanization and pandemic-related limitations have led to an increase in RT, providing a vital getaway for city inhabitants looking for amusement and a respite from the metropolis. It is important for marketers to highlight the esthetic value of rural places, which includes promoting and conserving rural buildings, landscapes, and agricultural environments. Promoting the educational elements of RT – such as regional handicrafts, agricultural goods, and conventional farming methods – can also make the experience more memorable and offer worthwhile educational opportunities. Long-term green consumption can be encouraged via educational events such as environmental knowledge competitions, which can strengthen tourists' awareness of green habits. Therefore, educators should focus on creating stimulating learning environments that highlight the unique characteristics of rural resources and promote a stronger connection to the natural world.

8.5 Conclusion

The preservation and prudent use of natural and cultural resources for the benefit of neighboring communities is a key component of a sustainable strategy for the expansion of RT. In India, sustainable methods are prioritized for the preservation and utilization of natural resources. An important idea is sustainable sharing, which encourages responsible use that benefits the neighborhood without diminishing resources. This strategy fosters understanding and appreciation of the uniqueness and fragility of rural assets, highlighting the significance of educational value in addition to recreational delight. Globalization and excessive tourism, in particular, pose threats to the preservation of cultural resources because, unlike natural resources, cultural practices are frequently not legally protected (Widawski *et al.*, 2023). Community pride and awareness are essential for the strength of cultural preservation, and education can help to foster these traits. RT benefits greatly from educational projects like educational homesteads, which combine learning with the natural and cultural surroundings. Having been a part of India's RT industry for more than 20 years, these programs provide immersive, interactive experiences that let visitors connect with the customs and natural surroundings of the area. In line with international trends, India's RT emphasizes sustainability, local involvement, and a balance between business and nature. The internet is utilized to distribute information and participate in the best possible way in order to promote eco-friendly, small-scale tourism. The area's distinctive natural and cultural assets must be effectively highlighted and used if sustainable RT development is to succeed. A conscious and active participation in the local community's molding of tourism is needed, presenting and updating information in real time with the use of contemporary technologies. In general, comprehensive management and education are essential for sustainable rural tourism in India, which benefits travelers as well as the environment and local populations.

References

Anderson, W.. (2015) Cultural tourism and poverty alleviation in rural Kilimanjaro, Tanzania. *Journal of Tourism and Cultural Change* 13(3), 208–224.

Andrei, D.R., Sandu, M., Gogonea, R.-M., Chirițescu, V. and Kruszlicika, M. (2012) Modeling of rural tourism towards sustainable development from the perspective of specifically organic food. *Scientific Papers Series Management, Economic Engineering in Agriculture and Rural Development* 12, 21–28.

Andriotis, K. and Vaughan, R.D. (2003) Urban residents' attitudes toward tourism development: The case of Crete. *Journal of Travel Research* 42(2), 172–185.

Apin, L. and Hassan, M.S. (2023) Operationalising sustainable development practice for area-based sustainability: Implementing UN-SDGs (2015–2030) at the local district area. *Russian Law Journal* 11(1S), 18–38.

Ataei, P., Karimi, H., Ghadermarzi, H. and Norouzi, A. (2020) A conceptual model of entrepreneurial competencies and their impacts on rural youth's intention to launch SMEs. *Journal of Rural Studies* 75, 185–195.

Ayhan, Ç.K., Özkök, F. and Tatlı, H. (2020) Land use suitability analysis of rural tourism activities: Yenice, Turkey. *Tourism Management* 76, 103949.

Berry, S. and Ladkin, A. (1997) Sustainable tourism: A regional perspective. *Tourism Management* 18(7), 433–440.

Bimonte, S. and Punzo, L.F. (2016) Tourist development and host–guest interaction: An economic exchange theory. *Annals of Tourism Research* 58, 128–139.

Blackstock, K.. (2005) A critical look at community based tourism. *Community Development Journal* 40(1), 39–49.

Blancas, F.J., Lozano-Oyola, M., Gonzalez, M., Guerrero, F.M. and Caballero, R. (2011) How to use sustainability indicators for tourism planning: The case of rural tourism in Andalusia (Spain). *Science of the Total Environment* 412, 28–45.

Bramwell, B. and Lane, B. (2012) Towards innovation in sustainable tourism research. *Journal of Sustainable Tourism* 20(1), 1–7.

Chin, C.H., Thian, S.S.-Z. and Lo, M.C. (2017) Community's experiential knowledge on the development of rural tourism competitive advantage: A study on kampung semadang–borneo heights, Sarawak. *Tourism Review* 72(2), 238–260.

Christou, P.A. (2020) Tourism experiences as the remedy to nostalgia: Conceptualizing the nostalgia and tourism nexus. *Current Issues in Tourism* 23(5), 612–625.

Croes, R. and Vanegas, M., Sr (2008) Cointegration and causality between tourism and poverty reduction. *Journal of Travel Research* 47(1), 94–103.

Du, Z.B. and Su, Q. (2011) Study on the relationship between the community participation of rural tourism, residents' perceived tourism impact and sense of community involvement—A case study of Anji rural tourism destination, Zhejiang Province. *Tourism Tribune* 26(11), 65–70.

Erokhin, Vasily. (2014) Approaches to sustainable rural development in a predominantly non-rural region. *Економика Пољопривреде* 61(2), 291–306.

Erokhin, V., Heijman, W. and Ivolga, A. (2014) Sustainable rural development in Russia through diversification: The case of the Stavropol Region. *Visegrad Journal on Bioeconomy and Sustainable Development* 3(1), 20–25.

Falak, S., Chiun, L.M. and Wee, A.Y. (2014) A repositioning strategy for rural tourism in malaysia-community's perspective. *Procedia-Social and Behavioral Sciences* 144, 412–415.

Franzoni, Simona. (2015) Measuring the sustainability performance of the tourism sector. *Tourism Management Perspectives* 16, 22–27.

Getz, Donald. (2008) Event tourism: Definition, evolution, and research. *Tourism Management* 29(3), 403–428.

Ghoddousi, S., Pintassilgo, P., Mendes, J., Ghoddousi, A. and Sequeira, B. (2018) Tourism and nature conservation: a case study in Golestan National Park, Iran. *Tourism Management Perspectives* 26, 20–27.

Gilbert, D. and Tung, L. (1990) Public organizations and rural marketing planning in england and wales. *Tourism Management* 11(2), 164–172.

Hampton, M.P. (2005) Heritage, local communities and economic development. *Annals of Tourism Research* 32(3), 735–759.

He, Wei. (2012) Sustainable rural tourism development and environment protection. *Advanced Materials Research* 347, 2893–2896.

Hegarty, C. and Przezbórska, L. (2005) Rural and agri-tourism as a tool for reorganising rural areas in old and new member states—a comparison study of Ireland and Poland. *International Journal of Tourism Research* 7(2), 63–77.

Higham, J.E. and Ritchie, B. (2001) The evolution of festivals and other events in rural southern New Zealand. *Event Management* 7(1), 39–49.

Huete-Alcocer, N. and Valero-Tévar, M.Á. (2021) Impact of information sources on promoting tourism in a rural region: The case of the roman villa of Noheda. *Sustainability* 13(14), 8038.

Ivona, Antonietta. (2021) Sustainability of rural tourism and promotion of local development. *Sustainability* 13(16), 8854.

Jamal, T. and Higham, J. (2021) *Justice and Ethics: Towards a New Platform for Tourism and Sustainability*. Routledge, pp. 1–15.

KC, A. and Parajuli, R.B.T. (2014) Tourism and its impact on livelihood in Manaslu conservation area, Nepal. *Environment, Development and Sustainability* 16, 1053–1063.

Lee, S., Kim, D., Park, S. and Lee, W. (2021) A study on the strategic decision making used in the revitalization of fishing village tourism: Using A'WOT analysis. *Sustainability* 13(13), 7472.

Lee, T.H. and Hsieh, H.-P. (2016) Indicators of sustainable tourism: A case study from a Taiwan's wetland. *Ecological Indicators* 67, 779–787.

Korani, Z. and Shafiei, Z. (2021) In search of traces of 'The Tourist Gaze'on locals: An ethnographic study in Garmeh village, Iran. *Journal of Tourism and Cultural Change* 19(5), 549–567.

Lewis, J.B. and Delisle, L. (2004) Tourism as economic self-development in rural Nebraska: A case study. *Tourism Analysis* 9(3), 153–166.

Marques, Helena. (2006) Searching for complementarities between agriculture and tourism—the demarcated wine-producing regions of northern Portugal. *Tourism Economics* 12(1), 147–155.

Lo, M.-C., Chin, C.-H. and Law, F.-Y. (2019) Tourists' perspectives on hard and soft services toward rural tourism destination competitiveness: Community support as a moderator. *Tourism and Hospitality Research* 19(2), 139–157.

Marques, C.P., Guedes, A. and Bento, R. (2022) Rural tourism recovery between two COVID-19 waves: The case of Portugal. *Current Issues in Tourism* 25(6), 857–863.

Mikhaylova, A.A., Wendt, J.A., Hvaley, D.V., Bógdał-Brzezińska, A. and Mikhaylov, A.S. (2022) Impact of cross-border tourism on the sustainable development of rural areas in the Russian–Polish and Russian–Kazakh borderlands. *Sustainability* 14(4), 2409.

Milman, A., Jones, F. and Bach, S. (1999) The impact of security devices on tourists' perceived safety: The central Florida example. *Journal of Hospitality & Tourism Research* 23(4), 371–386.

Mitchell, M. and Hall, D.R. (2005) Rural tourism as sustainable business: Key themes and issues. In: Hall, D., Kirkpatrick, I. and Mitchell, M. eds (eds) *Rural Tourism and Sustainable Business*. pp. 3–14.

Mutlu Danaci, H. and Eser, R.E. (2013) Ecotourism and vernacular architecture in Koprulu Canyon. *Zeitschrift Für Die Welt Der Türken* 5(1), 41–46.

Mvula, C.D. (2001) Fair trade in tourism to protected areas—a micro case study of wildlife tourism to South Luangwa National Park, Zambia. *International Journal of Tourism Research* 3(5), 393–405.

Palacios-Florencio, B., Santos-Roldán, L., Berbel-Pineda, J.M. and Castillo-Canalejo, A.M. (2021) Sustainable tourism as a driving force of the tourism industry in a post-COVID-19 scenario. *Social Indicators Research* 158(3), 991–1011.

Pechlaner, H., Bialk-Wolf, A. and Zacher, D. (2015) The role of cultural activities in tourism development: An urban–rural perspective. *Tourism Culture & Communication* 15(1), 47–58.

Polukhina, A., Sheresheva, M., Efremova, M., Suranova, O., Agalakova, O. *et al.* (2021) The concept of sustainable rural tourism development in the face of COVID-19 crisis: Evidence from Russia. *Journal of Risk and Financial Management* 14(1), 38.
Roberts, L., Hall, D. and Morag, M. (2017) *New Directions in Rural Tourism*. Routledge.
Saarinen, J. and Lenao, M. (2014) Integrating tourism to rural development and planning in the developing world. *Development Southern Africa* 31(3), 363–372.
Sharpley, Richard. (2003) Rural tourism and sustainability-a critique 38–53.
Serra Cantallops, A., Antoni, J.R.C. and Muntaner, R.E. (2015) Characteristics and peculiarities of rural tourism in the Balearic Islands. *Tourism Planning & Development* 12(2), 125–144.
Sharpley, R. and Roberts, L. (2004) Rural tourism—10 years on. *International Journal of Tourism Research* 6(3), 119–124.
Timothy, D.J. (1999) Cross-border partnership in tourism resource management: International parks along the US-Canada border.. *Journal of Sustainable Tourism* 7, 3–4.
Timothy, D.J. and Tosun, C. (2003) Tourists' perceptions of the Canada–USA border as a barrier to tourism at the International Peace Garden. *Tourism Management* 24(4), 411–421.
Tosun, Cevat. (2006) Expected nature of community participation in tourism development. *Tourism Management* 27(3), 493–504.
Trukhachev, Alexander. (2015) Methodology for evaluating the rural tourism potentials: A tool to ensure sustainable development of rural settlements. *Sustainability* 7(3), 3052–3070.
Utami, D.D., Dhewanto, W. and Lestari, Y.D. (2023) Rural tourism entrepreneurship success factors for sustainable tourism village: Evidence from Indonesia. *Cogent Business & Management* 10(1), 2180845.
Widawski, K., Krzemińska, A., Zaręba, A. and Dzikowska, A. (2023) A sustainable approach to tourism development in rural areas: The example of Poland. *Agriculture* 13, 10.
Wright, W. and Middendorf, G. (2010) *The Fight Over Food: Producers, Consumers, and Activists Challenge the Global Food System*. Penn State Press.
Yan, H. and Morpeth, N.D. (2015) Approaches to planning and tourism. In: *Planning for Tourism: Towards a Sustainable Future*. pp. 1–26.
Yun, H.Jeong. (2014) Spatial relationships of cultural amenities in rural tourism areas. *Tourism Planning & Development* 11(4), 452–462.
Zou, T., Huang, S. and Ding, P. (2014) Toward a community-driven development model of rural tourism: The Chinese experience. *International Journal of Tourism Research* 16(3), 261–271.
Slater, S.J., Christiana, R.W. and Gustat, J. (2020) Peer reviewed: Recommendations for keeping parks and green space accessible for mental and physical health during COVID-19 and other pandemics. *Preventing Chronic Disease* 17.
Štetić, S. (2012) Specific features of rural tourism destinations management. *Journal of Settlements and Spatial Planning* 1(Special Issue), 131–137.

9 Investigating Cultural Heritage and Tourism Dynamics in the Magical Town of Cholula, Mexico

Tahir Sufi[1]*, Anil Kumar Goyal[2] and Edgar Rojas-Rivas[1]
[1]Department of Tourism, School of Business and Economics, Universidad de las América's Puebla (UDLAP), Puebla, Mexico; [2]Institute of Hotel Management, Catering and Nutrition Pusa, New Delhi, India

9.1 Introduction

This chapter explores the concept of cultural heritage and its relationship with tourism in the "magical town" of Cholula. Specifically, it investigates how the unique heritage of Cholula is preserved, maintained, and integrated into tourism, further empowering the local community and preserving its cultural identity. Cultural heritage plays a vital role in promoting tourism by providing a unique experience that attracts visitors. The integration of cultural heritage assets not only helps preserve heritage but also enhances the tourism experience, contributing to the sustainable development of tourism (Fyall and Garrod, 1998). The interaction between heritage and tourism leads to the activation and protection of heritage, enriching the cultural diversity of tourism offerings (Lan *et al.*, 2021). Cultural heritage tourism combines aspects of ruralism with key cultural heritage elements, creating a distinctive appeal for tourists (Muangasame and Tan, 2022).

Tourists' appreciation of the esthetics of the environment and their engagement significantly influences the shaping of cultural identity within heritage tourism contexts (Yang *et al.*, 2022b). The strategic integration of intangible cultural heritage and the tourism industry ensures sustainable viability and enriches the overall tourism experience (Xiao, 2022). Sustainable cultural development involves the continuity of heritage, cultural tourism, and rural tourism, and emphasizes the interconnectedness of these elements (Jurėnienė and Radzevičius, 2022). The relationship between cultural heritage and tourism impacts entrepreneurs' financial outcomes in rural tourism, highlighting the economic benefits derived from leveraging cultural heritage assets (Starčević *et al.*, 2022). Utilizing cultural heritage for tourism purposes not only attracts tourists but also contributes to the preservation and conservation of heritage sites (Đukanović *et al.*, 2021). Tangible and intangible factors related to rural heritage, such as local cuisine and traditional architecture, are essential components of culture-based rural tourism (Petelca *et al.*, 2021).

The sustainable development of heritage tourism requires collaboration with community tourism, health tourism, and rural tourism, highlighting the significance of community culture in enhancing the tourism experience

*Corresponding author: tahir.sufi@gmail.com

DOI: 10.1079/9781836992646.0009

(Zhao, 2022). By adapting the interpretation and presentation of rural heritage, rural tourism in specific regions has been effectively promoted, providing visitors with a deeper engagement with local culture (Harun, 2018). Striking a balance between heritage protection and tourism development is crucial for ensuring the long-term sustainability of rural tourism initiatives (Ramón-Cardona and Sánchez-Fernández, 2022).

Despite the growing body of literature on cultural heritage tourism and its impact on rural development, there remains a gap in understanding how specific regions like Cholula, Mexico, navigate the challenges of integrating cultural heritage into sustainable tourism. Existing studies have focused on general heritage preservation or rural tourism but have not thoroughly explored the unique interplay between intangible cultural elements, such as architecture, dance, music, food, traditions, and festivals, in the context of Cholula. Additionally, the role of local communities in shaping tourism experiences and preserving cultural heritage through social media feedback remains under explored. To address these gaps, the current research seeks to answer the following questions.

1. How does the integration of Cholula's architectural, cultural, and religious heritage empower tourism development?
2. What are the challenges and opportunities associated with leveraging architectural, cultural, and religious heritage for sustainable tourism development in Cholula?
3. How do tourists' perceptions and interactions with cultural heritage sites influence the preservation and promotion of Cholula's identity?

To achieve these objectives, a qualitative research design was adopted to collect secondary data from social media platforms to study cultural heritage and rural tourism in Cholula. The information was collected based on reviews published on the Tripadvisor platform about the emblematic places of San Pedro Cholula, such as the Great Pyramid of Cholula, the regional museum, and architectural heritage such as churches and convents, among others. In addition, information was explored in publications about the cultural heritage of the destination, as well as the experiences associated with rural tourism in the region. The information was analyzed through descriptive qualitative analysis with the identification of themes and subthemes. Comments, perceptions, and reactions were standardized through a rigorous process, where words are standardized, errors are eliminated, and connectors are removed, with the emerging categories having their frequency calculated and cognitive importance determined (Symoneaux *et al.*, 2012).

9.2 Review of the Literature

9.2.1 Investigating cultural heritage and tourism dynamics in Cholula

Cultural heritage encompasses the tangible and intangible legacies of past societies, in contrast to a notion of being a collection of artifacts or monuments. The concept represents dynamism in the social process involving the heritagization of community practices, beliefs, and values inherited across several generations (Zhu and Liu, 2021). The concept has undergone a transition from being one that deals with the monuments and other physical aspects to a living heritage encompassing cultural and knowledge of communities, reflecting a new approach for a contemporary society (Poulios, 2014). Going beyond historical significance, this new version of the framework plays a crucial role in shaping identities of nations, communities, and their people. The elements of the framework are reflected among those who claim to own the cultural heritage in the form of a unique sense of belonging and identity (Tutchener, 2013). The connection between identity and heritage is evident in the context of intangible cultural heritage representing expressions and knowledge that communities refer to as their heritage (Wasino *et al.*, 2023). Therefore, heritage preservation and promotion are vital for forging a social cohesion based on shared heritage within communities (Vegheş, 2019).

Cultural heritage management has entered a phase of digitization, transformed its landscape and allowing broader access and engagement with cultural resources (Thekkum Kara, 2021), thereby opening new avenues for education, tourism, and community engagement. Additionally, this advanced framework enables interactive and participatory approaches to heritage conservation (Isa *et al.*, 2018). However, there is also a flip side to such an approach on account of questions raised about the authenticity of cultural heritage. Such questions find a basis due to rising heritage commercialization and commodification, diluting the essence and emotional value of heritage (Son, 2023).

9.2.2 Cultural heritage tourism theory

Cultural heritage tourism (CHT) and destination image theory are critical frameworks in understanding the dynamics of tourism, particularly in regions rich in cultural heritage. The interplay between these concepts is essential for developing effective tourism strategies that enhance visitor satisfaction and promote sustainable tourism practices. CHT offers experiences of historical tourist sites, representing the stories and people of the past, and provides opportunities of economic development to the locals and promotes local businesses, resulting in job creation (Huete-Alcocer and Ruiz, 2019). Therefore, this form of tourism is based on the perceived quality of the cultural experiences impacting tourist satisfaction and resulting in revisits of tourists to such locations (Abbasi *et al.*, 2021). To support this framework, destination image theory elaborates how tourist perceptions about destinations have a potential to influence visit intentions of tourists, and embraces various perceptual elements like cognitive, affective, and conative processes that shape such tourist perceptions about the destination (Yang *et al.*, 2022a). Cognitive function represents the destination beliefs of tourists, affective function captures how travelers feel, and conative function reflects tourist behavior in visiting the destination (da Silva Añaña *et al.*, 2018; Wang *et al.*, 2020). The tourist experience is a prime factor for establishing the destination image as such experience triggers the cognitive and affective factors responsible for tourist satisfaction and recommending the destination (Iordanova and Stylidis, 2019). Such a framework therefore demonstrates how immersive and engaging cultural experiences promote cultural heritage tourism (Huete-Alcocer and Ruiz, 2019).

Furthermore, the creditability of destination information addresses sustainability concerns by triggering responsible behavior among tourists (Qiu *et al.*, 2022), who follow sustainable practices when credible information is provided, highlighting the significance of effective communication strategies in promoting CHT. Word of mouth is yet another critical dimension shaping destination image, in addition to shaping tourists' behavior, by mediating the relationship between perceived destination image and tourists' intentions to visit (Xu *et al.*, 2020). This highlights how fostering positive tourist experiences can make them share their experiences, thus enhancing the destination's image in the process. The cognitive-affective framework establishes how cognitive evaluations and emotional responses create destination images and demonstrates how understanding the elements of destination image is vital in establishing effective marketing strategies (Amine and Semaoune, 2021; Yang *et al.*, 2022a; Aziz and Niazi, 2023). Community perception is another important factor for the sustainability of CHT as it offers unique insights about such sites and demonstrates the emotional bonds of the community engaging in tourism initiatives. The involvement of local communities therefore can lead to better tourists' experiences and improved destination image (Xue and Kerstetter, 2017; Stylidis, 2020). Finally, the concept of self-congruity, demonstrating a vital link between the tourist's self-image and the destination's image, can be crucial for forging destination loyalty as the tourist revisit intentions resonate with their self-identity. Such an understanding therefore highlights how crafting a destination image that appeals to various tourist segments is vital for a better destination image (Huang, 2018; Nguyen, 2023).

9.2.3 Cultural heritage and tourism

Apart from contributing to the identity of the communities, cultural heritage attracts tourists from international destinations, making an enormous economic contribution. Research into the economic impact of CHT has gained lots of attraction over the decades. Large numbers of studies report the positive economic contribution of heritage, and researchers therefore propose treating cultural variables as cultural capital that needs to be taken into account in formal behavioral models of economic growth and development (Benhamou, 2020). Some econometric studies explore its impact on growth directly or through tourism demand involving treating the heritage monuments of UNESCO's world heritage sites (WHS) as time-invariant structural variables (Hosseini *et al.*, 2021). Acknowledging this approach, the Organization for Economic Co-operation and Development (OECD), recommends treating such cultural assets as elements of territorial attractiveness that lead to the differentiation and competitiveness of destinations, thereby offering a link between cultural heritage and tourism (Mason and Ying, 2020). As a major economic driver, CHT provides lots of benefits to local communities (Lussetyowati, 2015). However, there is also a flip side to CHT. Over the years, there have been cases where residents of some important tourism destinations have complained of overtourism, resulting in the appropriation of local cultures (Diaz-Parra and Jover, 2021). As a result, there has been political and social concern across several countries, especially in Europe. Cities like Athens, Barcelona, and Rome have witnessed tremendous increases in tourism thereby causing socio-spatial issues. The term overtourism was coined in 2017, attracting the attention of a large number of research studies, including from the UN World Tourism Organization and the European Parliament (Koens and Klijs, 2022).

To safeguard heritage from the onslaught of overtourism, therefore, sustainable management practices are essential. Such a proactive approach ensures tourism in the context of cultural heritage has a positive contribution for preserving the heritage while, at the same time, offering benefits to the local communities (Bani Omar, 2022). Awareness and education additionally can foster a sense of cultural heritage preservation, responsibility, and stewardship (Winarni, 2018). Education and awareness is not just a reactive measure of overtourism, such proactive measures need to be adopted through imparting education through school curricula and community programs. Additionally, through campaigns aimed at local communities, a deeper appreciation of heritage can be cultivated for community-level participation in the preservation efforts (Batmaz, 2023). Management of cultural heritage can be a complex phenomenon as it encompasses a complex network of tangible and intangible elements. However, through an integrated approach, cultural heritage management can be sustainable in meeting the challenges arising out of overtourism. It can continue to shape community identities, ensure cohesion, and at the same time offer economic benefits. In the backdrop of globalization and technological advancement, cultural heritage preservation is essential for cultural diversity, in addition to ensuring that future generations inherit their cultural legacies and draw their identity from a well-preserved heritage.

9.2.4 Tourism in Cholula

The city of Cholula, officially known as Cholula de Rivadavia, is in the state of Puebla, Mexico and is known for its Great Pyramid, with the Iglesia de Nuestra Señora de los Remedios sanctuary on top. Over the centuries, it has emerged as a vibrant tourism destination on account of its historical significance and cultural heritage, thereby earning its name "Pueblo Mágico." The city attracts tourists on account of its pre-Hispanic and colonial history, creating a unique destination for tourists interested in culture and history. The Great Pyramid of Cholula is the largest pyramid in the world, showcasing its prestige of being part of glorious ancient civilizations. Gastronomy is also a factor influencing tourism here; the unique gastronomical culture of the city represents a fusion of indigenous and Spanish influences and appeals to experiential tourism of the town (Sosa *et al.*, 2021; Michaca-Rodríguez *et al.*, 2023). The gastronomical tourism aligns with the cultural tourism, as the gastronomy contributes to shaping overall tourist experiences and additionally promotes the destination image (Sosa *et al.*, 2021).

9.2.5 Destination marketing and tourism trends in Cholula

Cholula has witnessed an increase in tourism over the decades, leading to incremental economic growth attributed to numerous reasons. The most important push behind increasing tourist numbers is designating Cholula as a magical town. The development of new tourism infrastructure has been another contributing factor in the increase in tourist activities (Gutiérrez Juárez *et al.*, 2022). The development of new parks, museums, and transportation systems, mainly aimed at enhancing tourist experiences and making the destination attractive for local and international tourists, are examples of how policy makers aim to upgrade the infrastructure to support tourism (Ismael-Simental *et al.*, 2019; Gutiérrez Juárez *et al.*, 2022). In keeping up with the trend of promoting sustainable tourism, the destination should focus on preserving the area's natural and cultural resources (Durán-Díaz, 2023). Promoting positive aspects of the town or city through cultural attractions is a unique approach that can overcome security-related concerns among tourists and has proved to be an effective marketing strategy achieved through marketing campaigns, cultural events, improvement of public spaces, and promoting its culinary delights and festivals (Ismael-Simental *et al.*, 2019).

Cultural events like music, dance, and crafts offer immersive experiences of local culture to visitors. The landmark events of Mexico, Day of the Dead and Carnival of Cholula, highlight the community's vibrant cultural practices in addition to fostering the sense of belonging for local communities (Serrano, 2024). Additionally, such festivals attract tourists and contribute to safeguarding local traditions as well as promoting cultural heritage (Serrano, 2024). Stakeholder engagement, especially indigenous perspectives on tourism planning, is another example of best practice in tourism. Accommodating diverse perspectives allows authentic representation of the region's cultural heritage and has led to the development of unique tourism products that reflect the values and traditions of these communities (Espeso-Molinero *et al.*, 2019).

Such an approach, therefore, enhances tourist experience and ultimately provides marginalized communities economic opportunities and a platform to share their cultural narratives (Espeso-Molinero *et al.*, 2019).

The unique culture of Cholula has successfully attracted digital nomads and remote workers. In the backdrop of the Covid-19 pandemic, the accelerated trend for remote work has led workers to look for attractive destinations that can provide them with a good work–life balance, something that Cholula has to offer. Whether it is its peaceful and attractive atmosphere, affordable living costs, or well-established information and communication network, the destination is emerging as an attractive option for digital nomads. Content creators also find the destination a perfect place for leveraging the social media that play a significant role in promoting the destination. The user-generated content and influencer marketing, therefore, have emerged as effective marketing strategies for attracting millennials and Gen Z travelers.

9.2.6 Drawbacks of tourism in Cholula

Despite the efforts of sustainability, the rapid growth of tourism has resulted in concerns of residents about their heritage and environment (Gutiérrez Juárez *et al.*, 2022; Durán-Díaz, 2023). For sustainability of its cultural heritage, especially the Great Pyramid and colonial churches, careful planning and investment in conservation efforts is required. The stakeholder approach therefore is vital to ensure success in sustainability efforts that balances tourism promotion and economic development (Zamora, 2023). To maintain such a balance is a common challenge in almost all destinations. Another challenge is ensuring that the tourism benefits are equitably distributed among stakeholders, especially the local communities.

In this regard, tourism policies play a prominent role by ensuring effective policies and strategies that ensure the benefits of tourism development are equitably distributed. To achieve this goal, collaborative governance models based on stakeholder approach need to be put in place. By involving various stakeholders like government, non-governmental organizations, local businesses, and local community, a more inclusive tourism development framework can be ensured (Gutiérrez Juárez *et al.*, 2022; Schumacher *et al.*, 2022). These frameworks can guarantee sustainable tourism initiatives and promote social cohesion and community engagement. While the destination is the focal point of cultural tourism, the sustainable practices based on stakeholder approach (especially the indigenous communities) should enrich the tourist experience while fostering economic development.

The destination can be constantly evolving as a cultural tourism hub by adopting new trends and technology while preserving its inheritance; it can therefore serve as a model cultural destination that looks forward while balancing tourism growth and preserving its cultural identity. Planners face the challenges of balancing economic benefits with the sustainability of its cultural heritage through dialog and collaboration with the stakeholders. Additionally, an increase in tourist numbers often results in gentrification and many indigenous communities then face displacement (Gutiérrez Juárez *et al.*, 2022; Schumacher *et al.*, 2022) – yet another challenge for planners.

9.3 Methodology

The study this chapter is based on adopted a qualitative research design to explore CHT dynamics within the Cholula region of Mexico. Furthermore, it adopted thematic analysis to capture and analyze experiences of tourists and local community members to explore how the region's cultural aspects promote local tourism. There are several examples demonstrating how similar studies have adopted thematic analysis to explore the interaction between cultural heritage and tourism through qualitative data analysis from secondary sources. For example, Lee and Chang's (2023) study analyzed online reviews and visitor feedback to assess the role of heritage sites in shaping tourist experiences. Qualitative data from travel forums was utilized to identify key themes related to visitor satisfaction and heritage conservation. Similarly, Safaaa *et al.* (2017) study adopted thematic analysis to explore tourists' perceptions by analyzing online reviews to understand patterns in tourists' engagement, and Cenni and Goethals (2017) also used thematic analysis to analyze Tripadvisor reviews to explore tourists' emotional response to cultural heritage sites in Italy.

The present study adopted a six-phase thematic analysis that included studying the qualitative data, generating initial codes, identifying subthemes, grouping the codes into broader themes, reviewing and refining the themes, and finally defining and naming the themes as recommended by Labra *et al.* (2020). Qualitative data in the shape of online reviews about the Great Pyramid of Cholula, the Regional Museum of Cholula, and architectural heritage sites posted on Tripadvisor were used. Tripadvisor is a widely recognized platform where travelers share their experiences (Nicoli and Papadopoulou, 2017). Adopting such an approach allowed the capturing of diverse tourist perceptions on various factors that define the cultural dimensions of Cholula. The study also consulted blogs and news portals to gain a deeper understanding of how such factors contribute to cultural tourism in Cholula. Additionally, such sources assisted in contextualizing how the local community engages with cultural heritage and how tourism supports its preservation. The reviews were standardized by correcting grammatical errors and removing irrelevant words for accurate analysis and interpretation. The data were subsequently analyzed to identify recurring themes, patterns, and trends. Subsequently, the frequency of occurrence was computed and tabulated, as presented in Table 9.1.

The issues of credibility, transferability, dependability, and confirmability for qualitative research was taken care of as recommended by Nowell *et al.* (2017) for such research. The study ensured its credibility by utilizing multiple sources of data such as Tripadvisor reviews, online publications, and blogs. The diversity of data made the study more insightful and ensured that the insights generated are creditable. Transferability was ensured through detailed contextualization of Cholula's cultural heritage and allowed researchers to apply the findings to other similar regions. In a similar way, data standardization was achieved through removal of irrelevant content and elimination of technical errors. All such measures enhanced the study's dependability for being consistent and replicable. Finally, confirmability of the study was achieved by the objective handling of data. Since the study was based on secondary data, it additionally used systematic coding techniques to minimize bias. All such steps collectively therefore ensured data integrity and reliability of the research findings.

9.4 Results and Discussion

9.4.1 Demographic analysis of travelers

The demographic profile of the respondents shows a diverse group of travelers, with a predominance of female reviewers, particularly single female travelers. There is a balanced mix of male and female travelers across different travel groups, including solo travelers, families, and groups. Based on the profile pictures of the reviewers on Tripadvisor, the age range spans from 20 to 50 years, covering both millennials and Gen X. The respondents are regular travelers, familiar with travel norms, and they traveled either alone, with their families, or in groups, such as friends or multigenerational families. The reviews reflect a broad international presence, with travelers from various nationalities, including those from Asia, Europe, and North America. Although the reviews do not explicitly state the purpose of travel, it may be inferred that their visits were likely motivated by a mix of leisure, cultural exploration, business, and adventure.

9.4.2 Analysis of qualitative data

The data for this study were collected from Tripadvisor reviews posted by visitors to tourist attractions in the region, including the Great Pyramid of Cholula, the Regional Museum of Cholula, and several architectural heritage sites. A total of 120 reviews were gathered, with an equal distribution of 40 reviews from each site to ensure balanced representation. Reviews were double-checked by neutral moderators. The inductive method was used in the analysis because it was targeted at capturing emergent concepts. Open line-by-line coding was used to read and reread the reviews to understand and document the reviewers' experiences, behaviors, feelings, attitudes, and perspectives regarding their visits to the Cholula pyramids, tunnels, regional museum, covenants, and churches. Commonalities and differences between and among the reviews were looked for. In order to describe linkages, uncover underlying concepts, and establish categories and themes, the various stories were compared and contrasted to improve the codes and to map connections. Constant interaction with the data guaranteed that new ideas were rooted in the experiences of the reviewers. Analytical memoranda and notes were taken by the researcher at various points in the process to record their evolving understanding of key concepts and their interconnections. The theoretical underpinnings of the study were debated and documented, along with any new themes or notions that emerged.

9.4.3 Emergence of themes

As outlined in Table 9.1, thematic patterns were systematically examined within individual cases and groups, to identify recurring themes and their interrelationships. Through a detailed analysis of the data, a total of 25 subthemes emerged. These subthemes represented various nuanced aspects of the research objectives. Upon further refinement, these subthemes were synthesized into six broad themes, central to the study's objectives: architectural heritage, cultural heritage, religious heritage, gastronomic, tourist experiences, and the environment.

9.4.3.1 Architectural heritage and tourism development

The architectural heritage of Puebla, Mexico, is a testament to the region's rich history, blending indigenous and colonial influences. A centerpiece of this heritage is the Great Pyramid of

Table 9.1. Emergence of broad theme from the reviews.

Broad theme	Emergent subthemes	No. of reviews			
		Cholula pyramids	Regional museum	Churches and convents	Total
Architecture heritage	Pyramid size and significance	10			10
	Tunnels	8			8
	Architecture and esthetic	2	9	13	24
	Preservation and restoration efforts			2	2
Cultural heritage	Cultural and historical importance	5	6	6	17
	Cultural celebrations	9			9
	Emotional impact			4	4
	Dance and music			12	12
Religious heritage	Churches			16	16
	Catholics			4	4
	Religion			6	6
	Religious significance			7	7
Gastronomic	Food	3	2	4	9
	Bars/beverages	2	1	2	5
	Local food carts	3	2	4	11
	Restaurants	4	2	2	9
Tourist experiences	Tourism infrastructure and accessibility	4	7		11
	Disappointment due to closures	4			4
	Family and friends' experience	3			3
	Visitors' experience	6	7	5	18
Environment	Weather and sun exposure	2			2
	Surroundings		3		3
	View and scenery	6			6
	Volcanoes	2			2
	Valley	5			5

Cholula, the largest pyramid by volume in the world. Beneath its massive structure, a network of tunnels reveals the layers of civilization that flourished before the Spanish conquest (Bagnara Milan, 2022). The Church of Nuestra Señora de los Remedios, built atop the pyramid by Spanish colonizers, symbolizes the fusion of Mexican and Christian architecture. Other architectural marvels in Puebla include the Church of Santo Domingo, with its ornate Rosary Chapel, and the Puebla Cathedral, a baroque masterpiece in the city's historic center. Architectural highlights also extend to the regional museum, which showcases artifacts from the pre-Columbian era to the colonial period, preserving Puebla's layered history. The preservation and restoration of these sites have played a crucial role in the region's tourism development. Puebla's architectural beauty draws international visitors each year, contributing significantly to the local economy. Hence, heritage tourism fosters economic growth and reinforces the cultural identity of the region (Vegheş, 2019). Enjoyment of the architecture is evident from the following observation by one tourist visiting historical sites.

> "Baroque" doesn't begin to describe Cholula's stunning Templo San Francisco Acatep. Talavera tiles in varied patterns and colors (though mostly cobalt blue, white and yellow) lie so thick on the façade that the columns, niches and arches are difficult to discern. The heavy wooden front doors, the only feature unadorned, were locked when we arrived, but when we walked around to the courtyard to the right as we faced them, we found a sacristan who unlocked them. Inside, the church is heavily embellished with gilded carving and plasterwork, but the vaulted ceilings keep the effect from being overwhelming. Flanked by curving polychrome columns, the altarpiece features statues of saints that seem almost naïve by comparison.

In addition to providing positive reviews about the destinations, reviewers also encourage others to visit these places. This trend highlights the potential to boost tourism in these areas, which in turn can have a significant positive impact on the economic well-being of the historical sites.

> Entering the church gave me goosebumps. The grandeur of how this church was built was stunning. The pieces inside the church were made by the indigenous people. I loved this church and I think travelers should include this in their itinerary. Don't forget to thank God!

> This recently opened museum is educational, entertaining, and all-round impressive. Beginning with stories of nearby volcanoes, the exhibits address legend and vulcanology. Extensive exhibits of pre-Hispanic ceramic and stone sculptures tell the story of past religions, artistic achievement, and the significance of the area through Mexican history. There are historical and contemporary folk-art exhibits as well. I was particularly fascinated by two exhibits: an interactive room explaining the glyphs that can be found in so many pre-Hispanic sites, and the videos and digital reconstructions of stages of the Great Pyramid's construction (and sometimes destruction). The museum is beautifully designed and laid out, using traditional as well as progressive exhibition techniques. Staff are very helpful, and we had a friendly and invaluable discussion with a museum intern. We spent a whole afternoon there, but it could be seen in just an hour or two. The entrance fee is reasonable, and there are free admission days. Highly recommended – especially BEFORE visiting the Great Pyramid archaeological site!

9.4.3.2 Cultural heritage and tourism development

Puebla is rich in vibrant dance, music, festivals, and cultural traditions that attract visitors from around the world. One of the most iconic dance forms in Puebla is the "Danza de los Quetzales," a traditional performance that honours the ancient Aztec heritage. The dancers wear colorful feathered costumes, symbolizing the quetzal bird, and the dance is performed during religious and civic celebrations (Serrano, 2024). Puebla is also known for "Son Jarocho," a lively folk music genre that combines indigenous, Spanish, and African influences. Puebla's festivals are a big attraction for tourists, with "Cinco de Mayo" being the most famous one that celebrates Mexico's victory over the French in the Battle of Puebla in 1862. The attractions of the festival include military parades and cultural performances. "Festival Internacional de Puebla" is yet another major festival of the region highlighting the region's cultural heritage through cultural performances from different parts of Mexico, making Puebla a key cultural tourism destination (Espeso-Molinero *et al.*, 2019). All such elements collectively contribute to the sustainable development of tourism in addition to fostering economic growth. Capturing the sentiments of cultural aspects of the destination, one of the tourists posted on the Tripadvisor:

> This was an amazing experience! It is a must if you find yourself in Puebla. You get to climb parts of the pyramid which is not allowed in most authentic pyramids. Once you are on top of the pyramid, which was turned into a church by the Spanish conquistadors, you get a view of the whole city including the famous volcano Popocatepetl meaning "white smoke". If you are lucky, you will also get to see los Voladores which is an amazing ritual to the sun. There are also Aztec dancers who perform mostly during the weekends. I had the pleasure of meeting them and they welcomed me and told me history of Mexico while they shared homemade Pulque which is an alcoholic beverage derived from the Agave plant.

> Out of all the beautiful places that I've been to this experience was by far the best because I was given such a great hospitality by the locals that I felt like I lived there my whole life.
>
> Didn't have a whole lot of time here but when we arrived there were some local Indians doing a "goodbye to the sun" dance, it was pretty cool to watch, I then, got a blessing from them via smoke, very cool.

9.4.3.3 Religious heritage and tourism development

The content analysis of Tripadvisor reviews regarding the churches and convents in Cholula reveals religious significance as a theme. Catholicism has a deep connection with the religious history of the area. Visitors frequently highlight the spiritual importance of the region's churches, many of which serve as significant pilgrimage sites, reflecting the enduring Catholic heritage. The religious architecture, festivals, and cultural practices linked to these religious landmarks have preserved local traditions and driven tourism development (Vizcaino-Suárez, 2018.). By attracting both religious pilgrims and tourists interested in cultural and historical exploration, Cholula's religious significance plays a vital role in boosting local tourism, supporting the area's economy, and promoting the preservation of its heritage. This is reflected in the following reviews.

> They were having a service and it was amazing to see the faith and respect people have in this place, beyond admiring the place itself, even if you don't share it people gathering respecting something so deeply is amazing. Even children were quiet, the smell is unique to this old churches you will always remember it.
>
> This church was built on top of the great pyramid of Cholula, which is the largest one in the world volume-wise. The pyramid is partially uncovered, and serves as an attraction in this millenarian town. The church holds the statue of the Virgin of the Remedies, which is the most important religious image for people in Cholula. The ascent is trying, but well worth it since from the top you can see a large span of the Valley of Puebla, and weather permitting, the four volcanoes which delimit it.
>
> This pyramid is the biggest in Mexico and is located in a small town called Cholula which has 365 churches!! Is a very cute small town very close to the city. You have to see it.

9.4.3.4 Gastronomy and tourism development

The Cholula region is renowned for its rich and diverse culinary heritage, featuring a fusion of indigenous, Spanish, Arabian, and Asian influences. The unique blend of different cultures in its culinary traditions adds to it to being a unique gastronomic destination that influences tourism significantly (Acle-Mena *et al.*, 2020). The trend is in agreement with the literature, demonstrating how the quality and promotion of local cuisine are highly correlated with tourist satisfaction (Yasami *et al.*, 2020). Mexican cuisine is a part of the "Intangible Heritage of Humanity" as defined by UNESCO because of the traditional processes and seasonality of the ingredients, which have unique sensory characteristics (Fattacciu, 2019). Traditional dishes like mole, chile en nogada, carnitas, and cochinitapibil are some of the most popular dishes among locals and international tourists (Rojas-Rivas *et al.*, 2020). The state of Puebla is recognized as one of the most important gastronomic capitals of Mexico, and it combines the indigenous past with culinary innovation. In terms of tourism, it has been highlighted that its gastronomic offerings are considered a main attraction in the motivation of tourists to visit iconic locations in the state. Recent data have established that three out of every ten visitors or tourists to the city of Puebla choose this destination for its gastronomy (Acle-Mena *et al.*, 2020).

> There are tunnels underneath the pyramid and the food is really tasty!
>
> Very cool place to visit! Very affordable only about $5USD. Wide open and allows you to walk around and view everything at your own pace. Tons of food and beverages nearby as well. I highly recommend taking a look at the pyramid. An amazing view and beautiful church at the top of the hill. Highly recommend visiting if you are in Puebla Mexico.
>
> Truly unnecessary, not that great, not quite a pyramid. Poblanos may hate me for this, but it's not worth it. On the bright side, restaurants and most of the food around the pyramid is awesome!

9.4.3.5 Tourist experience and tourism development

The tourist experience in Cholula highlights a mix of positive and negative aspects. While many visitors appreciated the rich cultural heritage, vibrant markets, and unique archeological sites like the pyramid, tunnels, and churches, there was some disappointment due to unexpected closures of key attractions. Despite these setbacks, the overall experience often left a lasting impression. Positive experiences, such as seamless access to sites, local food, and well-maintained infrastructure, play a critical role in enhancing Cholula's tourism potential (Zhao, 2022). When visitors have fulfilling and enjoyable experiences they are more likely to recommend the destination to others, boosting tourism growth (Xiao, 2022). Developing accessible and reliable tourism infrastructure, minimizing disruptions, and catering to diverse visitor needs contribute to the region's sustainable tourism development (González-Herrera *et al.*, 2023). These views are exhibited in a few of the following reviews.

> Visited this church on a bus tour from Cholula, and even though the grounds turned out to be closed, the brief minutes we had there were incredible. Speaking only of the exterior files' façade, like many architectural sites in the area, it is phenomenal and unique.

> This is worth visiting if you go to Cholua. You see well installed the things that they dug up. There is a large shop, but it and the restaurant were closed. There was a free historic performance in the central courtyard.

> First things first – the tunnels underneath the pyramid are currently closed, so don't go there expecting to Indiana Jones your way underneath as we did. With that said, it's still an interesting experience that isn't as famous and generic as some of the other landmarks you might have on your Mexico list. We hired a guide, with limited results, as his English was spotty at best. However, what did come through was interesting, and provided context we never would have had just walking around.

9.4.3.6 Environment and tourism development

The natural beauty of Cholula, as reflected in visitor reviews, plays a significant role in shaping the overall tourist experience. The stunning views from the Great Pyramid, the scenic surroundings of the archeological sites, and the pleasant weather are frequently highlighted as key aspects that enhance the charm of the region. Visitors often comment on the beauty of the landscapes, which create a picturesque backdrop for exploring the historical and cultural landmarks. Elements like clear skies, panoramic views, and the unique blend of natural and man-made beauty make the experience more memorable (Yang *et al.*, 2022b). By promoting these natural attractions and integrating them into tourism development strategies, Cholula further boosts its appeal. Preserving the environment, improving access to scenic spots, and promoting eco-friendly tourism attract more visitors and enhance the region's tourism potential (Yang *et al.*, 2022b). This is evident from a few reviews posted on Tripadvisor.

> Going to Cholula and not visiting the Regional Museum is unforgivable. The Museum is a large enclosure excellently distributed by rooms, with modern and spacious gardens, the only bad thing is that it takes at least 3 hours to visit and appreciate its content if you go with limited time. It is also surrounded by gardens, as well as access to the pyramid, the entrance to the Sanctuary and finally the railway terminal towards Puebla. It is worth mentioning that this terminal, as well as the railway, are new, its facilities are modern and for the time being the service is free. Highly recommended.

> Although it looks like a hill with a colonial church on top, this is actually a series of pyramids built on top of each other. The tunnels make this site unique and are not for the claustrophobic. The outside sites are well labelled and quite interesting and the small museum has a wonderful display of what the pyramid actually looked like before the Spanish conquest. The church is lovely and the view fantastic. I only wish the weather would have permitted us to see the volcano.

9.5 Conclusion

The architectural heritage of Mexico, such as Great Pyramid of Cholula, the Church of Nuestra Señora de los Remedios, the Church of Santo Domingo, and the Puebla Cathedral, and tunnels, reflects a unique blend of Mexican and Spanish influences that play a vital role in enhancing tourism. These architectural marvels, along with the Regional Museum of Cholula, serve as major attractions for architectural heritage tourism. By integrating historical education into tourist experiences, Mexico has successfully positioned itself as a destination that showcases its rich historical legacy. These findings are similar to those found in the research by Tutchener (2013), Wasino *et al.* (2023), and Vegheș (2019), who all emphasize the importance of heritage tourism for preserving history while boosting the economy. However, Zhang *et al.* (2021) argued that too much commercialization of heritage sites may harm their historical value and reduce the quality of the visitor experience. The risk associated with mass tourism, especially causing damage to historic buildings, is an area explored by Pati and Husain (2023) and García-Ortiz *et al.* (2014), making it mandatory for planners to strategize offering protection to these sites while benefiting from tourism.

Puebla, as a tourist destination, stands out as a unique cultural destination due to its traditional dance, music, and its iconic cultural expressions such as the "Danza de los Quetzales" and "Son Jarocho," reflecting the deep influence of its indigenous past. Such historical footprints of its culture offer immersive experiences to tourists. Additionally, festivals contribute to the tourist's immersive experience resulting in enhanced destination satisfaction. Festivals like Cinco de Mayo and the Festival Internacional de Puebla are examples that underscore the region's cultural traditions and enhance its appeal as a key cultural destination. Such findings align with Abbasi *et al.* (2021) and Yang *et al.* (2022b), emphasizing how cultural heritage can promote tourist immersive experiences through a sustainable approach and while ensuring economic development. However, in contrast, Gallaga *et al.* (2022) argued how overemphasizing native cultural experiences at the cost of the broader cultural aspects of heritage tourism might lead to decline among diverse tourist groups.

The impact of religious tourism on the overall tourist experience is evident from the present study. The outcome is in sync with Vizcaino-Suárez (2018), emphasizing the prominent role of religion in preserving local culture and supporting economic development. However, Hung *et al.* (2016) argued that over-commercialization of pilgrimage sites may erode their spiritual authenticity.

The analysis also highlights the significant impact of Cholula's diverse culinary heritage on tourism, driven by the fusion of indigenous, Spanish, Arabian, and Asian influences. The findings align with existing literature that underscores the strong correlation between the quality and promotion of local cuisine and tourist satisfaction (Yasami *et al.*, 2020). Mexican cuisine, recognized by UNESCO as part of the "Intangible Heritage of Humanity," plays a pivotal role in attracting both local and international tourists. Puebla's status as a gastronomic capital, where culinary tradition meets innovation, further supports the region's appeal as a destination for food-driven tourism. Tripadvisor reviews reinforce that gastronomy remains a primary motivation for 30% of visitors to the city of Puebla, confirming the culinary sector's vital role in shaping tourist preferences (Acle-Mena *et al.*, 2020).

The tourist experience in Cholula reveals a blend of positive and negative elements. While visitors greatly value the region's rich cultural heritage, archeological wonders, and unique religious sites, there are some frustrations due to unexpected closures of key attractions. Nevertheless, the overall experience remains impactful for many visitors due to architecture, local cuisine, and cultural performances adding to the destination's appeal. Positive experiences lead to word of mouth recommendations, which drive tourism growth in the region. These findings are consistent with prior research by da Silva Añaña *et al.* (2018), Wang *et al.* (2020), and Yang *et al.* (2022b), which emphasize the importance of overall tourist satisfaction in fostering sustainable tourism growth. As demonstrated by Zhang *et al.* (2021), excessive focus on improving infrastructure might hinder the tourist's authentic experience. In support of this claim, Liang *et al.* (2021) also suggested how the disturbed schedule due to renovation and other activities for a long time could diminish the perceived value of heritage destinations. Cholula continues to charm tourists, and visitors are all in praise of its picturesque environment and its historical and cultural landmarks – all such attractions make the tourist experience memorable. Planners are therefore expected to integrate all such unique destination attractions before formulating their tourism development strategies (Yang *et al.*, 2022b).

9.5.1 Theoretical implications

The findings of the study offer valuable insight into and contribute to the existing literature on heritage tourism. Architecture, culture, religion, gastronomy, and natural beauty all add value in culture heritage tourism. Such a framework aligns with the heritage tourism framework proposed by Garrod and Fyall (2000) recommending that tourism preservation significantly contributes to sustainable tourism growth. Furthermore, the blend of Mexican and Spanish architectural influences reflected in the Great Pyramid of Cholula supports Urry and Larsen's (2011) concept of the tourist gaze. The framework demonstrates how the historical and cultural landmarks shape the tourist experience, something this study also hints at. The findings of this study also show how religious tourism can potentially reinforce the tourist–pilgrim continuum, where religious sites attract both spiritual seekers and cultural tourists, as highlighted by the study by Dimitrovski *et al.* (2020). At the same time, the present study offers insights into how overcommercialization can erode authentic tourist experiences (Zhang *et al.*, 2021).

9.5.2 Managerial implications

The findings of the study highlight the need for a balanced approach to tourism development by stressing the preservation of heritage while enhancing visitor experiences. Tourism managers are expected to prioritize sustainable practices, which includes considering limiting the commercialization of heritage sites to preserve their authenticity. To achieve these objectives, effective visitor management strategies, such as restricted access to fragile sites and making substantial improvements in infrastructure, might be needed to mitigate negative impacts arising from mass tourism. Educational initiatives aimed at enhancing tourist experiences and raising awareness of the region's history could be instrumental in sustainability efforts. Taking advantage of being a global leader in culinary arts due to its heritage status, the continued promotion and innovation in culinary traditions could make Cholula a global destination. The collaboration between public and private stakeholders, finally, could offer tremendous benefits to the local communities without compromising on sustainability.

9.5.3 Limitations of the research

The present study has certain limitations that future research studies could make a note of. First, the research study was geographically restricted to a single heritage site. Such a limited approach could be a challenge when applied to other heritage destinations. Different heritage sites are unique in their socio-cultural environments, so the unique characteristics of Cholula may not fully represent similar sites globally. Additionally, the study might have overlooked other critical factors influencing heritage tourism since it focused on the Cholula region only. Finally, the reliance on secondary data sources can be a limitation in summarizing findings and it is recommended that future studies try to incorporate real-time observations in visitor behavior.

9.5.4 Future scope of research

Future studies are expected to consider comparative studies across other heritage sites in Mexico and internationally to enhance the generalizability of the findings. Cross-cultural comparisons offer deeper insights into how different regions find a balance between preservation and tourism growth (Dimitrovski *et al.*, 2020). Surveys involving real-time observation could capture evolving trends in tourist behavior towards heritage tourism. The long-term impact of commercialization and infrastructure development on heritage authenticity is yet another area worth probing. Similarly, tracking the consequences of tourism management decisions over time could offer valuable insights into the sustainability of heritage tourism. As in every other field, the role of digital technology and virtual tourism could provide insights into tourist engagement strategies.

References

Abbasi, G.A., Kumaravelu, J., Goh, Y.-N. and Singh, K.S.D. (2021) Understanding the intention to revisit a destination by expanding the theory of planned behaviour (TPB). *Spanish Journal of Marketing - ESIC* 25(2), 282–311. DOI: 10.1108/sjme-12-2019-0109.

Acle-Mena, R.S., Santos-Díaz, J.Y. and Herrera-López, B. (2020) La gastronomía tradicional como atractivo turístico de la ciudad de Puebla, México. *Revista de Investigación, Desarrollo e Innovación* 10(2), 237–248. DOI: 10.19053/20278306.v10.n2.2020.

Amine, B. and Semaoune, K. (2021) An analysis of tourist's behavioural intention in the digital era: Using a modified model of the reasoned action theory. *International Journal of Marketing, Communication and New Media*. DOI: 10.54663/2182-9306.2021.v9n17.4-26.

Aziz, S. and Niazi, M.A. (2023) Understanding environmentally responsible behavior of tourists at coastal tourist destinations. *Social Responsibility Journal* 19(10), 1952–1977. DOI: 10.1108/srj-10-2022-0412.

Bagnara Milan, S. (2022) The interconnectedness of Pueblo arts and architecture and its significance in the sustainable regeneration of northern New Mexico's cultural landscape. *Advances in Science, Technology & Innovation* 217–224. DOI: 10.1007/978-3-030-99480-8_16.

Bani Omar, S.Y. (2022) The significance of cultural heritage in fostering economic viability. *Global Journal of Economics and Business* 12(3), 414–424. DOI: 10.31559/gjeb2022.12.3.9.

Batmaz, O. (2023) Examination of primary school teachers' opinions on cultural heritage education in life science lesson. *Journal of Computer and Education Research* 11(22), 429–446. DOI: 10.18009/%20jcer.1241337.

Benhamou, F. (2020) Heritage. In: *Handbook of Cultural Economics*, 3rd edn. DOI: 10.4337/9781788975803.00037.

Cenni, I. and Goethals, P. (2017) Negative hotel reviews on tripadvisor: A cross-linguistic analysis. *Discourse, Context & Media* 16, 22–30. DOI: 10.1016/j.dcm.2017.01.004.

da Silva Añaña, E., dos Anjos, F.A. and de Pereira, M. (2018) Touristic destination image in light of the service dominant logic of marketing. *Tourism & Management Studies* 14(3), 7–18. DOI: 10.18089/%20tms.2018.14301.

Diaz-Parra, I. and Jover, J. (2021) Overtourism, place alienation and the right to the city: Insights from the historic centre of Seville, Spain. *Justice and Tourism* 16–33. DOI: 10.4324/9781003143055-2.

Dimitrovski, D., Ioannides, D. and Nikolaou, K. (2020) The tourist-pilgrim continuum in consumer behaviour: The case of international visitors to Kykkos Monastery, Cyprus. *Journal of Tourism and Cultural Change* 19(5), 568–586. DOI: 10.1080/14766825.2020.1828437.

Durán-Díaz, P. (2023) Sustainable land governance for water–energy–food systems: A framework for rural and peri-urban revitalisation. *Land* 12(10), 1828. DOI: 10.3390/land12101828.

Espeso-Molinero, P., Carlisle, S. and Pastor-Alfonso, M.J. (2019) Knowledge dialogue through indigenous tourism product design: A collaborative research process with the Lacandon of Chiapas, Mexico. *Sustainable Tourism and Indigenous Peoples* 265–283. DOI: 10.4324/9781315112053-16.

Fattacciu, I. (2019) Seed of unity or seed of change? Maize, memory and identity in the making of Mexican cuisine as intangible heritage. *Quaderni Storici: Rivista Quadrimestrale* 317–352. DOI: 10.1408/96902.

Fyall, A. and Garrod, B. (1998) Heritage tourism: At what price. *Managing Leisure* 3(4), 213–228. DOI: 10.1080/136067198375996.

Gallaga, E., Trujillo, J. and Andrlić, B. (2022) Archaeological attractions marketing: Some current thoughts on heritage tourism in Mexico. *Heritage* 5(1), 567–580. DOI: 10.3390/heritage5010032.

García-Ortiz, E., Fuertes-Gutiérrez, I. and Fernández-Martínez, E. (2014) Concepts and terminology for the risk of degradation of geological heritage sites: Fragility and natural vulnerability, a case study. *Proceedings of the Geologists' Association* 125(4), 463–479. DOI: 10.1016/%20j.pgeola.2014.06.003.

Garrod, B. and Fyall, A. (2000) Managing heritage tourism. *Annals of Tourism Research* 27(3), 682–708. DOI: 10.1016/s0160-7383(99)00094-8.

González-Herrera, K.C., Moran-Bravo, L.D.C. and Ortega-Pérez, E. (2023) Sustainable tourism. Puebla and other entities. In: *Proceedings T-1. Ecorfan*. DOI: 10.35429/p.2023.3.1.105.

Gutiérrez Juárez, E., Kurjenoja, A.K., Schumacher, M., Villalvazo, M.G., Meza, E.G. *et al.* (2022) Neoliberal urban development vs. rural communities: Land management challenges in San Andrés Cholula, Mexico. *Land* 11(7), 1058. DOI: 10.3390/%20land11071058.

Harun, S.N. (2018) Promoting rural tourism in Perak Tengah District through the expanding of heritage interpretation and presentation. *Planning Malaysia Journal* 16, 7. DOI: 10.21837/pmjournal.

Hosseini, K., Stefaniec, A. and Hosseini, S.P. (2021) World heritage sites in developing countries: Assessing impacts and handling complexities toward sustainable tourism. *Journal of Destination Marketing & Management* 20, 100616. DOI: 10.1016/%20j.%20jdmm.2021.100616.

Huang, S. (2018) The moderation of gender and generation in the effects of perceived destination image on tourist attitude and visit intention: A study of potential Chinese visitors to Australia. *Journal of Vacation Marketing* 25(3), 375–389. DOI: 10.1177/135676671.

Huete-Alcocer, N. and Ruiz, V.R.L. (2019) The role of destination image in tourist satisfaction: The case of a heritage site. *Economic Research-Ekonomskalstraživanja* 33(1), 2444–2461. DOI: 10.1080/1331677%20x.2019.1654399.

Hung, K., Yang, X., Wassler, P., Wang, D., Lin, P. *et al.* (2016) Contesting the commercialization and sanctity of religious tourism in the Shaolin Monastery, China. *International Journal of Tourism Research* 19(2), 145–159. DOI: 10.1002/jtr.2093.

Iordanova, E. and Stylidis, D. (2019) The impact of visitors' experience intensity on in-situ destination image formation. *Tourism Review* 74(4), 841–860. DOI: 10.1108/tr-12-2018-0178.

Isa, W.M.W., Zin, N.A.M., Rosdi, F. and Sarim, H.M. (2018) Digital preservation of intangible cultural heritage. *Indonesian Journal of Electrical Engineering and Computer Science* 12(3), 1373–1379. DOI: 10.11591/ijeecs.%20v12.i3.

Ismael-Simental, E., Kurjenoja, A., Cuenca, A.L. and Medina, L.R. (2019) Building the city through culture: Puebla's cultural urban assemblage (1987-2017). *Social & Cultural Geography* 23(1), 101–119. DOI: 10.1080/14649365.2019.

Jurėnienė, V. and Radzevičius, M. (2022) Peculiarities of sustainable cultural development: A case of dark tourism in Lithuania. *Journal of Risk and Financial Management* 15(6), 264. DOI: 10.3390/jrfm15060264.

Koens, K. and Klijs, J. (2022) Overtourism – Identifying the underlying causes and tensions in European tourism destinations. *A Research Agenda for Urban Tourism*. DOI: 10.4337/9781789907407.00020.

Labra, O., Castro, C., Wright, R. and Chamblas, I. (2020) Thematic analysis in social work: A case study. *Global Social Work-Cutting Edge Issues and Critical Reflections* 10(6), 1–20.

Lan, T., Zheng, Z., Zhang, R., Law, R. and Zhang, M. (2021) Resident-tourist value co-creation in the intangible cultural heritage tourism context: The role of residents' perception of tourism development and emotional solidarity. *Sustainability* 13(3), 1369. DOI: 10.3390/su13031369.

Lee, K.-C. and Chang, M.-H. (2023) Discussion on accessible tourism through online reviews of accessible tourism destinations: Focused on TripAdvisor reviews. *Journal of Tourism Studies* 35(3), 25–50. DOI: 10.21581/%20jts.2023.8.35.3.25.

Liang, S., Li, C., Li, H. and Cheng, H. (2021) How do you feel about crowding at destinations? An exploration based on user-generated content. *Journal of Destination Marketing & Management* 20, 100606. DOI: 10.1016/j.jdmm.

Lussetyowati, T. (2015) Preservation and conservation through cultural heritage tourism: Case study of Musi Riverside, Palembang. In: *Procedia - Social and Behavioral Sciences*, Vol. 184. pp. 401–406. DOI: 10.1016/%20j.%20sbspro.2015.05.109.

Mason, A.R. and Ying, M. (2020) Evaluating standards for private-sector financial institutions and the management of cultural heritage. *Advances in Archaeological Practice* 8(1), 1–14. DOI: 10.1017/aap.2019.44.

Michaca-Rodríguez, J., Edgar, H.-C. and de dios-Feria, Z.E. (2023) Gastronomy as a factor influencing tourism in the magical towns: The case of Las Cholulas, Puebla. *Sustainable Tourism. Puebla and Other Entities* 53–64. DOI: 10.35429/p.2023.3.53.54.

Muangasame, K. and Tan, E. (2022) Phygital rural cultural heritage: A digitalisation approach for destination recovery and resilience. *Worldwide Hospitality and Tourism Themes* 15(1), 8–17. DOI: 10.1108/whatt-08-2022-0096.

Nguyen, Q.V. (2023) Impact of self – congruity and destination image on tourist loyalty: Evidence from recreation farm tourism. *International Journal of Professional Business Review* 8(4). DOI: 10.26668/businessreview/%202023.v8i4.156.

Nicoli, N. and Papadopoulou, E. (2017) TripAdvisor and reputation: A case study of the hotel industry in Cyprus. *EuroMed Journal of Business* 12(3), 316–334.

Nowell, L.S., Norris, J.M., White, D.E. and Moules, N.J. (2017) *Thematic Analysis*. SAGE Publications Inc.

Pati, A. and Husain, M. (2023) People's perspectives on heritage conservation and tourism development: A case study of Varanasi. *Built Heritage* 7, 1. DOI: 10.1186/s43238-023-00098-w.

Petelca, O., Ciprian-Constantin, P. and Iurie, B. (2021) The multifunctionality and sustainability of small and medium enterprises in tourism-based rural communities. *The Annals of the University of Oradea: Economic Sciences* 30(2), 123–131. DOI: 10.47535/1991auoes30(2)013.

Poulios, I. (2014) *Past in the Present: A Living Heritage Approach - Meteora, Greece*. Available at: https://doi.org/10.5334/bak

Qiu, H., Wang, X., Wu, M.-Y., Wei, W., Morrison, A.M. *et al.* (2022) The effect of destination source credibility on tourist environmentally responsible behavior: An application of stimulus-organism-response theory. *Journal of Sustainable Tourism* 31(8), 1797–1817. DOI: 10.1080/%2009669582.2022.2067167.

Ramón-Cardona, J. and Sánchez-Fernández, M.D. (2022) Tangible and intangible heritage of Ibiza (Spain) and its potential to be valued as a tourist offer. *Heritage* 5(3), 1350–1369. DOI: 10.3390/heritage5030070.

Rojas-Rivas, E., Rendón-Domínguez, A., Felipe-Salinas, J.A. and Cuffia, F. (2020) What is gastronomy? An exploratory study of social representation of gastronomy and Mexican cuisine among experts and consumers using a qualitative approach. *Food Quality and Preference* 83, 103930. DOI: 10.1016/j.foodqual.2020.103930.

Safaaa, L., Housni, K.E. and Bédard, F. (2017) Authenticity and tourism: What TripAdvisor reviews reveal about authentic travel to Marrakech. In: *Information and Communication Technologies in Tourism 2017*. pp. 595–606. DOI: 10.1007/%20978-3-319-51168-9_43.

Schumacher, M., Villalvazo, M.G., Kurjenoja, A.K. and Durán-Díaz, P. (2022) The writ of amparo and indigenous consultation as instruments to enforce inclusive land management in San Andrés Cholula, Mexico. *Land* 12(1), 9. DOI: 10.3390/land12010009.

Serrano, J. (2024) El futuro en común: Las comunidades indígenas en las ciudades del bajo Río Negro, Norpatagonia, Argentina. *Encartes* 7(14), 75–101. DOI: 10.29340/en.v7n14.369.

Son, C.M. (2023) Exploring the impact of cultural heritage on destination branding and tourist experiences: Perspective from South Korea. *Journal of Hospitality & Tourism Management* 6(1), 1–10. DOI: 10.53819/81018102%20t4138.

Sosa, M., Aulet, S. and Mundet, L. (2021) Community-based tourism through food: A proposal of sustainable tourism indicators for isolated and rural destinations in Mexico. *Sustainability* 13(12), 6693. DOI: 10.3390/%20su13126693.

Starčević, K., Vojinović, Z. and Glavaš, J. (2022) Intangible cultural heritage and their influence on financial results in rural tourism. *Ekonomikapoljoprivrede* 69(2), 483–496. DOI: 10.5937/ekopolj2202483s.

Stylidis, D. (2020) Exploring resident–tourist interaction and its impact on tourists' destination image. *Journal of Travel Research* 61(1), 186–201. DOI: 10.1177/0047287520969861.

Symoneaux, R., Galmarini, M.V. and Mehinagic, E. (2012) Comment analysis of consumer's likes and dislikes as an alternative tool to preference mapping. A case study on apples. *Food Quality and Preference* 24(1), 59–66. DOI: 10.1016/j.foodqual.2011.08.013.

Thekkum Kara, G.K. (2021) Developing a sustainable cultural heritage information system. *Library Hi Tech News* 38(6), 17–20. DOI: 10.1108/lhtn-08-2021-0053.

Tutchener, David. (2013) Collective identity and heritage: Ebb and flow at national and global levels. *Journal of Globalization Studies* 4(2). Available at: https://cyberleninka.ru/article/n/collective-identity-and-heritage-ebb-and-flow-at-national-and-global-levels.pdf

Urry, J. and Larsen, J. (2011) *The Tourist Gaze 3.0*. SAGE.

Vegheș, C. (2019) Cultural heritage: The forgotten resource for marketing and sustainable development of the local communities. *European Journal of Sustainable Development* 8(3), 41. DOI: 10.14207/%20ejsd.2019.%20v8n3p41.

Vizcaino-Suárez, L.P. (2018) Tourism and indigenous heritage in Latin America: As observed through Mexico's magical village Cuetzalan. *Journal of Heritage Tourism* 14(5–6), 579–580. DOI: 10.1080/1743873%20x.2018.1556834.

Wang, J., Li, Y., Wu, B. and Wang, Y. (2020) Tourism destination image based on tourism user generated content on internet. *Tourism Review* 76(1), 125–137. DOI: 10.1108/tr-04-2019-0132.

Wasino, W., Lim, C. and Dewayani, E. (2023) User interface design of West Java's intangible cultural heritage website using user-centered design. *International Journal of Application on Sciences, Technology and Engineering* 1(2), 421–432. DOI: 10.24912/ijaste.v1.i2.421-432.

Winarni, F. (2018) Aspek Hukum Peran Serta Masyarakat Dalam Pelestarian Cagar Budaya. *Mimbar Hukum - Fakultas Hukum Universitas Gadjah Mada* 30(1), 94. DOI: 10.22146/jmh.29160.

Xu, F., Niu, W., Li, S. and Bai, Y. (2020) The mechanism of word-of-mouth for tourist destinations in crisis. *SAGE Open* 10(2), 215824402091949. DOI: 10.1177/2158244020919491.

Xiao, L. (2022) Intangible cultural heritage reproduction and revitalization: Value feedback, practice, and exploration based on the IPA model. *Computational Intelligence and Neuroscience* 1–13. DOI: 10.1155/%202022/8411999.

Xue, L. and Kerstetter, D. (2017) Discourse and power relations in community tourism. *Journal of Travel Research* 57(6), 757–768. DOI: 10.1177/%200047287517714908.

Yang, S., Isa, S.M., Yao, Y., Xia, J. and Liu, D. (2022a) Cognitive image, affective image, cultural dimensions, and conative image: A new conceptual framework. *Frontiers in Psychology* 13. DOI: 10.3389/fpsyg.2022.935814.

Yang, W., Chen, Q., Huang, X., Xie, M. and Guo, Q. (2022b) How do aesthetics and tourist involvement influence cultural identity in heritage tourism? The mediating role of mental experience. *Frontiers in Psychology* 13. DOI: 10.3389/fpsyg.2022.990030.

Yasami, M., Promsivapallop, P. and Kannaovakun, P. (2020) Food image and loyalty intentions: Chinese tourists' destination food satisfaction. *Journal of China Tourism Research* 17(4), 592–612. DOI: 10.1080/%2019388160.2020.1784814.

Zamora, C.R. (2023) The importance of preserving and protecting natural resources for the future of tourist destinations in Mexico. *Journal of Management and Sustainability* 13(2), 39. DOI: 10.5539/jms.%20v13n2p39.

Zhang, T., Yin, P. and Peng, Y. (2021) Effect of commercialization on tourists' perceived authenticity and satisfaction in the cultural heritage tourism context: Case study of Langzhong Ancient City. *Sustainability* 13(12), 6847. DOI: 10.3390/su13126847.

Zhao, K. (2022) Industrial heritage tourism sustainable development. *Advances in Economics, Business and Management Research*. DOI: 10.2991%20/aebmr.k.220307.518.

Zhu, Y. and Liu, J. (2021) Cultural reproduction: Ethnic festivals as intangible cultural heritage in China. *Loisir et Société / Society and Leisure* 44(3), 415–428. DOI: 10.1080/07053436.2021.1999090.

Đukanović, Z., Živković, J., Radosavljević, U., Lalović, K. and Jovanović, P. (2021) Participatory urban design for touristic presentation of cultural heritage sites: The case of Negotinske Pivnice (wine cellars) in Serbia. *Sustainability* 13(18), 10039. DOI: 10.3390/su131810039.

10 Unearthing Treasures: Leveraging Cultural Heritage for Sustainable Rural Tourism – A Comparative Study

Komal Sandhu[1], Ishan Atrey[2*] and Navajyoti Samanta[3]
[1]Department of Law, Gurugram University, Gurugram, India; [2]Integrated Programme in Law, Indian Institute of Management Rohtak, Rohtak, India; [3]Leicester Law School, University of Leicester, Leicester, UK

10.1 Introduction

Rural tourism encapsulates travel experiences that immerse visitors in the authentic essence of rural areas worldwide. It provides interactions with regional customs, traditions, and lifestyles that go beyond simple tourism. The rich tapestry of cultural heritage, which consists of both tangible and intangible elements passed down through the years, is its fundamental component. Everywhere in the world, rural communities have distinct cultural identities that are intricately entwined with their environment (UNWTO, 2017). The soul of rural areas can be glimpsed through many legacy assets, such as colorful folklore, local crafts, historic landmarks, and ancient archeological sites. This rich cultural legacy is celebrated and preserved by rural tourism, which welcomes visitors to discover centuries-old customs, sample local cuisine, take part in local celebrations, and interact with craftspeople practising age-old trades. Rural areas open up new business options, generate employment, and boost local economies by embracing and exhibiting their cultural history. Besides, rural tourism protects cultural traditions from the globalization wave by fostering a sense of pride and community among locals (Sudheer, 2021). Because of its complementarity with other economic activities, gross domestic product (GDP) contribution, job creation, and ability to encourage the spreading of demand across time (overcoming the effects of seasonality) and throughout a greater territory, rural tourism has a significant potential to stimulate social transformation and local economic progress (Ministry of Tourism, Government of India, 2021). Rural tourism acts as a link between individuals from all origins, encouraging tolerance for cultural diversity and aiding in the preservation of our common human history. Any form of tourism aimed at promoting rural life, art, culture, and heritage while also enhancing the local economy and social cohesion is referred to as rural tourism, as discussed by Kapur (2016). It improves the quality of the entire tourism experience by facilitating meaningful contact between visitors and residents. With an emphasis on active engagement in rural lifestyle, this type of tourism allows tourists to fully immerse themselves in the customs and culture of the

*Corresponding author: ishan.atrey@iimrohtak.ac.in

DOI: 10.1079/9781836992646.0010

area, participate in day-to-day village activities, and even choose to stay overnight to get a more in-depth understanding of village life. Agricultural tourism, cultural tourism, natural tourism, adventure tourism, and eco-tourism are only a few of the diverse aspects that make up rural tourism, and they are all interconnected.

Globally, rural areas are experiencing depopulation, economic loss, and cultural destruction. Jobs are lost during economic downturns, and populations decline as a result of depopulation brought on by young people moving into cities (analyzed by Gocer *et al.*, 2024). Globalization simultaneously threatens cultural heritage, posing a challenge to the survival of customs and identity. Using cultural heritage for sustainable rural tourism is crucial in addressing these issues. This strategy seeks to maintain population growth, revitalize economies, and safeguard cultural identity. Communities can promote economic growth, resilience, and cultural preservation by examining the potential of cultural heritage in rural tourism (UNESCO, 2008). This provides a means of achieving inclusive and sustainable rural development.

Rural tourism is a significant aspect in the propulsion of rural revitalization. Moreover it exerts a profound impact on the development of the country as it has a strong influence on the industrial growth, ecological balance, cultural upliftment, and effective leadership and management as well as the well-being of people. Rosalina *et al.* (2021) argued that there have been notable changes in the psychology and behavior of tourists over recent years. Tourism demands have changed from searching for "beautiful scenery" to striving for a "good life." The prevailing tourist choice today is a rather low-density travel experience, which is slowly moving towards short-stay trips. Tourists also look for safety and comfort during their travels. Rural tourism, characterized by local culture and environmentally friendly practices, is therefore gaining more popularity among the public compared with traveling to cities, well-known scenic areas, or other crowded places (Bhuiyan and Darda, 2021). The development of rural tourism offers two benefits: protection of rural ecology and local culture and the promotion of rural economic growth alongside rural revitalization.

The promotion of ethical, sustainable, and broadly accessible tourism with an eye on achieving the 2030 Agenda for Sustainable Development and the UN's sustainable development goals (SDGs) falls under the purview of the UN World Tourism Organization (UNWTO). The UNWTO advocates for responsible tourism and promotes tourism as a catalyst for equitable development, economic growth, and environmental sustainability. It also provides leadership and support to the tourism industry in improving knowledge and tourism policies globally. Singh *et al.* (2024) highlighted the UNWTO, which has 159 member nations as of 2025, and promotes the application of the Global Code of Ethics in Tourism to optimize the socio-economic benefits of tourism while reducing any potential drawbacks. Guidelines for the development and management of sustainable tourism are applicable to all types of tourism in all locations, including mass tourism and the many niche tourism markets. In order to ensure the long-term sustainability of tourism growth, a sufficient balance between these dimensions must be maintained. Sustainability principles pertain to the environmental, economic, and socio-cultural components of tourism development. Thus, sustainable tourism should utilize the available resources optimally, respect the host communities' socio-cultural authenticity, and guarantee sustainable economic activities that yield equitable socio-economic gains for all parties involved. In order to promote widespread engagement and consensus building, sustainable tourism development necessitates the informed participation of all pertinent stakeholders in addition to strong political leadership. Achieving sustainable tourism is an ongoing process that calls for ongoing impact monitoring along with the occasional introduction of preventive or corrective actions (Sudheer, 2021). In addition to providing visitors with a meaningful experience and maintaining a high level of satisfaction, sustainable tourism should also increase visitors' understanding of sustainability issues and encourage the adoption of sustainable travel habits.

10.2 Transboundary Perspectives

The term "world heritage" refers to the shared, irreplaceable wealth of humanity as determined by the World Heritage Committee of the UN Educational, Scientific, and Cultural Organization (UNESCO), and it is acknowledged by all people as a historic site, cultural artefact, or natural landscape of exceptional significance and universal value. Three components comprise world heritage: natural heritage, mixed cultural and natural heritage properties, and cultural heritage including cultural landscapes (as stated by the UNESCO World Heritage Convention). Travelers can experience the genuine attractiveness of rural landscapes, customs, and civilizations outside of their own boundaries through the unique chances provided by rural tourism in other countries. For adventurous visitors looking to get in touch with the natural world and explore local culture, rural regions abroad offer a diverse range of activities, from gorgeous towns and rolling hills to indigenous populations and historic landmarks. Rural tourism creates economic opportunities for local populations and protects natural and cultural assets in many foreign nations, acting as a catalyst for sustainable development. Hiking, animal viewing, cultural exchanges, and agricultural experiences are just a few of the activities available to visitors. Lane and Kastenholz (2015) discussed these activities as not only making unforgettable experiences but also helping sustain rural communities' livelihoods and conservation efforts. New Zealand, for instance, provides experiences for rural tourists that revolve around its breath-taking scenery, the Maori way of life, and agritourism offerings like farm stays and wine tours (Mackay *et al.*, 2019). Or, in Japan, travelers can discover ancient villages, take part in workshops for traditional crafts, and enjoy the peace of rural areas, such as the Japanese Alps or the island of Shikoku (Mackay *et al.*, 2019). In nations like India, which have a rich cultural past, rural tourism provides a window into the many traditions and customs of the nation by allowing visitors to explore historic temples, outlying marketplaces, and artisanal craft communities (Vikramaditya, 2024). Similar experiences may be found in European nations like France and Italy, which provide agritourism stays in quaint rural houses and farms, as well as vineyard tours and culinary customs.

Pristine rainforests, indigenous villages, and sustainable agriculture initiatives can be discovered through ecotourism and community-based tourist experiences for example in rural regions of Latin American nations like Ecuador and Costa Rica. It can be said that rural tourism abroad gives visitors an opportunity to get away from the bustle of the city and engage with the natural beauty, cultural diversity, and history of rural landscapes everywhere (Gunter and Cedia, 2021). Rural tourism offers fascinating experiences that promote a deeper appreciation for the landscapes and cultures of foreign nations, whether one is looking for adventure, relaxation, or cultural immersion.

10.2.1 Economic benefits: job creation and income generation for local communities

The conservation of cultural beauty is highly economically rewarding, triggering positive impacts such as economic growth, social development, and improving standards of living for the local populace. This is because involving local people in the process of preserving certain structures also has several positive effects; the foremost of these is that it helps in the creation of both economic benefits as well as better management of cultural conservation. The engagement of locals in the conservation of the facilities creates awareness and makes people to take on responsibilities, hence enhancing the overall management of heritage facilities. This involvement is essential in the expansion of economic returns following the implementation of conservation measures since it creates jobs and generates income (Sudheer, 2021). The redevelopment of historical sites and consideration of agricultural landscapes can increase tourist traffic, which can affect the development of local economies. Heritage conservation programmes thus bring about higher tourism levels, which offers employment opportunities to people and yields income to businesses. These initiatives also help to affirm and enhance social and cultural life in communities by contributing to the public sense of the self and the ongoing process of culture in general. People's engagement in the management of various heritage programs is therefore crucial to the success of these programs and so that benefits are evenly shared among all stakeholders, as well as making sure that the heritage conservation programs are economically sustainable in the long run. The involvement of communities creates an opportunity for local economic benefits while at the same time conserving the history of the nation for the benefits of the generations to come (Kumar and Asthana, 2020). Furthermore, the preservation of cultural assets has a positive economic impact on communities, demonstrating that it fosters the development of economic stimuli that contribute positively to society and enhance people's quality of life.

This is one way in which heritage conservation is beneficial as it provides chances for employment and income generation disregarding the constraints of specific sites or cultures. Measures to preserve cultural and historical landscapes and agricultural areas may positively influence income generation and tourism and, therefore, influence life in a community (Madandola and Boussaa, 2023). In the broader social context, the active participation of communities in the management of cultural heritage contributes to local revenues and assists in the protection of such sites. Gould and Burtenshaw (2014) considered that economic incentives, such as job creation and income generation, are pivotal in upholding the sustainable preservation of cultural heritage, ensuring its enduring safeguarding for the welfare of current and future generations.

10.2.2 Social benefits: preservation of cultural identity and community pride

Sustainable development can be effectively correlated with social aspects of saving and preserving cultural heritage as reference to the identity and cultural values of communities is culturally significant. Endere *et al.* (2024) stated that the maintenance of culture and traditions plays an essential role in supporting and celebrating the individuality of people's histories. Communities are able to preserve their history, traditions, and rituals when these are protected, which promotes unity and a sense of belongingness for people in that community (Nilson and Thorell, 2018). Thus, the protection of cultural heritage as an element of cultural identity is an important factor in enhancing the quality of social relations within a community. In as much as people appreciate a cause, idea, or policy championing cultural harmony, those who are aware of each other's culture and embrace it have better rapport, leading to better social relations. The unity of such a population is important when responding to external tensions and threats, and provides a means by which such populations are sustainable. The main function of the protection of cultural property is as a measure against the future loss of the values and structures of the culture, thus promoting the sustainable development and permanency of the community (Kandavel, 2024).

In addition, this preservation of cultural heritage inspires people, especially those of the cultural group, to have pride and be proud of their cultural assets. Anyone who sees that their ethnicity or assets are being respected, preserved, and valued develops dignity as a result. Such appreciation leads to the cultivation of a spirit of stewardship for these assets, hence encouraging the community's active participation in the preservation of their cultural endowment. Hoang (2021) highlighted that effective participation in these practices supports not only the protection of cultural heritage but also indicates the community's desire to support its language, beliefs, and traditions.

Pride in cultural heritage can go a long way in empowering communities and even help in regeneration. When citizens are proud of their customs and traditions, they will willingly introduce them to the other people and guests (Kapur, 2016). Sharing contributes to the exchange of cultures and, consequently, increases the desirability of the community as a cultural attraction. The emotional attachment to the values of heritage conservation can catalyze tourism and related economic growth within the community, thus leading to economic upliftment and stability in the area.

In other words, it is impossible to address the question of sustainable development without understanding the importance of preserving the cultural background as the major source of people's pride (Wu *et al.*, 2016). These social impact issues are basic to the enhancement of societal capital, social capability, and personal social adjustment within various societies. Through the enhancement and implementation of focal heritage areas, key players can be assured of the continued sustenance of culture and communal practices.

Other relevant parties, such as government agencies, non-governmental organizations (NGOs), and companies, have to realize that such values exist and are significant. In this way, it is possible to direct the funding and backing for projects that contribute to the protection of cultural assets. This support can be in the form of financial assistance in projects that are aimed at the preservation of the cultural property, education programs that foster the recognition of the importance of cultural resources, and political measures aimed at preventing the adverse impacts of such factors as, for example, urbanization and climate change on the cultural property (Jangra *et al.*, 2023). Stakeholder involvement and activities done in cooperation with the locals are essential for sustainability of the heritage preservation tasks. Involving the community in the planning and implementation of such programs guarantees that the initiatives address the needs and desires of the community, hence boosting the success rate of the preservation process. This also means that communities get to participate in the process and own some form of protection of their cultural assets (Oloidi and Okonkwo, 2020). It is also crucial to consider cultural tourism as a part of sustainable development because the impact of cultural resources on social values is very high. In other words, heritage preservation sustaining ethnic culture and reinforcing the sense of belonging contributes to positive community development and reduces instances of illness in society (Mayaram, 2024). The managers and people involved in heritage management need to enhance their competence and focus on financial commitment towards heritage management to ensure the continuity of social and cultural endowment. In this sense, cultural cooperation can result in the preservation of cultural heritage which, in return, can lead to the development of various communities and to making ways of living more viable, for people of all cultures (Bansal and Choudhary, 2024).

10.2.3 Environmental benefits: conservation of natural and built heritage

Non-renewable resources and cultural identity and values for future generations are some of the benefits of heritage conservation. Environmental protection supports the existing animal and plant species by increasing their numbers and improving the ecological systems. A preservation of cultural heritage can lower the societal implications of climate change to a society. Admitting that the future is a progressive flow that adds new features and changes the existing ones is essential in engaging in the processes of conserving and sponsoring heritage for people today and tomorrow (Kandavel, 2024). Heritage conservation services give service providers competitive advantages in marketing and management pertinent to certain unique features, high standards of service, and appropriate prices. In circumstances where tourism is one of the economic priorities, the sustainable development of tourism together with the integration of cultural values becomes a concern. Almeida and Machado (2020) discussed how preservation of natural and cultural heritage, received from the ancestors, and their rational, efficient, and sparing utilization will become essential for the implementation of the national paradigm of tourism development. This approach ensures that these resources are conserved and used in a manner so that the current population as well as future generations can benefit from them.

In this context it is crucial to define success not only in terms of financial profits but also as sustainable profits for future generations. The key to accumulating economic benefits in the present and in the future lies in preserving heritage along with consistency in the achievement of holistic objectives such as environmental conservation, cultural sustainability, and coping mechanisms against the ravages of climate change. Heritage protection is beneficial in multiple ways. It retains the community's character, tangible history, and non-consumable assets for generations to come (Hariram *et al.*, 2023). Conservation improves biological and thus ecological diversity and can also reduce the effects of climate change on society. Since heritage conservation programs have to be planned considering possible future benefits, such projects should enable competitive advantages in tourism through their unique features, standards of service, and appropriate prices. The theory of sustainable tourism development compatible with preserving cultural heritage is critical for economic settings focusing on tourism (Kandavel, 2024). The national tourism development paradigm is that perceived success in terms of apositive outcome within the present generation has to enable positive outcomes in generations to come.

Therefore, for the purpose of sustainability of limited resources, commercial plans should be in harmony with cultural goals. For investment operations and economic development programs to be implemented with success, relevant regulations need to strike a balance between short-term economic requirements and long-term cultural values (Rosalina *et al.*, 2021). The importance of culture and the conserving natural as well as built heritage greatly impacts the environment. Cultural sites and the environment are two fundamental features that need to be preserved to sustain the future of later generations. By elaborating the regulations that consider both economic and cultural factors simultaneously in the decision-making process, balanced results could be achieved that will serve both today's economy and the resource-limited culture and environment of the future (Semwal *et al.*, 2024). This combined strategy can secure the longevity of our natural and cultural asset base, and build more sustainable and resilient communities.

10.3 Cornerstones of Progress: How Preserving Cultural Heritage Paves the Way for Sustainable Development

Across the globe, the narrative for rural communities often paints a bleak picture – depopulation, economic stagnation, and a slow erosion of cultural identity (Hamid, 2024). But within the heart of these communities lies a potent force for positive change: their cultural heritage. Preserving this heritage is not just about dusty artefacts and forgotten narratives. It is about igniting a spark of pride and belonging within the community. This sense of connection to their roots fosters a spirit of unity and purpose, combating depopulation and isolation. The benefits extend far beyond the village. Cultural heritage also acts as a magnet for tourists seeking authentic experiences (Hariram *et al.*, 2023). Vibrant marketplaces showcase the work of local artisans, captivating festivals draw visitors from afar, and historic sites connect with the past. This influx of tourism injects lifeblood into local economies, creating jobs and revitalizing businesses. Restaurants can offer local delicacies, homestays showcase traditional architecture, and workshops teach age-old crafts – all become pillars of a thriving rural economy (Childs, n.d.).

The wisdom embedded within cultural heritage extends to the very environment itself. Traditional practices often hold the key to sustainable resource management. From time-tested methods of farming that nurture the land to a deep respect for nature ingrained in cultural beliefs, these practices offer a path towards environmental sustainability (Bansal and Choudhary, 2024). Protecting heritage sites often means protecting ecosystems like sacred groves or traditional water-harvesting systems, ensuring a healthy future for generations to come. Furthermore, cultural heritage preservation fosters a sense of responsibility towards the environment. Traditional knowledge systems emphasize living in harmony with nature, promoting practices like organic farming, biodiversity conservation, and water management (Kandari *et al.*, 2014). This creates a crucial link between cultural identity and environmental stewardship, encouraging communities to become custodians of their land.

10.4 Role of Technology in Rural Tourism: Relevant Experiences in India

The incorporation of technology into rural tourism has enormous potential to protect India's cultural legacy and to advance sustainable development. Innovative ways to improve visitor experiences and increase the visibility of rural places can be found in digital technologies including internet platforms, mobile apps, and virtual reality. In India, the government initially sets up the infrastructure and later charges fees for the visitors. Apps specifically designed for rural tourism can function as all-inclusive guides, giving users details about nearby landmarks, historical places, cultural events, and customs. To enhance the visiting experience and make exploring rural areas easier, these apps can have interactive maps, audio instructions, and augmented reality capabilities (Mayaram, 2024). Mobile apps can also include features for making reservations for lodgings, transportation, and guided tours, which will help local businesses and guests by streamlining travel logistics. Virtual reality technology enables the whole world to explore rural locations and cultural heritage places. Virtual reality experience can take users to distant villages, historic sites, and scenic fields, allowing them to experience the rich cultural diversity of rural India from the comfort of their own homes. Shrivastava *et al.* (2023) stated that it will stimulate fascination and curiosity about rural areas, resulting in a visitation boost from individuals who want to learn more about the cultural assets by providing immersive interactive storytelling and virtual tours. In this case, online platforms are vital for the promotion of rural tourism and the conservation of cultural treasures since they help to draw attention and spread information about obscure localities. Websites, social media, and travel blogs contain curated information about rural experiences, which may include unique traditions, handcrafted souvenirs, traditional dances and songs, and local cuisines. Online platforms will enable the education of potential tourists on the real India while drawing their attention to the country's rural areas and conveying the spirit of Indian villages (D'Souza and Nirupama, 2021).

Technology-enabled programs can strengthen communities by offering opportunities for capacity building and economic engagement. For example, remote workshops on skill development, entrepreneurial training, and digital literacy may give rural residents the knowledge and skills needed to use technology to promote tourism. If rural tourism projects can be technology-driven, they can improve the socio-economic environments of rural communities and create sustainable livelihoods through encouraging innovation and entrepreneurship (Hariram *et al.*, 2023). The potential use of digital technologies such as virtual reality, online platforms, and mobile apps in rural India has the unique potential to contribute to sustainable development and cultural heritage. Technology-enabled tourism projects can significantly expand the socio-economic and cultural potential of rural places to empower local populations, expand market opportunities, and increase guest satisfaction.

Influencer marketing compounds this by a thousand. Their massive followings put eyes on under-represented locations in rural areas that only an influencer could bring to the forefront. By helping local infrastructure and improving essential services, rural tourism sites can be marketed to a wider public, increasing awareness of unique characteristics and cultural events, which would otherwise go unnoticed. Influencers can help rural places find new audiences and cultures to engage, increasing tourism interest in those rural locations (Serra and Gretzel, 2020). As a whole, the integration of digital marketing and social media strategy can significantly promote rural tourism development, and thus promote economic and cultural revitalization. The use of digital marketing and social media is essential for augmenting the prominence and allure of rural tourism places. Digital marketing strategies like pay-per-click (PPC) advertising and search engine optimization (SEO) work together to efficiently promote rural areas to a worldwide audience through platforms like Facebook, Instagram, and X (originally Twitter). These tactics raise interest in and awareness of rural locations, improving their accessibility and allure to potential tourists. Content created by users greatly enhances marketing campaigns. Travelers who post about their experiences on social media offer genuine recommendations that appeal to their followers and frequently inspire others to visit these remote areas (Bhattacharya *et al.*, 2023). Because of the visually engaging content and the personal recommendations made by the promoter, this organic promotion may result in more interest and visits from tourists. Special mention can be made of the various village cooking channels from South Asia that enhance the creative economy as well as providing a steady source of advertising revenue.

Eco-friendly technologies can be used to foster sustainable tourism practices, particularly in the environment. Such technologies include those oriented towards renewable energy, waste reduction, and water conservation, all of which are essential in the execution of tourism activities in a sustainable manner (Gunter and Cedia, 2021). The development of these technologies helps the responsible management of natural resources and reduces the impact of tourism in rural areas. Sustainable transport options can be promoted, such as establishing electric vehicle (EV) charging stations and providing e-bike rentals which will encourage sustainable travel throughout rural areas. Pardo-Bosch *et al.* (2021) discussed how these programs help to lessen the carbon footprint connected to tourism by encouraging the use of low emission forms of transportation. The long-term sustainability of rural tourism depends on the incorporation of these environmentally friendly technologies and transportation options. They are in line with larger initiatives to encourage ethical tourist practices in addition to improving the environmental sustainability of rural travel destinations. In order to ensure that rural areas can continue to draw in and welcome tourists, while maintaining their natural and cultural heritage, these initiatives encourage the creation of a more resilient and sustainable rural tourism infrastructure (Sudheer, 2021).

E-commerce websites create demand for regional products to be purchased online by consumers shopping for authentic, local products. These are essentially online marketplaces, providing the opportunity for local businesses to extend their reach and tap into the rising desire for unique, locally produced goods. They also allow travelers to buy local items before they even get to where they are going. This is designed to assist tourists in general for their journey planning and to bring in more excitement. Familiarity of local offerings can help travelers appreciate and plan their travels and make them easier to access, which ultimately can lead to enhancing the overall travel experience (Serra and Gretzel, 2020).

10.5 Case Study of the Homestay Sector in Darjeeling

The homestay sector in the Darjeeling district of West Bengal provides a good case study of how rural tourism or micro tourism may boost the local economy and provide a wholesome experience for the tourists. The Darjeeling district is situated in the Himalayan region of the state of West Bengal, India and is bordered by Nepal and Bhutan. Darjeeling town is its main urban center and was popularized as a hill station during the British era by its officers who wanted to periodically escape the humidity of Kolkata, which is located about 650 km to the south. Darjeeling district is also famous for its tea, which was recognized as the first "geographical indication" (GI) in India. Darjeeling draws in a large number of tourists each year, although until the 2000s this was concentrated around the urban center of Darjeeling town. Rural homestays were used by trekking and mountaineering enthusiasts who would venture out of the normal tourist circuits. With the percolation of internet connectivity, many of the rest stops for trekkers transformed to the first homestays. The benefits are obvious: in an area where there are low employment opportunities a single homestay provides employment opportunities for about three or four people (Sharma and Chettri, 2023). The government of West Bengal stated that there were 248 registered licensed homestays in 2024 (Department of Tourism, Government of West Bengal, n.d); the Himalayan Hotel Owners Association (HHOA) gave a number of around 450 for the same year; however, unofficially, the actual number is estimated to be more than double this. Empirical studies have shown that homestays boost the income of the owner of the residence as well as those associated with it (Bhattacharya *et al.*, 2023). Tourists are also more likely to return to the same or similar host.

Tourists are becoming more concerned about the ethical impact of tourism and are more likely to choose sustainable homestay tourism over chain hotels in urban areas. Research also shows that tourists gain a deeper understanding of the cultural and natural heritage of the area and that the tourism is more ecologically sustainable when homestays are used.

Table 10.1. Number of registered homestays in each state in India in March 2023. From Government of India, Ministry of Tourism (2023).

State/union territories	Number
Andaman and Nicobar Islands	20
Andhra Pradesh	2
Arunachal Pradesh	107
Assam	4
Bihar	1
Chandigarh	3
Chhattisgarh	4
Dadra and Nagar Haveli and Daman and Diu	1
Delhi	33
Goa	46
Gujarat	7
Haryana	33
Himachal Pradesh	307
Jammu and Kashmir	12
Jharkhand	2
Karnataka	22
Kerala	35
Ladakh	14
Madhya Pradesh	75
Maharashtra	35
Manipur	2
Meghalaya	8
Nagaland	4
Odisha	15
Puducherry	2
Punjab	3
Rajasthan	39
Tamil Nadu	245
Telangana	3
Uttarakhand	57
Uttar Pradesh	102
West Bengal	607
Total	1850

Table 10.1 shows a clear divide in the promotion of homestays in each state: while West Bengal, Himachal Pradesh and Tamil Nadu lead with the number of homestays, only two other states in India have more than 100 homestays. This is an area of concern as homestays form the foundation of rural tourism. As has been discussed elsewhere in this chapter, the main focus of rural tourism is to impart a sustainable, wholesome, and memorable experience for the guests, and that homestays provide an immersive experience which is unique and as close to experiencing the "real lives" of hosts as possible. The National Strategy for Promotion of Rural Homestays – An Initiative towards Atmanirbhar Bharat (Ministry of Tourism, Government of India, 2022) focuses on how homestays can improve the rural economy and enrich travelers, however the low number of homestays especially in tourist hotspots and emerging areas like in the north-eastern states is worrying. Around the world sustainable rural tourism is based on homestays and host–guest interactions; this is the missing link in Indian rural tourism which can only addressed by concerted efforts of the Indian states. More states should be encouraged to emulate the relative success of homestays in West Bengal, which cluster around Darjeeling and Sunderbans, both having multiple UNESCO tags (UNESCO, 2024a). Homestays around UNESCO heritage sites, which may provide sufficient activities to have an overnight stay, should be the main goal of states looking to strategically increase the number of homestays.

However, preserving cultural heritage is not just about safeguarding the past; it is about embracing its potential for the future. Technology can be a powerful tool in this process. Online platforms can showcase traditional crafts and connect artisans with a global audience. Digital storytelling can breathe life into historical narratives, captivating younger generations and fostering a sense of pride in their heritage. Mobile apps can guide tourists to hidden gems of heritage sites, promoting responsible tourism practices. The benefits are not limited to rural communities. Preserving cultural heritage fosters a sense of global understanding and tolerance (Rana, 2024). As visitors immerse themselves in diverse traditions, they develop a deeper appreciation for different ways of life. This intercultural dialog creates a more inclusive and vibrant world, fostering peaceful coexistence and mutual respect (Lopes and Hiray, 2024).

Moving forward, a multipronged approach is crucial for successful cultural heritage preservation and sustainable development. This includes:

- *Community empowerment.* Local communities can be empowered to make decisions regarding the preservation and utilization of their heritage. Training programs can equip them with the skills and knowledge to manage heritage sites and develop tourism initiatives (Bhattacharya *et al.*, 2023).
- *Government support.* Governments can offer financial assistance for heritage restoration projects, provide tax incentives for businesses promoting local heritage, and invest in infrastructure development that supports sustainable tourism in rural areas.
- *Public–private partnerships.* Collaboration between the public and private sectors can leverage expertise and resources for heritage preservation initiatives. This can involve tourism development projects undertaken by private entities in partnership with local communities (Serra and Gretzel, 2020).
- *Education and awareness.* Educational programs can instill a sense of value for cultural heritage among younger generations. This can include incorporating local history and traditions into school curriculums and organizing workshops that showcase traditional crafts.

Preserving cultural heritage is not a nostalgic indulgence, it is a strategic investment in a sustainable future. By nurturing the traditions, knowledge, and wisdom embedded in their heritage, rural communities can build a vibrant future based on economic prosperity, environmental sustainability, and a deep sense of identity. Cultural heritage preservation should be considered to be a cornerstone of sustainable development, ensuring that the vibrant tapestry of our past continues to pave the way for a thriving future.

10.5.1 Data analysis

The lynchpin of structured tourism in India should be focused on the UNESCO world heritage sites, both cultural and natural. India has 43 properties on the world heritage list (Table 10.2). Each should have a list of rural tourism hotspots ascribed to it. Places like Bhimbetka, Shantiniketan, Hampi, and all the national parks are ripe for homestays and rural tourism with little government support.

Many of the world-renowned heritage sites in India are close to rural or semi-rural areas. It is easy to harness the intimate knowledge of the local population to create another layer of tourist experience. The move should be away from big ticket tourism to a more "live like the locals" experience. For rural heritage tourism to be boosted, the focus should be on more discerning tourists, those who want to learn and enjoy the experience. Thus the marketing for this type of experience should be focused on the appropriate demographics.

India has 15 intangible cultural heritages listed by UNESCO (Table 10.3), and efforts could be made by respective state governments to incorporate them into the rural tourism sector, where workshops and performances can be melded to produce a better guest experience. The central government of India has focused on highlighting yoga worldwide, but there needs to be a concerted effort by the state governments to popularize it. More rural populations should be encouraged to provide professional yoga courses and use digital media to reach wider audiences. Similarly, festival-based intangible cultural heritages like Garba and Durga Puja should be highlighted to foreign and domestic tourists to attract more visitors during these festivals. Some of the intangible cultural heritages can only be adequately practiced in rural settings. Dedicated rural tourism hotspots with a night stay should be developed in each of these circuits with the rural halt as the baseline for cultural immersion. Similarly, the Kubh-Mela should be more professionally managed with the entry of private players and other stakeholders.

The list given in Table 10.4 of places submitted by India for UNESCO status provides a wide range of places where rural tourism and homestay infrastructure can be developed. With proper guidance and governmental help, the total spend per person on internal tourism could be greatly increased which would lead to knock-on economic effects in rural areas. Several of the destinations do not have adequate accommodation facilities, which leads to tourists going there for day trips from the

Table 10.2. Some properties in India on the UNESCO world heritage list. From UNESCO (2024b).

Properties on the world heritage list	State	Year
Agra Fort	Uttar Pradesh	1983
Ajanta Caves	Maharashtra	1983
Ellora Caves	Maharashtra	1983
Group of Monuments at Mahabalipuram	Tamil Nadu	1984
Sun Temple	Konark	1984
Churches and Convents of Goa	Goa	1986
Fatehpur Sikri	Uttar Pradesh	1986
Group of Monuments at Hampi	Karnataka	1986
Khajuraho Group of Monuments	Madhya Pradesh	1986
Elephanta Caves	Maharashtra	1987
Group of Monuments at Pattadakal	Karnataka	1987
Great Living Chola Temples	Tamil Nadu	1987, 2004
Buddhist Monuments at Sanchi	Madhya Pradesh	1989
Humayun's Tomb	New Delhi	1993
Mahabodhi Temple	Bihar	2002
Champaner-Pavagadh Archaeological Park	Gujarat	2004
Chhatrapati Shivaji Terminus (formerly Victoria Terminus)	Maharashtra	2004
Jantar Mantar	Rajasthan	2010
Hill Forts of Rajasthan	Rajasthan	2013
Archaeological Site of Nalanda Mahavihara at Nalanda, Bihar	Bihar	2016
Historic City of Ahmadabad	Gujrat	2017
Victorian Gothic and Art Deco Ensembles of Mumbai	Mumbai	2018
Dholavira: a Harappan City	Gujarat	2021
Kakatiya Rudreshwara (Ramappa) Temple	Telangana	2021
Santiniketan	West Bengal	2023
Moidams: the Mound-Burial System of the Ahom Dynasty	Assam	2024

nearest big city. The focus should be on ensuring that tourists stay the night at the destination as this would lead to more development of both infrastructure and human resources in the rural area. The state government should strategize with local stakeholders and develop a 2-day overnight itinerary for each of the destinations. For example, rock art sites of the Chambal Valley can be explored while staying *in situ* without having to travel from Bhopal on a day trip. These circuits should be developed through public–private partnerships similar to the pilgrimage tourism circuits like the Char Dham Yatra.

10.6 Challenges Faced by Rural Communities

A significant portion of the global tourism market is made up of rural travel. Numerous locations, from strolling through verdant jungles to scaling mountains, depend on rural tourism to generate much-needed income for the regional economy. Any tourism that occurs in non-urbanized areas is referred to as rural tourism. Given that rural tourism is intrinsically associated with green places and typically environmentally beneficial travel activities like

Table 10.3. Intangible cultural heritage of India recognized by UNESCO. From UNESCO (2024c).

Intangible heritage	State	Year
Ramlila, the traditional performance of the Ramayana	India (nationwide)	2008
Tradition of Vedic chanting	India (nationwide)	2008
Kutiyattam, Sanskrit theater	Kerala	2008
Ramman, religious festival and ritual theatre of the Garhwal Himalayas	Uttarakhand	2009
Mudiyettu, ritual theatre and dance drama of Kerala	Kerala	2010
Kalbelia folk songs and dances of Rajasthan	Rajasthan	2010
Chhau dance	West Bengal, Jharkhand, Odisha	2010
Buddhist chanting of Ladakh: recitation of sacred Buddhist texts in the trans-Himalayan Ladakh region	Ladakh	2012
Sankirtana, ritual singing, drumming, and dancing of Manipur	Manipur	2013
Traditional brass and copper craft of utensil making among the Thatheras of Jandiala Guru, Punjab	Punjab	2014
Yoga	India (nationwide)	2016
Nawrouz	India (nationwide)	2016
Kumbh Mela	Uttar Pradesh, Uttarakhand, Madhya Pradesh, Maharashtra	2017
Durga Puja in Kolkata	West Bengal	2021
Garba of Gujarat	Gujarat	2023

hiking or camping, it is directly related to the notion of sustainable tourism.

10.6.1 Economic decline

Rural communities' ability to thrive is seriously hampered by economic loss. A lot of rural communities mainly depend on traditional sectors like mining, forestry, and agriculture, which are prone to changes in the market, environmental conditions, and consumer demand. Economic stagnation as a whole, job losses, and lower earnings can result from the demise of these industries (Lankauskienė and Raudonė, 2024). Increasing the difficulty of maintaining livelihoods is the fact that rural populations are especially susceptible to economic downturns due to the lack of variety in their economic activity.

10.6.2 Depopulation

Another urgent problem that rural communities around the world are dealing with is depopulation. The population of rural areas has decreased as a result of youth migration to metropolitan areas in quest of better employment, education, and living prospects. In addition to decreasing the labor force available for nearby industry, this demographic change accelerates the average age of rural populations. The viability and vibrancy of these

Table 10.4. List of places whose applications are pending for induction into the UNESCO heritage site list. From UNESCO (2024d).

Tentative list	State	Year
Monuments and Forts of the Deccan Sultanate	Karnataka, Maharashtra, Andhra Pradesh	2014
Cold Desert Cultural Landscape of India	Himachal Pradesh, Ladakh	2015
Keibul Lamjao Conservation Area	Manipur	2016
Garo Hills Conservation Area (GHCA)	Meghalaya	2018
Historic Ensemble of Orchha	Madhya Pradesh	2019
Hire Benkal Megalithic Site	Karnataka	2021
Sri Veerabhadra Temple and Monolithic Bull (Nandi), Lepakshi	Andhra Pradesh	2021
Iconic Riverfront of the Historic City of Varanasi	Uttar Pradesh	2021
Temples of Kanchipuram	Tamil Nadu	2021
Vadnagar: MultiLayered Historic Town	Gujarat	2022
Geoglyphs of Konkan Region of India	Maharashtra	2022
JingkiengJri: Living Root Bridge Cultural Landscapes	Meghalaya	2022
Sri Veerabhadra Temple and Monolithic Bull (Nandi), Lepakshi	Andhra Pradesh	2021
Iconic Riverfront of the Historic City of Varanasi	Uttar Pradesh	2021
Temples of Kanchipuram	Tamil Nadu	2021
Hire Benkal Megalithic Site	Karnataka	2021
Gwalior Fort	Madhya Pradesh	2024
KhooniBhandara, Burhanpur	Madhya Pradesh	2024
Rock Art Sites of the Chambal Valley	Madhya Pradesh	2024
Bhojeshwar Mahadev Temple, Bhojpur	Madhya Pradesh	2024
Gond Monuments of Ramnagar, Mandla	Madhya Pradesh	2024

communities are threatened as fewer young people choose to live in rural locations, where they struggle to preserve infrastructure, social cohesion, and basic services (Inoue *et al.*, 2021).

10.6.3 Cultural destruction

Rural communities' identity and legacy are seriously threatened by cultural erosion. Traditional customs, values, and cultural manifestations are being lost as a result of globalization, rapid urbanization, and shifting social standards. Intangible cultural legacy could be lost if younger generations move into metropolitan areas and disturb the transmission of cultural knowledge across generations (Giliberto and Labadi, 2021). Furthermore, the distinctive character and allure of rural communities may be further undermined by the dissolution of authentic cultural identities brought about by the commodity of culture and the influence of mainstream media.

10.6.4 Natural calamities

The resilience and capacity of rural people to maintain their way of life are further hampered by environmental deterioration and natural disasters. Droughts, floods, and other extreme weather events are examples of climate change-related phenomena that can destroy agricultural output, up-end local economies, and jeopardize the physical infrastructure of rural communities. The difficulties that rural people experience are exacerbated by inadequate access to basic amenities like healthcare, education, and transportation, which widens the gap between rural and urban areas (Rana, 2024).

10.6.5 Accessibility

Global rural tourism is a daunting task due to the absence of reliable internet connectivity and poor infrastructure for e-facilities. Rural tourism growth depends on the availability and use of digital tools such as desktops, laptops, and smartphones. In addition, many rural societies have low levels of digital literacy which affects their ability to interact with online platforms like virtual reality and phone apps. This illiteracy in technology not only curtails the potential that digital tools hold in promoting rural tourism but also extends the gap between urban and rural areas as far as technology is concerned (Correa and Pavez, 2016). As a result, these obstacles restrict rural communities from fully participating in and benefiting from tech-driven advancements in tourism, thereby impeding attempts towards sustainability promotion within their territories. It is therefore important that some of these issues are addressed so as to improve the uptake and efficiency of digital tourism resources in different parts of the world, especially in those which lack good connectivity. Rural areas can maximize the use of technology through improving the infrastructure and enhancing digital literacy so that they can develop their tourism industry, hence increasing economic opportunities while narrowing down the disparity between people who have access to information and communications technology and those who do not. They will be able to tap into global tourism markets, actively becoming players at a time when these regions are working towards sustainable development (Ferrari *et al.*, 2022).

10.6.6 Authenticity

Currently, the cultural integrity of rural tourism experiences needs to be maintained to effectively market the rural culture to tourists and to protect cultural assets. When it comes to people from rural areas, they have their own cultural backgrounds and ways of life. It is also important to understand that rural tourism has its own unique selling point – it is the simple life with limited resources and originality of the rural culture (Rana, 2024). So rural tourism should be portrayed in a way that supplements the positive experience to be provided to tourists, rather than in a way that detracts from the experiences they are interested in. To provide a memorable experience to the tourist, rural people may start altering the customs of the village to meet the tourists' expectations. There is a need to ensure that authenticity of the rural culture is not compromised and that representations of their culture should be achieved in a manner that matches the real cultural practices and surrounding environment in order to strike the right balance and authenticity of the rural tourism (Gocer *et al.*, 2024). In addition, changes in integration should be made in such a way that does not forcibly displace the culture of a place but instead strengthens it. This means a rigorous process of decision making and consultation with the local peoples to ensure that their identity and culture is captured appropriately. It is vital that new policies are put in place that will act as a guide to the use of the digital tools in encouraging rural tourism with the intention of conserving and promoting culture. In this way, it is possible to further develop positive aspects associated with technological advancements without compromising the genuineness of rural tourism experiences and preserving cultural heterogeneity, thereby promoting the sustainable tourism concept.

10.6.7 Sustainability

There is a range of factors that must be taken into consideration when rural tourism is being developed. One of the foremost factors that has to be considered is the proper utilization of resources in such a way that the ecology of the rural area is not adversely affected. The development of tourism can cause environment issues for the local communities (D'Souza and Nirupama, 2021). Sometimes the economic opportunities provided may be less than the negative impacts on environmental sustainability. As tourism increases in rural areas, more locales need to be found to accommodate the increased number of visitors. This can inflict environmental damage, erase the authenticity of a culture, and in turn reduce the quality of life for local residents. More visitors coming to rural areas may potentially harm the environment. Many rural areas have delicate ecosystems that can be easily disturbed by mass tourism. When not managed sustainably, waste management, water pollution, and deforestation can result in a loss of biodiversity López-Sanz *et al.* (2021). Finally, the attractions that rural tourism chooses to capitalize on could be damaged by detrimental changes in the future due climate change. The use of digital technologies may backfire by worsening the effects of environment degradation due to high energy consumption by the digital gadgets themselves and the escalating levels of e-waste. Further, while the use of technology can improve productivity and efficiency, a lack of policies and progressive distribution of benefits derived from its adoption may deepen pre-existing gaps in socio-economic status (Russo and Palla, 2023). Therefore, it is important to recommend and enforce sustainable technologies, a fair distribution of gains to those involved, and to engage the locals in technology decisions.

10.6.8 Infrastructure

Various difficulties may be encountered as rural areas often do not have sufficient infrastructure for technological advancement of the tourism sector, starting with transport networks, through accommodation facilities, and ending with basic services. Haldar (2018) discusses the problem of a lack of infrastructure development in relation to tourist visits, which may cause discomfort and therefore affect the quality of the experience. Inadequate infrastructure – for example poor access roads, poor accommodation, and other related facilities – poses a threat to the viability of rural tourism enterprises. These constraints relate to post-modern theory where tourism infrastructure is central; to encourage the positive integration of information technology into rural tourism, these infrastructure chasms must be closed. This means that adequate measures have to be taken to ensure that strong and optimum transportation systems are provided, good-quality accommodation is available, and adequate amenities are provided. In this way, rural areas can improve attractions and sightseeing experiences, making tourists' visits memorable and facilitating a sustainable economic basis for tourism companies. Hence, the need for targeted investments and strategic planning in the improvement of infrastructure cannot be overemphasized, as this is pivotal to the effective deployment of technology-induced tourism development making balanced growth and development a reality in rural regions.

10.6.9 Cultural sensitivities

Any efforts that encourage people to travel should be done respecting the ethical and cultural values in the relevant countries. The use of technology in conveying cultural identity should be done so that it has a positive impact on areas. When portraying cultural elements, special attention should be paid to depicting them as close to reality as is possible, and, at the least, not offending the local population. If cultural taboos are violated there may be repercussions from the local community as they may be offended, leading to a deterioration in their relationship with visitors (Lee *et al.*, 2013). A failure to consider such sensitivities is detrimental to the reputation of specific tourism projects, and, by extension, reduces their accomplishments. Consequently, it is vital for the promoters of such strategies to approach communities and involve them in planning and decision making. Cooperating this way will ensure that cultural identity remains protected and the tourist and the locals have a positive interplay, thus promoting sustainability and a successful implementation of technology-enhanced tourism (Government of India, 2022).

10.7 Policy Recommendations and Best Practices

Culture and heritage protection must be effectively integrated into sustainable urban environments by monitoring the environmental impacts caused by rural tourism. As a result of this process, cultural heritage is preserved while urban sustainability is promoted. The active participation of the local community in tourism planning is essential in guaranteeing that their viewpoints and interests are adequately considered and addressed. Such involvement is a fundamental aspect of inclusive and comprehensive planning processes. Facilitating proactive engagement in tourism development and cultural preservation initiatives is crucial for leveraging economic prospects. It is possible to align these initiatives with the community's economic and cultural objectives, thereby fostering sustainable economic growth (Giliberto and Labadi, 2021). Proactive engagement not only ensures the preservation of cultural heritage but also enhances the socio-economic fabric of the community by creating job opportunities and promoting local businesses. Moreover, the objective of infrastructure improvements in rural areas is to significantly elevate the tourism experience. This is achieved by augmenting accessibility, connectivity, and visitor amenities. In order to make rural areas more accessible and appealing to tourists, improvements in infrastructure, such as improved transportation networks, better communication systems, and upgraded visitor facilities, are being made (Rahman and Baddam, 2021). By attracting more tourists and encouraging longer stays, these improvements foster a more seamless and enjoyable visitor experience, stimulating local economies.

The need to protect cultural historical assets cannot be overemphasized because of the dangers associated with commercialization of such resources for tourist attraction. Within the scope of cultural heritage management, the use of natural essential oils is considered as a key strategy to carrying out proper restoration procedures that are friendly to the environment and do not harm the artwork in any way (Russo and Palla, 2023). Apart from preserving the environment, this method also retains the structure and appearance of cultural artefacts for many years hence. The preparation and promotion of cultural and historical resources has to be treated as a top priority (Bhuiyan and Darda, 2021). For example, overcommercialization can have negative impacts since it can cause the decline of culture identity and history. These assets, by preserving the authenticity of the artistic value, can be preserved for future generations and remain significant for historical and cultural study. Utilizing cultural resources is considered to be efficient for marketing sustainable tourism in rural areas. Such strategies should be developed in a way that calls attention to the multicultural aspect of rural regions and the history of the country, so as to encourage responsible tourism. In this way, tourism can be promoted that avoids erasing cultural differences to favor the economic richness of certain multinational companies.

The continuous renewal of past initiatives is encouraged through an active policy of involving communities and stakeholders in the decision-making process. By integrating various perspectives and encouraging teamwork, these activities encourage rounded ways of dealing with heritage issues that are consistent with sustainability values. In this sense, capacity building programs have a significant part to play. They enable communities to gain the necessary skills and knowledge that are required for participating in tourist-based businesses, which is a way of creating job opportunities as well as entrepreneurship development (D'Souza and Nirupama, 2021). They also promote cultural property conservation efforts by preserving heritage sites for future generations. Through local people's participation in such projects, there is not only a sense of belonging but also equal sharing of benefits derived from heritage conservation programs. Additionally, the integration of sustainability principles into proactive geosciences, interdisciplinary methodologies, and technical advances is vital in conserving cultural resources from physical calamities (Bhuiyan and Darda, 2021). Proactive geosciences involve utilizing geological knowledge to anticipate and manage natural hazards. The integration of this scientific method with interdisciplinary approaches can enable heritage initiatives to develop comprehensive strategies, for example addressing the complex issues related to subterranean heritage conservation. Zolotarev *et al.* (2023) conducted a comprehensive assessment of technology advancement such as remote sensing, geographic information systems (GIS), and 3D modeling, that can facilitate detailed documentation and monitoring of heritage sites.

The same tools also help in the early identification of possible threats, hence allowing targeted action that may reduce risks. Technology has also become a way of enhancing community participation by establishing information-sharing platforms and collaborative decision-making portals. Capacity building programs are essential in order to empower communities for tourism involvement, enterprise development, especially at the grassroots level, and cultural preservation exercises. By fostering collaboration and leveraging innovative approaches, heritage initiatives can achieve long-term sustainability and contribute to the preservation of cultural heritage for future generations.

Tourism development plans should be oriented towards the achievement of sustainable development goals and the emphasis of social purpose and stability. These strategies have to be planned and implemented in such a manner that the development of communities and the sustainable use of culture and nature resources will be economically beneficial in the long run. Local people, NGOs, other business investors, and government organizations must come together to conserve the resources used in rural tourism. Such cooperation is essential as it enables the spread of undertakings and, most importantly, the harnessing of the capacities and resources possessed by specific stakeholders (Bhuiyan and Darda, 2021). Bhuiyan and Darda underscored that this framework strengthens the effectiveness of sustainable tourism programmes and actions. Another important factor is the involvement of all parties, particularly at the formulation and decision-making stages. Different participants will experience different impacts hence the importance of these inclusive stages. This is necessary not only for the legitimacy of the planning process but it also reduces the likelihood of failure in its implementation. Coordinating different standpoints and experience helps in establishing more secure and enduring polices on tourism. Local communities contribute increased understanding of community needs and cultural/contextual issues; NGOs contribute the databases and lessons in sustainable practices and corporate governance accountability; businesses contribute the capital and management expertise; and, lastly, governmental organizations contribute legal policies/frameworks and penalties. A combination of the multiple stakeholders is most effective in addressing the complex processes of sustainable tourism. Zolotarev *et al.* (2023) discuss that through the support of multistakeholders it is possible to identify and exclude most potential risks, outline the correct recommendation, as well as define the conditions for further controlling and guarding the tactics needed to develop socially accountable and sustainable tourism.

Existing modes of transport need to be enhanced for the masses, additional transportation infrastructure has to be built, and direct links to first-tier transport networks developed. Transportation improvements decrease the time required to get somewhere and improve connectivity, thus more people can visit rural destinations (Semwal *et al.*, 2024). With the increasing accessibility of rural tourist destinations by cars and other modes of transport, the caliber of the rural tourism experienced is enhanced and the satisfaction of tourists increased. An improvement in infrastructure equates to a consistent development and promotion of activities of rural tourism enterprises, ensuring the promotion of economic activities and the retention of rural traditions. The emphasis placed on infrastructural investments also recommends an integrated strategy approach that gives due regard to cultural authenticity coupled with functionality as a key success factor of tourism. Furthermore, expanding the range of lodging options in rural areas guarantees that guests will have convenient and comfortable places to stay when visiting. Creating environmentally friendly lodges, guesthouses, homestays, and campsites that complement the local landscape while offering contemporary conveniences can be one way to do this. Rural locations can attract a wider spectrum of visitors and can extend visitor stays by providing a variety of sustainable lodging options, which will boost local economies and assist local communities (UNESCO, 2024d).

Further improving amenities and facilities at cultural heritage sites prolongs visitor stays and improves the overall visitor experience. This could involve building visitor centers, interpretive centers, rest places, and dining establishments. In addition to increasing tourist enjoyment, providing sufficient facilities guarantees the maintenance and safeguarding of cultural heritage places by managing visitor impacts and promoting responsible tourism practices. In rural locations, infrastructure investment also promotes economic growth and community development. Infrastructure improvements boost local economies, generate employment opportunities, and draw private capital to tourism-related enterprises (Liu, 2022). Furthermore, enhanced infrastructure has the potential to increase the value of cultural heritage assets, resulting in higher tourism earnings and sustainable community development. Funding environmentally friendly infrastructure advances environmental preservation objectives and strengthens the general resilience of rural areas. In order to minimize environmental effects and maintain the natural beauty of rural settings, infrastructure projects may incorporate green building practices, renewable energy sources, and water conservation measures. Leveraging cultural assets for sustainable rural tourism requires significant investments in infrastructure, lodging, and transportation. In order to successfully integrate cultural heritage into tourism policies and ensure the long-term success of rural tourism efforts, infrastructure development plays a critical role in promoting environmental sustainability, boosting accessibility, enhancing visitor experience, and supporting local economies (Chen *et al*., 2024).

10.8 Conclusion

This chapter highlights the immense potential for leveraging cultural heritage to promote sustainable rural tourism in India. It contends that authentic yet curated lived experiences should be offered via homestays by showcasing the customs, traditions, and lifestyles of varied rural communities of India enmeshed around a natural, cultural, or archeological site. Cultural heritage-centric rural homestay-based tourism not only preserves the cultural heritage but also enhances social cohesion, economic growth, and environmental sustainability. However, challenges like the digital divide, poor human resources and infrastructure, population decline, cultural degradation and convergence, threats of overcommercialization, and economic downturns threaten the nascent rural tourism sector in India outside a few hotspots. To realize the full potential of this kind of rural tourism it is essential to empower local communities, foster public–private partnerships, and promote awareness and training. Rural tourism initiatives should follow a bottom-up approach and actively engage with all relevant stakeholders to balance sustainable and equitable economic development with cultural preservation. Technology and marketing support to rural entrepreneurs is essential for the success of homestays; targeted human resource training should be provided to allow hosts to offer innovative ways to enhance visitor experiences and support rural communities while maintaining authenticity. These strategies should be tailored to the unique cultural and environmental features of each region as adaptability to local context is key to a successful rural tourism experience. It is thus imperative to build a collaborative framework ensuring that rural tourism continues to benefit both present and future generations.

References

Almeida, A. and Machado, L. (2020) Heritage as a source of competitive advantage: Lessons from Madeira. In: *Multilevel Approach to Competitiveness in the Global Tourism Industry*. IGI Global, pp. 177–188.

Bansal, N. and Choudhary, H. (2024) Fostering digital equity: Evaluating impact of digital literacy training on internet outcomes in rural marginalized communities in India. *International Journal of Life Long Education* 43(5), 473–493.

Bhattacharya, P., Mukhopadhyay, A., Haldar, S. and Saha, J. (2023) Commercialization of home through homestay tourism: A study on Chatakpur of Darjeeling district (India) in commensurate to ASEAN standard and revisit intention. *Global Social Welfare* 3–14.

Bhuiyan, M.H. and Darda, M. (2021) Tourism for sustainable development goals (sdgs) achievement in bangladesh. *Bangladesh Journal of Public Administration* 29(2), 54–62.

Chen, L., Hu, Y., Wang, R. and Li, X. (2024) Green building practices to integrate renewable energy in the construction sector: A review. *Environmental Chemistry Letters* 22(2), 752–882.

Childs, C. (n.d.) How Culture and Heritage Tourism Boosts More than a Visitor Economy. *My Travel Research*. Available at: https://www.mytravelresearch.com/culture-and-heritage-tourism-boosts-visitor-economy (accessed 13 August 2024).

Correa, T. and Pavez, I. (2016) Digital inclusion in rural areas: A qualitative exploration of challenges faced by people from isolated communities. *Journal of Computer-Mediated Communication* 21(3), 247–263.

Department of Tourism, Government of West Bengal (n.d). Available at: https://wbtourism.gov.in/home-stay-page/309 (accessed 12 August 2024).

D'Souza, M. and Nirupama, P. (2021) Impacts of social media on the growth of rural tourism in India. *International Research Journal of Modernization in Engineering Technology and Science* 3, 574–582.

Endere, M.L., Cantar, N.M. and Zulaica, M.L. (2024) Heritage and sustainability. In: *Encyclopedia of Archaeology*. Academic Press, pp. 589–596. DOI: 10.1016/B978-0-323-90799-6.00040-9.

Ferrari, A., Bacco, M., Gaber, K. and Brunori, G. (2022) Drivers, barriers, and impacts of digitalisation in rural areas from the viewpoint of experts. *Information and Software Technology* 145, 106816.

Giliberto, F. and Labadi, S. (2021) Harnessing cultural heritage for sustainable development: An analysis of three internationally funded projects in MENA countries. *International Journal of Heritage Studies* 28(2), 134–145.

Gocer, O., Boyacioglu, D. and Karahan, E.E. (2024) Cultural tourism and rural community resilience: A framework and its application. *Journal of Rural Studies* 107(3), 3–11.

Gould, P. and Burtenshaw, P. (2014) Heritage sites: Economic incentives, impacts, and commercialization. In: Smith, C. (ed.) *Encyclopedia of Global Archaeology*. Springer Link, pp. 3326–3331.

Government of India (2022) *Report on National Digital Tourism Mission*. Government of India, New Delhi.

Government of India, Ministry of Tourism (2023) Statement in reply to part (a) of Rajya Sabha unstarred question no. 1903 answered on 16.03.2023 regarding homestay policy in India, Government of India. Available at: https://tourism.gov.in/sites/default/files/2023-03/usq.1903%20for%2016.03.2023.pdf (accessed 21 November 2025).

Gunter, U. and Cedia, M. (2021) Can indigenous and community-based ecotourism serve as a catalyst for land sparing in Latin America? *Journal of Travel Research* 60(7), 1566–1580.

Haldar, P. (2018) Rural tourism – challenges and opportunities. In: *International Conference on Marketing and Society*. Noida.

Hamid, S. (2024, June 14) The simple joys of village life. *Rising Kashmir*.

Hariram, N.P., Mekha, K.B., Suganthan, V. and Sudhakar, K. (2023) Sustainalism: An integrated socio-economic-environmental model to address sustainable development and sustainability. *Sustainability* 15(13), 10682. Available at: https://www.mdpi.com/2071-1050/15/13/10682

Hoang, K.V. (2021) The benefits of preserving and promoting cultural heritage values for the sustainable development of the country. *E3S Web of Conferences* 234, Article 00076. Available at: https://www.e3sconferences.org/articles/e3sconf/abs/2021/10/e3sconf_icies2020_00076/e3sconf_icies2020_00076.html

Inoue, T., Koike, S. and Yamauchi, M. (2021) Exploring the impact of depopulation on a country's population geography: Lessons learned from Japan. *Population, Space and Place* 28, e2543.

Jangra, R., Kaushik, S., Singh, E., Kumar, P. and Jangra, P. (2023) The role of transportation in developing the tourism sector at high altitude destination, Kinnaur. *Environment, Development and Sustainability* 26(4), 9369–9395.

Kandari, L.S., Bisht, V.K., Bhardwaj, M. and Thakur, A.K. (2014) Conservation and management of sacred groves, myths, and beliefs of tribal communities: A case study from North India. *Environmental Systems Research* 3–11.

Kandavel, S. (2024) Village cooking channel team roped in to showcase creative economy of T.N. at Global Investors Meet 2024. *The Hindu*. Available at: https://www.thehindu.com/news/national/tamil-nadu/village-cooking-channel-team-roped-in-to-showcase-creative-economy-of-tn-at-global-investors-meet-2024/article67677422.ece (accessed 13 August 2024).

Kapur, S. (2016, October) Rural tourism in india: Relevance, prospects and promotional strategies. *International Journal of Tourism and Travel* 9(1 & 2), 40–49.

Kumar, S. and Asthana, S. (2020) Technology and innovation: Changing concept of rural tourism – a systematic review. *Open Geosciences* 12(1), 738–750.

Lane, B. and Kastenholz, E. (2015) Rural tourism: The evolution of practice and research approaches – towards a new generation concept? *Journal of Sustainable Tourism* 23, 1133–1156.

Lankauskienė, R. and Raudonė, Ž. (2024) Challenges and prospects for the determinants of socioeconomic development in rural areas: Evidence from Lithuania. *European Countryside* 16(1), 11–24.

Lee, S.G., Trimi, S. and Kim, C. (2013) The impact of cultural differences on technology adoption. *Journal of World Business* 48(1), 21–28.

Liu, Y. (2022) Application of digital technology in intangible cultural heritage protection. *Mobile Information Systems* 1, 3–11.

Lopes, R. and Hiray, D. (2024) Impacts of cultural events and festivals on cultural tourism. *Journal of Advanced Zoology* 45(S4), 174–179.

López-Sanz, J., Penelas-Leguía, A. and Rodríguez, P. (2021) Rural tourism and the sustainable development goals: A study of the variables that most influence the behavior of the tourist. *Original Research* 19, 2–10.

Mackay, M., Nelson, T. and Perkins, H.C. (2019) Agritourism and the adaptive re-use of farm buildings in New Zealand. *Open Agriculture* 4(1), 466–474.

Madandola, M. and Boussaa, D. (2023) Cultural heritage tourism as a catalyst for sustainable development: The case of old oyo town in Nigeria. *International Journal of Heritage Studies* 29, 22–37.

Mayaram, A. (2024) Tourism infrastructure can yield gains for rural India. *Economic Times*. Available at: https://economictimes.indiatimes.com/policy/tourism-infrastructure-can-yield-gains-for-rural-india/articleshow/4782118.cms?from=mdr (accessed 28 July 2024).

Ministry of Tourism, Government of India (2021) *National Strategy & Roadmap for Development of Rural Tourism in India*. Ministry of Tourism, Government of India, New Delhi.

Ministry of Tourism, Government of India (2022) *The National Strategy for Promotion of Rural Homestays – An Initiative towards Atmanirbhar Bharat*.

Nilson, T. and Thorell, K. (2018) *Cultural Heritage Preservation: The Past, the Present and the Future*. Högskolani Halmstad: Halmstad University Press, Halmstad.

Oloidi, A J. and Okonkwo, E.C. (2020) Revitalisation of cultural authenticity for tourism and community development in selected heritage sites. *Journal of Tourism and Heritage Studies* 9(2), 59–84.

Pardo-Bosch, F., Pujadas, P., Morton, C. and Cervera, C. (2021) Sustainable deployment of an electric vehicle public charging infrastructure network from a city business model perspective. *Sustainable Cities and Societies* 71, 3–10.

Rahman, S.S. and Baddam, P.R. (2021) Community engagement in Southeast Asia's tourism industry: Empowering local economies. *Global Disclosure of Economics and Business* 76–89. Available at: https://pdfs.semanticscholar.org/76a4/736614e018ddb513a551cc08ab720e3199c1.pdf

Rana, N. (2024) Digital preservation of traditional crafts. *International Journal of Research Culture Society* 8, 7–10.

Rosalina, P.D., Dupre, K. and Wang, Y. (2021) Rural tourism: A systematic literature review on definitions and challenges. *Journal of Hospitality and Tourism Management* 47, 135–148.

Russo, R. and Palla, F. (2023) Plant essential oils as biocides in sustainable strategies for the conservation of cultural heritage. *Sustainability* 170, 3–14.

Semwal, R., Tripathi, N., Tyagi, P.K., Pandey, U.K. and Dafouti, B.S. (2024) Revitalizing rural tourism in India: A comprehensive framework for AI integration. In: *International Conference on Innovation and Regenerative Trends in Tourism and Hospitality Industry (Advances in Economics, Business and Management Research)*, pp. 250–264.

Serra, F.F. and Gretzel, U. (2020) Influencer marketing for tourism destinations: Lessons from a mature destination. In: *Information and Communication Technologies in Tourism*. Springer, Switzerland, pp. 66–77.

Sharma, S. and Chettri, R. (2023) An analytical study on the growth of home stays in darjeeling hills. *International Journal for Multidisciplinary Research* 5(3), 3–8.

Shrivastava, A., Jodhana, L.S. and Chourasia, S. (2023) Exploring the potential of augmented reality and virtual reality on indian tourism industry. *Gurugram University Business Review* 3(2), 39–48.

Singh, P., Seal, M. and Sharma, D. (2024) Analysing the role and contribution of tourism in achieving sustainable development goals. In: *Harmonizing Global Efforts in Meeting Sustainable Development Goals*, IGI Global, pp. 183–191.

Sudheer, Arya. (2021) Integrated village tourism for rural sustainability and development: A review of village tourism and its impact on the environment and society of Kumbalangi model tourism village. *JournalNX: A Multidisciplinary Peer Reviewed Journal* 7(5), 256–266.

UNESCO (2008) *Safeguarding Intangible Heritage and Sustainable Cultural Tourism: Opportunities and Challenges*. UNESCO, Bangkok.

UNESCO (2024a) World Heritage and Sustainable Development. Available at: https://whc.unesco.org/en/sustainabledevelopment (accessed 11 August 2024).

UNESCO (2024b) Properties Inscribed on the World Heritage List. Available at: https://whc.unesco.org/en/statesparties/in (accessed 11 July 2024).

UNESCO (2024c) Elements on the Lists of Intangible Cultural Heritage. Available at: https://ich.unesco.org/en/state/india-IN?info=elements-on-the-lists (accessed 10 July 2024).

UNESCO (2024d) Tentative List for Homestays. Available at: https://whc.unesco.org/en/tentativelists/?action=listtentative&state=in&order=states (accessed 10 August 2024).

UNWTO (UN Tourism Organization) and Huzhou City (2017) *International Rural Tourism Development – An Asia-Pacific Perspective*. UNWTO, Madrid.

Vikramaditya, R. (2024) Rural tourism: A driver of the rural economy in India. *Times of India*. Available at: https://timesofindia.indiatimes.com/readersblog/black-ink/rural-tourism-a-driver-of-the-rural-economy-in-india-51053 (accessed 12 August 2024).

Wu, S.R., Fan, P. and Chen, J. (2016) Incorporating culture into sustainable development: A cultural sustainability index framework for green buildings. *Sustainable Development* 24(1), 64–76.

Zolotarev, S., Kusakina, O. and Ryazantsev, I. (2023) Transport accessibility assessment of rural tourism facilities. *E3S Web of Conferences* 376, 1–8.

Further Reading

Bhaskar, P. (2023) Rural tourism development in Rajasthan: An empirical study of factors affecting tourist satisfaction in rural Rajasthan. *Pacific Business Review* 15, 3–11.

Ghosh, S. (2024) Cultural erosion in India: Cultural loss and a threat to the nation's traditions and ethnicity. *International Journal of Social Science and Economic Research* 9, 1092–1099.

Kannegieser, I. (2015) A home in the hills: Examining the socioeconomic benefits of homestay tourism on rural women and their communities in the Darjeeling district. *SIT Digital Collections* 11–35.

Labadi, S. (2014) UNESCO world heritage convention (1972). In: Smith, C. (ed.) *Encyclopedia of Global Archaeology*. Springer Link, pp. 7434–7439.

Mekonnen, H., Bires, Z. and Berhanu, K. (2022) Practices and challenges of cultural heritage conservation in historical and religious heritage sites: Evidence from North Shoa Zone, Amhara Region, Ethiopia. *Heritage Science* 10, 172–177.

Mukhopadhyay, D. (2024) Homestay-tourism: A viable alternative to the perils of overtourism in the Darjeeling Hills of West Bengal, India. *Environment Conservation Journal* 25, 517–526.

11 Factors Influencing Rural Youth Engagement in India: A Comprehensive Analysis

Deepika S.R.[1*], Sumathi Annamalai[1] and Angappa Gunasekaran[2]
[1]School of Business and Management, Christ University, Bangalore, India; [2]School of Business Administration, Penn State Harrisburg, Middletown, Pennsylvania, USA

11.1 Introduction

India, as one of the world's fastest growing economies, boasts a youthful demographic, with nearly half of its population under the age of 25 years. This demographic dividend presents a unique opportunity and challenge, particularly in rural areas where youth engagement is crucial for national development. The rural youth in India face a complex array of issues, including limited access to quality education, employment opportunities, and infrastructure, all of which influence their ability to contribute meaningfully to their communities and the broader economy. The chapter begins by exploring the factors that shape rural youth involvement in India, encompassing educational opportunities, economic prospects, social and community engagement, technological access, and government support. The discussion is enriched by insights from a focused group discussion conducted among a group of students and faculty members from a college situated in a rural area, surrounded by agricultural communities. By integrating empirical findings and practical observations, this chapter provides a comprehensive understanding of rural youth engagement in India. It aims to shed light on the interventions needed to empower rural youth, ensuring their active participation in the nation's growth trajectory. The proposed research framework and data collection instrument are designed to facilitate further exploration and targeted action in this vital area of development.

11.2 Literature Review

India is a fast-growing economies and is home to a significant portion of the world's youth. To fully harness the growth potential, it is essential to invest in human capital and address the specific needs and aspirations of rural youth across economic, social, technological, and ecological domains (Bharat Cares, n.d.). The factors affecting rural youth engagement are multifaceted, encompassing social, economic, and environmental dimensions. Identifying factors influencing rural youth involvement in India is crucial for understanding and enhancing their participation in economic, social, and community activities. Rural youth in India face unique challenges, including limited access to education, employment opportunities, and

*Corresponding author: deepika.sr@christuniversity.in

 Agripreneurship, Digital Inclusion and Sustainability: Rural Innovation and Growth in the Global South (eds Dutta and Trivedi)
DOI: 10.1079/9781836992646.0011

infrastructure, which affects their engagement. Socio-economic status, cultural norms, availability of resources, and governmental support are key variables that shape their involvement. Understanding these factors is essential for developing targeted interventions that can empower rural youth and integrate them more fully into the nation's growth and development.

11.2.1 Education and rural youth engagement

The engagement of rural youth in India is critically dependent on both education and skill development. The research indicates that there is a pressing need to raise awareness about the value of higher education among rural youth to enhance their participation (Undale, 2021). Concurrently, the Skill India initiative has been pivotal in equipping rural youth with vocational skills, fostering entrepreneurship, and instilling self-reliance, which has led to a positive shift in societal attitudes towards skill-based careers (Akther, 2023). However, there are contradictions and challenges to be addressed. While the Skill India initiative has made significant strides, there is still a need for specialized skill development programs tailored to regional demands and addressing gender disparities. Additionally, the decline in interest towards agriculture among rural youth poses a challenge in engaging them in traditional sectors (Haldar *et al.*, 2022). Initiatives like Skill India have shown positive outcomes, but there is a need for more targeted and inclusive programs. Addressing the challenges of regional specificity, gender disparities, and changing attitudes towards agriculture will be crucial for sustainable engagement of rural youth in the Indian economy (Sangameshwara and Mallikarjuna Swamy, 2007; Undale, 2021; Haldar *et al.*, 2022; Akther, 2023). Guo *et al.* (2024) explored the significance of extracurricular and co-curricular activities in fostering entrepreneurship among rural youth. Their study highlighted that such activities – like school clubs, community projects, and mentorship programs – are crucial in developing essential entrepreneurial skills such as creativity and problem-solving. Engaging in these activities positively influences rural youth's entrepreneurial intentions by boosting their confidence and interest in starting businesses. However, the study also identified challenges such as limited resources and inadequate infrastructure that hinder the effectiveness of these programs. The authors recommended increased investment in educational infrastructure, expanded mentorship opportunities, and enhanced partnerships between schools, businesses, and government agencies to better support rural youth in entrepreneurship.

11.2.2 Employment and economic opportunities for rural youth

Employment and economic opportunities for rural youth in India are critical for the nation's development and the reduction of rural poverty. The literature suggests that while agriculture remains a principal economic activity, it is insufficient in providing employment due to various challenges, including post-harvest losses and lack of value addition (Tripathi and Singh, 2017). The rural sector has been somewhat neglected in the discourse on "smart cities," with a need to integrate "smart village" policies that leverage information and communications technology (ICT) for employment opportunities (Fennell *et al.*, 2018). Government initiatives are in place to enhance skills and create jobs in rural areas, yet unemployment persists, indicating the need for more effective strategies (Kumar *et al.*, 2022). Contradictions arise in the effectiveness of current government schemes and the actual improvement in employment rates. While some papers suggest that government programs are ongoing (Kumar *et al.*, 2022), others imply that these may not be sufficiently addressing the complexities of rural youth unemployment (Hrumak *et al.*, 2018). Additionally, the potential for agripreneurship and non-agricultural employment through traditional farms and ICT initiatives is highlighted as a means to engage rural youth (Tripathi and Singh, 2017; Fennell *et al.*, 2018). The perception of agriculture as a profitable industry and the impact of skills training programs on youth engagement in agribusiness are significant (Boye *et al.*, 2024). In summary, there is a consensus on the need for diversified employment opportunities for rural youth in India, including the promotion of agripreneurship, skill development, and ICT-enabled services.

The literature underscores the importance of government support, innovation, and smart integration of rural areas into broader economic frameworks to create sustainable employment opportunities (Kume, 2013; Tripathi and Singh, 2017; Fennell *et al.*, 2018; Kumar *et al.*, 2022). As the labor market evolves, upskilling becomes essential for rural youth to remain competitive. Modern skill development, particularly in agriculture, industry, and services, equips them to navigate emerging challenges and seize new opportunities (The Hindu, 2024). The government's focus on introducing relevant training programs is a critical step in enabling rural youth to adapt and thrive in a changing economy. The literature on rural youth aspirations reveals a significant preference for staying in villages despite ongoing urbanization trends. With farming being the primary source of livelihood, a shift in the rural economy is evident as many farmers leave agriculture for non-farming jobs. This occupational transition, driven by factors like the agrarian crisis, underscores the necessity for upskilling rural youth to make agriculture an attractive vocation and to create alternative employment opportunities (The Hindu, 2024). Future efforts should focus on the effective implementation of policies and programs that address the multifaceted barriers to rural youth employment and capitalize on the untapped potential.

11.2.3 Social and community engagement of rural youth

Social and community engagement are critical for enhancing rural youth engagement in India. Studies have shown that community leadership practices and the understanding of factors influencing youth engagement are essential for developing interventions to improve youth participation and health behaviors (Majee *et al.*, 2019). However, there are challenges such as the impact of family structure changes on youth engagement, with community leaders perceiving non-nuclear families as less conducive to youth involvement in community life (Anakwe *et al.*, 2020). Educational initiatives that integrate community engagement into social work education have been found to promote social justice and empowerment, which are key to engaging youth in rural areas (Bhagwan, 2017). Interestingly, while community engagement strategies must be tailored to local contexts, common principles such as trust, transparency, communication, and commitment are universally important (Morrel-Samuels *et al.*, 2016). Youth leadership is also highlighted as a mechanism for positive youth development and social justice youth development, emphasizing the importance of youth-oriented, collaborative research processes (Iwasaki, 2016). Furthermore, university-community engagement programs have been shown to benefit both the youth and the institutions involved, suggesting a broader role for educational institutions in promoting rural youth engagement (Chile and Black, 2015). In summary, effective social and community engagement for better rural youth engagement in India requires a multifaceted approach that includes understanding local leadership practices, addressing family and structural challenges, integrating community engagement into education, overcoming practical barriers, and adhering to universal principles of engagement. Educational institutions can play a significant role in this process, and youth leadership should be fostered to ensure that youth voices are heard and valued in the development of community interventions (Chile and Black, 2015; Iwasaki, 2016; Morrel-Samuels *et al.*, 2016; Bhagwan, 2017; Majee *et al.*, 2019; Anakwe *et al.*, 2020; Chavda and Prakash, 2023).

11.2.4 Technology and infrastructure

Infrastructure and technology play a crucial role in enhancing youth engagement in rural India. The development of robust infrastructure, such as roads, electricity, and internet connectivity, serves as the backbone for technological advancements that can significantly impact rural communities. With improved access to technology, rural youth can gain better educational opportunities, access to information, and employment prospects. The integration of technology in sectors like agriculture, education, and healthcare not only empowers young people but also encourages them to actively participate in their communities' growth and development. As highlighted by Invest India, leveraging technology for rural development is essential for creating sustainable livelihoods and bridging the urban–rural divide, thereby fostering a more engaged and empowered rural youth population (India Invest, 2022).

The NIIT Foundation (2024) emphasizes that initiatives aimed at improving digital infrastructure, such as expanding internet access and providing digital literacy programs, can significantly enhance the engagement of rural youth. These advancements enable them to participate more actively in the digital economy, access information, and improve their socio-economic status. By leveraging technology, rural youth can overcome geographic limitations, connect with broader markets, and contribute more effectively to their communities' development, leading to overall socio-economic growth in rural areas.

11.2.5 Government policies and support

Government policies and support for rural youth in India are multifaceted, aiming to address the diverse needs of this demographic. The Skill India initiative is a prime example, focusing on education and employment by providing training in various sectors and offering financial aid and mentorship to young entrepreneurs (Akther, 2023). Meanwhile, the potential for ICT initiatives to foster "smart villages" is being explored, with the aim of creating employment opportunities through digital inclusion and leveraging mobile telephony to access services and information (Fennell *et al.*, 2018). Contradictorily, even where these efforts are in place, there is a noted decline in the interest of rural youth towards traditional agricultural occupations, despite agriculture being the primary occupation in rural India (Haldar *et al.*, 2022). This disinterest may be exacerbated by the lack of youth involvement in policy making, as government initiatives often do not consult young people, potentially leading to ineffective programs. Furthermore, the proportion of rural youth is expected to decrease, posing a challenge to agriculture and the potential loss of youthful innovation (IFAD , 2018).

In summary, the Indian government has implemented policies like the Skill India initiative to support rural youth through education, training, and entrepreneurship (Akther, 2023) but there is a need for more inclusive policy making that engages rural youth in the process to ensure the relevance and effectiveness of these initiatives. Additionally, the shift away from agriculture among rural youth and the anticipated demographic changes pose challenges that require the ongoing attention and adaptation of policies to sustain rural development and youth engagement (IFAD (International Fund for Agricultural Development), 2018; Haldar *et al.*, 2022).

11.2.6 Health and well-being of rural youth

One study focused on adolescent girls in rural eastern India highlighted the critical state of nutrition and physical health among rural youth (Rose-Clarke *et al.*, 2019). The research found that stunting and undernutrition are prevalent due to inadequate dietary intake, compounded by socio-economic factors such as poverty and lack of access to nutritious food. The study emphasized the need for targeted interventions, particularly for adolescent girls who are at risk of long-term health issues due to poor nutrition during crucial developmental years.

The mental health of rural youth in India is another area of concern. Studies reveal that mental health issues among rural adolescents are often overlooked due to stigma, lack of awareness, and limited access to mental health services. A descriptive study highlighted that depression, anxiety, and stress are prevalent among rural youth, yet there are few resources available to address these challenges effectively. The use of culturally adapted mental health assessment tools is necessary to identify and address these issues (Banerjee, 2023).

The lifestyle choices of rural youth in India are influenced by their socio-economic environment. Research using the simple lifestyle indicator questionnaire (SLIQ) found that rural youth often engage in unhealthy behaviors, such as tobacco use and lack of physical activity, which contribute to their overall poor health outcomes. The study called for comprehensive lifestyle interventions to promote healthy habits among rural adolescents (Chhajer and Hira, 2024).

Access to healthcare in rural areas remains a significant barrier to the well-being of rural youth. The healthcare infrastructure in these regions is often underdeveloped, with a shortage of medical professionals and facilities. Additionally, cultural barriers and lack of education contribute to the underutilization of available healthcare services. Initiatives aimed at improving healthcare access and education in rural areas are crucial for enhancing the health and well-being of rural youth (SMCGH, 2024).

Community-based programs play a vital role in improving the health and well-being of rural youth. The Rural Experiential Learning (REL) program, for instance, emphasizes the importance of engaging rural youth in community development activities. Such programs not only provide practical skills but also promote mental and social well-being by fostering a sense of purpose and community belonging (TISS (Tata Institute of Social Sciences), 2021).

11.2.7 Aspirations of rural youth

With a significant portion of India's population residing in rural regions, the aspirations of these youth are crucial for shaping the nation's future. Understanding their career aspirations and job preferences, and the barriers they face in achieving these goals, is essential for creating targeted policies and interventions that can help bridge gaps in education, skill development, and employment. Rural youth face several barriers in achieving their aspirations, particularly in the areas of education and employment. One study highlighted that lack of guidance in identifying suitable jobs, limited work experience, and a lack of relevant skills were the most prevalent barriers faced by rural youth (NSDC (National Skill Development Corporation), 2023). Moreover, the lack of financial resources and inadequate access to vocational training further hindered their ability to pursue their career goals. These barriers are more pronounced among females, who also face additional challenges such as household responsibilities and social constraints.

Rural youth in India are increasingly aspiring to diverse career paths beyond traditional government jobs, driven by greater internet access and exposure to technology. They seek high-paying jobs, self-employment, and opportunities in fields like sports and the arts. To meet these aspirations, there is a growing need for skill development in digital and green technologies, as well as entrepreneurship. The focus on equipping rural youth with relevant skills is crucial for making them employable, skilled, and self-reliant (Screwvala, 2022). This article on India's rural youth highlighted the contrast between their aspirations and the economic challenges they face. Despite being the backbone of the nation, rural communities struggle with limited resources, education, and employment opportunities. Rural youth aspire to a better future, but are often constrained by socio-economic barriers. Addressing these challenges requires targeted interventions in education, skill development, and infrastructure to unlock the potential of this dynamic group and to drive India's growth (Das, 2023). The aspirations of rural youth in India are shifting towards greater self-reliance and diversification of career opportunities. The skilling ecosystem must evolve to meet the changing demands of the job market, incorporating both digital and green skills. By empowering rural youth through targeted skill development and entrepreneurial support, India can harness the potential of its young workforce to drive national growth and development.

11.3 Methodology

Primary and secondary data were collected for writing this chapter. First, extensive literature reviews were done with the key words "youth engagement," "youth involvement," and "youth in India" in leading journals, and a secondary analysis was done on the previous literatures. Second, four focused group discussions (FGDs) were conducted in three government colleges in Tamil Nadu. The colleges are situated in predominantly rural areas, surrounded by agricultural and weaving communities. The institutions cater primarily to students from nearby villages and small towns, many of whom are first-generation learners. These students often face challenges such as limited access to educational resources and economic constraints, making the college a crucial gateway to higher education and personal development for rural youth in the regions. Each FGD had 8–14 students hailing from nearby villages and a total of 38 students were covered for primary data collection. QDA Miner Lite was used to code the data collected from the FGDs. The discussion aimed to gain insights into youth engagement practices, and the aspirations, challenges, and opportunities faced by the students, particularly in relation to their academic and professional pursuits.

11.4 Involvement of Youth in Empowerment, Upskilling, and Rural Development Practices

11.4.1 Planting trees

Planting trees and taking care of nature and the surroundings, in a village with less rainfall and a limited water source, is a task that came out as one of the prominent practices of youth engagement, in focused group discussion 1 (FGD1). Arun (name changed upon request) and his team have emerged as local change makers, driving youth involvement in planting trees as a key strategy for empowerment and rural development in their village near Punjai Puliampatti in Erode district. Their initiative not only focuses on environmental sustainability but also on upskilling young people by involving them in every step of the tree-planting process. Through this, young people in the village gain practical knowledge and a sense of responsibility towards their environment. Recognizing the potential of their efforts, Arun's team successfully influenced the local panchayat office to support and integrate tree planting into their development agenda. This collaboration has led to the allocation of resources for large-scale tree plantations, creating new employment opportunities for the village youth. The panchayat's involvement has also brought greater awareness about the importance of green cover in combating climate change and ensuring long-term agricultural productivity. By engaging with the local governance structure, Arun and his team have helped bridge the gap between government policies and grassroots action. Their work has inspired other villages in the region to adopt similar practices, highlighting the role of youth in driving meaningful change. This initiative has not only enhanced the green cover in the area but also fostered a stronger sense of community and collective purpose among the villagers. Arun's leadership exemplifies how youth-led movements can contribute to rural development and environmental conservation, creating a model that can be replicated across other rural areas.

11.4.2 Temple festivals, cultural celebrations, and fests

In FGD2, another prominent practice that the participants' narrated was thorough involvement in temple festivals, local fests, and rural sports activities. These events provide a platform for young people to connect with their cultural heritage while also fostering a sense of community. Temple festivals, in particular, offer opportunities for youth to take on leadership roles in organizing and managing the events, which helps develop their organizational and interpersonal skills. The inclusion of traditional rural sports in these festivals has further increased participation, as these activities resonate with the local culture and appeal to younger generations. Such engagement not only keeps traditions alive but also creates a sense of pride and belongingness among the youth. Additionally, these events often serve as a unifying force, bringing together different generations to celebrate their shared heritage. The discussion revealed that these practices are instrumental in maintaining cultural continuity while also adapting to the evolving interests of the younger population. One FGD4 participant narrated that in their village they have regular programs like vallikummi, oyilattam, and arichandrakummietc, where youths organize and participate. Kummi and oyilattam are forms of folk dances in Tamil Nadu.

11.4.3 Part-time employment and livelihood activities

All the participants in all FGDs acknowledged that a significant number (more than half) of students engage in part-time jobs, primarily in catering services, delivery activities, loading and transporting agricultural produce, and supporting weaving processes, while simultaneously pursuing their degrees. A significant number of students are also involved in livestock and cattle rearing as a source of income. This engagement highlights the financial needs and entrepreneurial spirit among the students.

11.4.4 Internet access and digital literacy

One positive aspect noted in all four FGDs was that the vast majority of students have access to the internet. This connectivity enables them to stay updated on current affairs and access educational content through platforms like YouTube and other web sources. Some students have even successfully ventured into content creation, demonstrating their adaptability and initiative in the digital space. They also provide mentoring and guidelines for juniors and peer members in their villages in enhancing digital literacy.

11.4.5 Sports activities

The students are actively involved in sports, with participation in activities such as marathons and cycling being quite common. This involvement in sports not only contributes to their physical well-being but also fosters a sense of community and teamwork. FDG3 participants indicated that their sports club encourages students to participate in various activities, and coaches and mentors them.

11.5 Challenges Faced

The discussion focused on the major challenges rural youths face.

11.5.1 Language proficiency and employment challenges

A major concern identified during the discussions was the students' struggle with English language proficiency. All the four FGDs highlighted this challenge. This limitation poses a significant barrier to securing employment opportunities in urban areas and also affects their confidence levels. The lack of fluency in English hinders their ability to express themselves effectively, which is crucial for career advancement. Though there are digital applications for improving English language proficiency such as spoken English platforms on the internet, the youth participants felt that these are not helping them and not keeping them engaged.

11.5.2 Skill development programs

While the government offers skill development programs, there is a noticeable lack of awareness among the students regarding this employment-specific skills training. The discussion also highlighted the presence of value-added courses in tailoring, particularly aimed at female students. However, these courses are perceived as lacking focus and effectiveness.

Overall, the FGDs provided valuable insights into the diverse experiences and challenges faced by rural students. While there are opportunities for academic and professional growth, language barriers and limited awareness of government-provided skill development programs are significant challenges that need to be addressed. Enhancing English language proficiency, increasing awareness of skills-based employment programs, and improving the focus of value-added courses could significantly benefit the students and better prepare them for their future careers. Also, the FGDs revealed that a small proportion of graduating students at the college opt to pursue postgraduate studies, while a similar number choose to enroll in professional courses like chartered accountancy. Additionally, only a few students showed interest in applying for government examinations, such as the Tamil Nadu Public Service Commission (TNPSC) group 1 and 2 exams.

11.6 Proposed Research Framework

After conducting an extensive literature review and four FGDs, a research model wsa conceptualized that synthesized existing empirical findings and the insights gained while interacting with the participants. By integrating the insights, the model aims to offer a comprehensive understanding of the topic.

11.6.1 Dependent variable

1. *Youth engagement (YE)*. The extent to which youth are actively involved in education, employment, community activities, and other developmental areas.

11.6.2 Independent variables

1. *Education and skill development (ESD)*:

   Access to education.
   Perceived quality of education.
   Participation in skill development programs.

2. *Economic opportunities (EO)*:

   Availability of employment opportunities.
   Entrepreneurship support.
   Satisfaction with local job opportunities.
   Migration intentions.

3. *Infrastructure and technology (IT)*:

   Internet connectivity.
   Use of digital tools for education/work.
   Transportation connectivity.

4. *Social and cultural factors (SCF)*:

   Community support for youth development.
   Influence of traditional gender roles.
   Family support for education and employment.
   Involvement in community activities.

5. *Government policies and programs (GPP)*:

   Awareness of government programs.
   Benefiting from government initiatives.
   Perceived effectiveness of government programs.

6. *Health and well-being (HW)*:

   Access to healthcare.
   Impact of substance abuse.
   Participation in physical/sports activities.

7. *Future aspirations and challenges (FAC)*:

   Aspirations for higher education, employment, business, etc.
   Perceived challenges in achieving goals.
   Optimism about the future in rural areas.

Figure 11.1 presents the framework which can serve as a valuable tool for future researchers seeking a comprehensive understanding of youth engagement. It can also be utilized for the collection of primary data. It is important to note that this is a proposed conceptual framework, developed based on the literature review and insights gathered from interactions with participants in the FGDs. The validity and reliability of the framework and its instruments still need to be tested.

11.6.3 Proposed instrument for data collection

The questionnaire prepared and presented in Appendix 11.1 can be adapted based on specific research needs and local contexts. The open-ended questions allow for deeper insights into the perspectives of rural youth.

11.7 Conclusion

The engagement of rural youth in India is crucial for the country's overall development, requiring a multifaceted approach that addresses education, employment, social engagement, and infrastructure. While initiatives like Skill India and government programs have made strides in empowering rural youth, significant challenges remain, particularly in regional specificity, gender disparities, and the declining interest in agriculture. To sustain and enhance rural youth engagement, it is essential to focus on targeted skill development, improving access

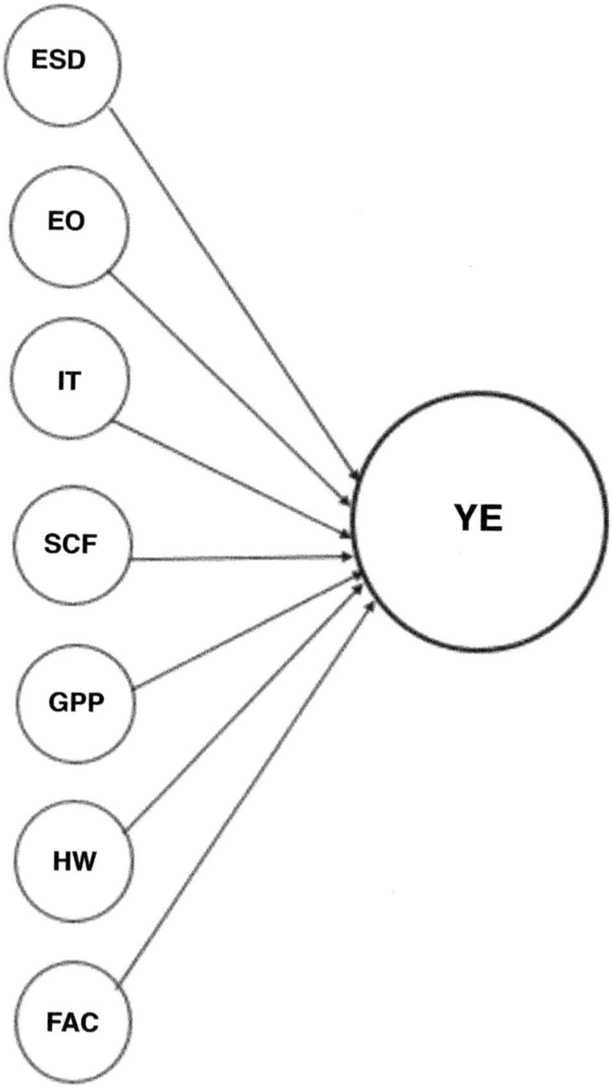

Fig. 11.1. Proposed research work. For abbreviations, see text.

to technology, and fostering community involvement. The case study of a rural college highlights the importance of addressing language barriers and increasing awareness of government-provided skills programs. By implementing inclusive policies and improving educational infrastructure, India can better harness the potential of its rural youth, ensuring their active participation in the nation's economic and social progress.

References

Akther, F. (2023) Role of skill India programs in fostering entrepreneurship among rural youth in India. *Formosa Journal of Science and Technology* 2(10), 2891–2902.

Anakwe, A., Majee, W., Keller, K.J.M. and Jooste, K. (2020) Family structure diversity: Views from rural community leaders and implications for youth engagement. *Community Development* 51(2), 157–171.

Banerjee, M. (2023, August 29) Navigating mental health challenges among India's Rural youth. LinkedIn Post. Available at: https://www.linkedin.com/pulse/navigating-mental-health-challenges-indias-rural-youth-manas-banerjee-nvi0f (accessed 17 September 2024).

Bhagwan, R. (2017) Community engagement within a social work programme in rural India. *Social Work* 53(3), 315–329.

Bharat Cares (n.d.) The needs and wants of the youth of rural India. *BharatCares*. Available at: https://bharatcares.org/The_Needs_and_Wants_of_the_Youth_of_Rural_India (accessed 29 August 2024).

Boye, M., Ghafoor, A., Wudil, A.H., Usman, M., Prus, P. *et al.* (2024) Youth engagement in agribusiness: Perception, constraints, and skill training interventions in Africa: A systematic review. *Sustainability* 16(3), 1096.

Chavda, K. and Prakash, V. (2023) Transport use limitations and its association with social participation among patients with stroke living in rural India. *Disability and Rehabilitation* 1–5.

Chhajer, R. and Hira, N. (2024) Exploring positive psychology intervention and mindfulness-based intervention in nature: Impact on well-being of school students in India. *Frontiers in Public Health* 12, 1297610.

Chile, L.M. and Black, X.M. (2015) University–community engagement: Case study of university social responsibility. *Education, Citizenship and Social Justice* 10(3), 234–253.

Das, A. (2023, August 11) International Youth Day: Understanding the Economic Realities of India's Rural Youth. IndiaTimes. Available at: https://www.indiatimes.com/explainers/news/international-youth-day-understanding-the-economic-realities-of-indias-rural-youth-611805.html

Fennell, S., Kaur, P., Jhunjhunwala, A., Narayanan, D., Loyola, C. *et al.* (2018) Examining linkages between smart villages and smart cities: Learning from rural youth accessing the internet in India. *Telecommunications Policy* 42(10), 810–823.

Guo, J., Khatibi, A. and Tham, J. (2024) The role of innovation and entrepreneurship education in promoting the growth of students in higher vocational colleges. *Library of Progress-Library Science, Information Technology & Computer* 44, 3.

Haldar, J., Prasad, J.N. and Singh, Y.K. (2022) To estimate the rural youth with their extent of participation in agricultural activities in block-Bhanupratappur (Chhattisgarh). *Small* 23, 32–85.

Hrumak, O.Y., Vovk, M.V. and Kindrat, O.V. (2018) Unemployment of rural youth: Causes and ways to solve. *Scientific Messenger of LNU of Veterinary Medicine and Biotechnology* 20(91), 57–61.

IFAD (International Fund for Agricultural Development) (2018) The youth advantage: Engaging young people in green growth. *ECG/IFAD Advantage Series*.

India Invest (2022) Technology and Rural Development. Invest India. Available at: https://www.investindia.gov.in/team-india-blogs/technology-and-rural-development

Iwasaki, Y. (2016) The role of youth engagement in positive youth development and social justice youth development for high-risk, marginalised youth. *International Journal of Adolescence and Youth* 21(3), 267–278.

Kumar, M.S., Lad, Y.A. and Mahera, A.B. (2022) An overview of employment opportunities in rural India. *Asian Journal of Agricultural Extension, Economics & Sociology* 40(12), 42–47.

Kume, E. (2013) On employment opportunities for albanian young people in rural areas through sustainable development of traditional farm. *Academic Journal of Interdisciplinary Studies* 2, 83–87.

Majee, W., Jooste, K., Aziato, L. and Anakwe, A. (2019) Scars of disengagement: Perspectives on community leadership and youth engagement in rural South Africa. *Global Health Promotion* 26(3), 5–14.

Morrel-Samuels, S., Bacallao, M., Brown, S., Bower, M. and Zimmerman, M. (2016) Community engagement in youth violence prevention: Crafting methods to context. *The Journal of Primary Prevention* 37, 189–207.

NIIT Foundation (2024) Bridging the Digital Divide: Empowering Rural India. NIIT Foundation. Available at: https://niitfoundation.org/bridging-the-digital-divide-empowering-rural-india (accessed 29 August 2024).

NSDC (National Skill Development Corporation) (2023) *Career Aspirations of Indian Youth*. Skill India, Delhi.

Rose-Clarke, K., Pradhan, H., Rath, S., Rath, S., Samal, S. *et al*. (2019) Adolescent girls' health, nutrition and wellbeing in rural eastern India: A descriptive, cross-sectional community-based study. *BMC Public Health* 19(1), 1–11.

Sangameshwara, B.G. and Mallikarjuna Swamy, U.M. (2007) Education for rural sector using knowledge network. In: *Intelligent Information Processing III: IFIP TC12 International Conference on Intelligent Information Processing (IIP 2006), September 20–23, Adelaide, Australia 3*, Springer US, pp. 145–150.

Screwvala, Z. (2022, July 14) Making our rural Indian youth employable, skilled and atmanirbhar. Times of India Blogs.

SMCGH (2024) Challenges and Solutions for Healthcare in Rural Communities. Available at: https://smcgh.edu.in/2024/02/12/challenges-and-solutions-for-healthcare-in-rural-communities (accessed 12 February 2024).

The Hindu (2024) Why is Upskilling Necessary for the Rural Youth: Explained. Available at: https://www.thehindu.com/news/national/why-is-upskilling-necessary-for-the-rural-youth-explained/article67703496.ece (accessed 16 August 2024).

TISS (Tata Institute of Social Sciences) (2021) Rural Experiential Learning (REL) Program. Available at: https://tiss.edu/uploads/files/Rural_Experiential_Learning_REL-2021_Final_compressed___.pdf (accessed August 2024).

Tripathi, P. and Singh, N. (2017) Promoting rural entrepreneurship through skill development for decent livelihood: A review. *International Journal of Current Research and Review* 9(15), 21–25.

Undale, P. S. (2021) Enhancing rural youths' employability with skill development. *Towards Excellence: An Indexed, Refereed & Peer Reviewed Journal of Higher Education* 13, 702–708.

Appendix 11.1

The following questionnaire is designed to assess various aspects of youth engagement in rural India. The questions are structured to cover key variables like education, economic opportunities, social and cultural factors, infrastructure, government policies, health and well-being, future aspirations and challenges faced by rural youth in India.

Section 1: Demographic information

1. Age:
   [ ] 15–18 years
   [ ] 19–22 years
   [ ] 23–26 years
   [ ] 27–30 years
2. Gender:
   [ ] Male
   [ ] Female
   [ ] Other
3. Educational qualification:
   [ ] Below 10th grade
   [ ] 10th grade
   [ ] 12th grade
   [ ] Diploma
   [ ] Bachelor's degree
   [ ] Master's degree
   [ ] Vocational training
4. Occupation:
   [ ] Student
   [ ] Farmer
   [ ] Employed in a business or company
   [ ] Self-employed/entrepreneur
   [ ] Unemployed
   [ ] Other (please specify): ____________

Section 2: Education and skill development

5. What is the highest level of education available in your village/town?
   [ ] Primary school
   [ ] Secondary school
   [ ] Higher secondary school
   [ ] College
   [ ] None
6. Are there any vocational training centers available in your area?
   [ ] Yes
   [ ] No
   [ ] Don't know

7. Have you participated in any skill development programs?
   [ ] Yes
   [ ] No
   [ ] If yes, please specify the type of program: ___________
8. How would you rate the quality of education available in your area?
   [ ] Very poor
   [ ] Poor
   [ ] Average
   [ ] Good
   [ ] Very good

Section 3: Economic opportunities

9. What type of employment opportunities are available in your area?
   [ ] Agriculture
   [ ] Small industries
   [ ] Service sector
   [ ] Government jobs
   [ ] None
10. Have you received any support for starting your own business?
   [ ] Yes
   [ ] No
   [ ] If yes, please describe the type of support: ___________
11. How satisfied are you with the employment opportunities available in your area?
   [ ] Very unsatisfied
   [ ] Unsatisfied
   [ ] Neutral
   [ ] Satisfied
   [ ] Very satisfied
12. Would you consider migrating to an urban area for better opportunities?
   [ ] Yes
   [ ] No
   [ ] Maybe

Section 4: Infrastructure and technology

13. Does your village/town have reliable internet access?
   [ ] Yes
   [ ] No
14. How often do you use digital tools (e.g. smartphone, computer) for educational or work purposes?
   [ ] Daily
   [ ] Weekly
   [ ] Monthly
   [ ] Rarely
   [ ] Never
15. Is there good transportation connectivity in your area?
   [ ] Yes
   [ ] No
   [ ] Somewhat

Section 5: Social and cultural factors

16. Does your community actively support youth development programs?
[ ] Yes
[ ] No
[ ] Not sure

17. How influential are traditional gender roles in your area regarding education and employment?
[ ] Very influential
[ ] Influential
[ ] Neutral
[ ] Little influence
[ ] No influence

18. Do your family members encourage you to pursue higher education or employment?
[ ] Yes
[ ] No
[ ] Sometimes

19. How involved are you in community activities (e.g. youth clubs, local governance)?
[ ] Very involved
[ ] Involved
[ ] Neutral
[ ] Rarely involved
[ ] Not involved

Section 6: Government policies and programs

20. Are you aware of any government programs aimed at supporting rural youth?
[ ] Yes
[ ] No
[ ] If yes, please name them: ___________

21. Have you benefited from any government program?
[ ] Yes
[ ] No
[ ] If yes, please specify: ___________

22. How would you rate the effectiveness of government programs in your area?
[ ] Very ineffective
[ ] Ineffective
[ ] Neutral
[ ] Effective
[ ] Very effective

Section 7: Health and well-being

23. How accessible is healthcare in your area?
[ ] Very accessible
[ ] Accessible
[ ] Neutral
[ ] Inaccessible
[ ] Very inaccessible

24. Have you or someone you know been affected by substance abuse?
[ ] Yes
[ ] No
[ ] Prefer not to say

25. How often do you participate in physical or sports activities?
[ ] Daily
[ ] Weekly
[ ] Monthly
[ ] Rarely
[ ] Never

Section 8: Future aspirations and challenges

26. What are your primary aspirations for the future? (Check all that apply)
[ ] Higher education
[ ] Stable employment
[ ] Starting a business
[ ] Migration to urban area
[ ] Community leadership
[ ] Other (please specify): ____________

27. What are the biggest challenges you face in achieving your goals? (Check all that apply)
[ ] Lack of education opportunities
[ ] Unemployment
[ ] Financial constraints
[ ] Lack of family support
[ ] Poor infrastructure
[ ] Other (please specify): ____________

28. Do you feel optimistic about your future in your rural area?
[ ] Yes
[ ] No
[ ] Not sure

Section 9: Open-ended questions

29. What changes would you like to see in your village/town to improve youth engagement?

____

30. Do you have any additional comments or suggestions regarding youth engagement in rural areas?

____

Section 10: General engagement level

31. Overall, I am actively involved in education, employment, community, and personal development activities.
[ ] Strongly disagree
[ ] Disagree
[ ] Neither agree nor disagree
[ ] Agree
[ ] Strongly agree

12 Education and Skill Development in Rural Areas

Nigar K. Mujawar[1*], Rekha L. Nirmale[2], Vidyanand S. Khandagale[3] and Wilko Reichwein[4]

[1]Department of Pharmaceutics, Womens College of Pharmacy, Peth Vadgaon, Kolhapur, India; [2]College of Education, Peth Vadgaon, Kolhapur, India; [3]Department of Education, Shivaji University, Kolhapur, India; [4]Hamburg Institute for Vocational Education and Training (HIBB), Hamburg, Germany

12.1 Introduction

Access to education and skill development are critical steps in the sustainable advancement of rural territories but there are often serious barriers at this level, including poverty, a high unemployment rate, and the absence of resources (McGrath and Powell, 2016). Education enlightens and trains the mind, transforming lives, and skill development trains and enables people to get secure employment, thus helping them to stand independently and escape poverty (DeJaeghere, 2013). In rural areas, where there is limited access to quality education and better paid employment opportunities, improvement in these becomes critical for economic enabling and social integration. Through education and training, rural populations have an increased chance of eliminating socio-economic constraints and can be productive players in the larger economy.

Five primary objectives in education and skill development in rural areas were examined in the study this chapter is based on (Fig. 12.1) (Trivelli and Morel, 2021).

The first objective was to analyze the current state of education and skills training in rural regions (Atchoarena *et al.*, 2003). Second, it sought to identify rural populations' challenges in accessing quality education and skill development programs. (Kraak *et al.*, 2013). Third, the study examined successful models and best practices that have proven effective in enhancing education and skills training in these areas (Hargreaves, 2014). Fourth, it assessed the role played by government policies, non-governmental organizations (NGOs), and private sector initiatives in promoting these crucial elements (Holveck *et al.*, 2007). Lastly, the study evaluated the impact of education and skill development on poverty reduction, economic growth, and social empowerment within rural communities (King and Palmer, 2006).

Education in many areas remains challenging, especially regarding infrastructure, resources, and geographic constraints (Correa and Pavez, 2016). Years of neglect by many states have also seen school infrastructural facilities such as proper classrooms, toilets, electricity, etc., being a preserve of the few rural schools that are fortunate to have them (Moulton, 2001). Also, the scarcity of

*Corresponding author: nigarwcop2017@gmail.com

DOI: 10.1079/9781836992646.0012

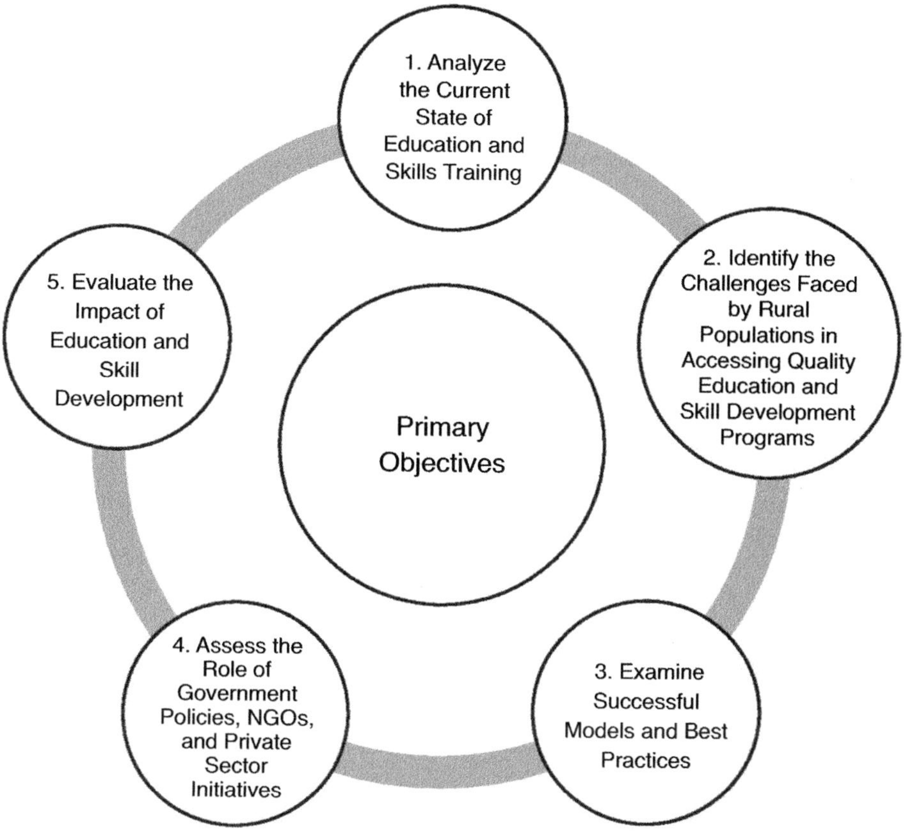

Fig. 12.1. Five primary objectives of education and skill development. Adapted from Trivelli and Morel (2021).

qualified teachers threatens the learning achievements in these areas even further. Besides, many children drop out of school, especially girls, due to poverty and social and cultural beliefs, meaning most children do not complete primary education. To overcome these barriers, there is a need to increase spending on school buildings, qualified teachers, and the community to enhance the belief in rural education (Shahidul and Zehadul Karim, 2015).

For development in the area of employment opportunities for the rural populace, skills development programs play very important roles in addressing the economic issues in affected regions, with particular emphasis on the agricultural value chain, small-scale industries, and service industries (Ra *et al.*, 2019). Vocational training programs offered by vocational and technical schools and colleges, namely carpentry, tailoring, and agro-processing, help the rural youth become more employable or to start up their own businesses. In addition, with technology being taken to rural areas, it has been noted that there is a need to conduct literacy training as this helps people access other markets and expand their outreach through computers. Realizing skills development for people working in rural settings can help spread sources of income and, at the same time, improve the economic base (Lubowa *et al.*, 2004). Nevertheless, significant hurdles still need to be overcome regarding the opportunities to deliver quality learning and skills in rural contexts. Lack of financial resources, human and material resource scarcities, and community constraints and perceptions, especially those of gender and culture, are significant challenges.

As for the future, it will be impossible to solve these problems without continuing investment in education and training facilities, without proper quality control in educational programs, and without stimulating the participation of communities to fight cultural prejudices (Dunbar, 2013). Examples from non-educational areas have shown that achieving a lasting positive shift is possible through the collective approach of governmental organizations, NGOs, and private sector representatives. More policy support, grassroots engagement with stakeholders, and policy innovative approaches are required to expand received learning achievements in the future, and subsequently to face the challenges in educating peasants and rural young people in the process of human resource development (Sanginga *et al.*, 2007).

12.2 Current State of Rural Education

As is well understood, the provision of rural education is poor compared with that of urban areas, with many challenges affecting education accessibility and quality. Limitations of infrastructure and facilities in the schools, including quality teachers and adequately built classrooms, with insufficient funding and support, compounded by the equally pervasive socio-economic constraints like poverty and gender disparity, make it hard for people in rural areas to get quality education (Fig. 12.2). The eradication of these problems is essential to enable all children to attain an equal education. The current state of rural education is briefly discussed here (Echazarra and Radinger, 2019).

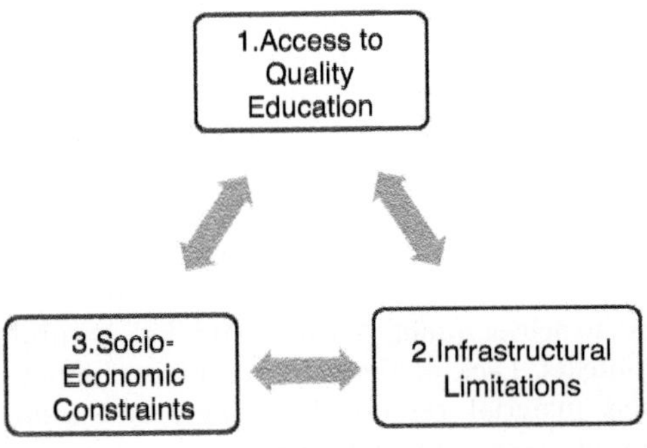

Fig. 12.2. Current state of rural education (Echazarra and Radinger, 2019).

12.2.1 Access to quality education

Education is one of the most essential and fundamental of human rights, but is still a dream to many, particularly in rural regions. That is due to the geographic remoteness of the regions and the shortage of schools and teachers who can provide quality education regularly. Transportation is a challenge since most students from rural areas are forced to travel long distances to school, making them attend school sporadically, especially the girls. The lack of secondary and higher education also limits the chances of rural young people gaining academic promotions, resulting in high dropout rates and low literacy (James *et al.*, 1999).

12.2.2 Infrastructural limitations

Lack of infrastructure is one of the most significant factor hindering rural schools. Some of the challenges faced in rural schools include inadequate school facilities like electricity, classrooms, toilets, and clean water. All these shortcomings are not just a problem in the learning environment but also in student retention. Computer and other physical teaching aids are lacking in most rural schools. Hence, there is a lack of modem teaching aids and technological-reinforced education. The lack of a proper road network and proper means of transport aggravates the problem by keeping schools physically unreachable for many students (Salemink *et al.*, 2017; Kaiser and Barstow, 2022).

12.2.3 Socio-economic constraints

Socio-economic barriers to providing education in rural areas is another concern. Due to poverty, child labor, and early responsibilities where children are forced to work to bring income for their families, they often cannot afford to go to school. In many rural areas, traditional culture hinders girls' education while, on the other hand, boys are forced to drop out of school to work. Other barriers are financial, such as the cost of school fees and other expenses (e.g. senior school fees, stationary, uniforms), among other necessities. Socio-economic barriers lower enrollment prospects and escalate dropout rates primarily in secondary education (Shabaya and Konadu-Agyemang, 2004; Sattar *et al.*, 2011).

12.3 Skill Development Initiatives

Skill development programs in rural areas form the basis for effectively enabling individuals to gain vocational skills that improve their living standards and contribute to overall economic productivity. Vocational training can result in more jobs while also developing business skills for economically independent societies and enhancing agricultural practices. Such initiatives are crucial in tackling unemployment and enhancing the rural economy for growth and development, as discussed in Table 12.1 (Collett and Gale, 2009).

All these programs aim to improve the living standards of the rural populace, empower them, fight poverty, and spur economic growth. Education, business skills, and knowledge of farming today are among the requirements that should be emphasized in the quest for enhanced rural development (Hartl, 2009).

12.4 Innovative Educational Models

Innovative educational models are transforming rural education by overcoming geographic and resource barriers. Initiatives such as mobile classrooms, community learning centers, and distance education programs provide flexible, accessible learning opportunities for students in remote areas. These models are essential to extend access to education for students in need and enhance learning achievement in underrepresented groups, as discussed in Table 12.2 (Simonson *et al.*, 2019).

12.5 Challenges Faced

Rural education faces numerous challenges that hinder its effectiveness and accessibility. Insufficient funding limits resources for infrastructure and learning materials, while the lack of qualified teachers hampers the quality of instruction. Additionally, cultural barriers often restrict educational participation, particularly for girls and marginalized groups, further exacerbating inequality in rural education systems (Monk, 2007).

12.5.1 Insufficient funding

Insufficient funding is a significant challenge affecting education and skill development in rural areas. Many countries, particularly in rural areas, have limited funds, which hampers their capacity to procure effective structures and educational items. Education in these regions entails challenges such as schools, infrastructure, and resources, with a lack of classrooms, toilets, libraries, and technology, among other basic needs (Shonhe, 2019).

Such a scarcity of facilities can result in what may be deemed a substandard learning situation in which the learners have very few resources regarding learning materials and aids. In addition, inadequate funding affects the teachers' remuneration, hence poor teacher attrition and high teacher turnover rates. Lack of sufficient funding means that rural schools cannot afford the routine maintenance and upgrades to their facilities that would improve the quality of education the children receive and deepens existing inequalities, creating cycles of poverty and stagnation (Du Plessis, 2014).

12.5.2 Lack of qualified teachers

The lack of qualified teachers is another pressing issue in rural education. Many rural areas face a shortage of trained educators due to low salaries, limited professional development opportunities, and geographic isolation (Monk, 2007). The idea of teaching in remote

Table 12.1. Skill development initiatives in rural areas (Collett and Gale, 2009; Teare, 2013; Rao, 2018; Olayiwola and Omitoyin, 2022; Begazo *et al*., 2023; Holleman, 2023; Adekunle, 2024).

Skill development initiatives	Description	Examples	Key benefits
Vocational training programs	Vocational training programs are necessary for promoting skill development and improving economic lives among rural people. They give hands-on experience in, for example, making a blackout, fixing pipes, and sewing; in other words, they empower employability. These programs are provided by centers for the community, NGOs, and even government projects, and they help satisfy the skill demands of the regions, enhance the employment rate, and encourage people to start businesses. They promote sustainable livelihoods and strengthen poverty-related economic capacities by making training demand-based. Since vocational skills are acquired that improve economic capabilities of individuals, such programs are helpful in the development of the community	1. Carpentry training centers in rural India, offering hands-on skills in construction and furniture making 2. Tailoring schools in Kenya, providing sewing and garment production training to women and young people	Enhances employability, empowering individuals with technical skills for local industries and self-employment opportunities (Adekunle, 2024)
Entrepreneurship skills	There is a need to integrate entrepreneurship skills to supplement government efforts to promote economic growth and self-employment by developing businesses among people in rural areas. Entrepreneurship training programs include business management, financial skills, and marketing skills. Business management training prepares people to carry out all business planning, organizing and strategic direction. Personal financial management ability is a financial competence majoring in budgeting, accounting, and financial planning. Marketing aptitude training demonstrates how to maintain product understanding, discover customer wants, and utilize social media and the web for marketing. Some practicalities are business incubation for early-stage development, business coaching for guidance, and business financing to establish new businesses. Such programs assist entrepreneurs in rural areas in addressing local demands for products and services, bringing about economic change, and generating employment, thus improving the quality of people's well-being and that of communities	1. Start-up India initiative, offering mentorship and financial support to rural entrepreneurs 2. She Trades initiative by ITC Ltd, helping rural women entrepreneurs gain market access and business skills	Promotes self-employment, encourages the creation of small businesses, and fosters financial independence in rural communities (Rao, 2018; Olayiwola and Omitoyin, 2022)

Continued

Table 12.1. Continued

Skill development initiatives	Description	Examples	Key benefits
Agricultural techniques	Various strategies that enable productivity and efficiency of farming practices in agriculture are also helpful in enhancing farming in rural regions. Precision agriculture uses GPS and controlled drip irrigation to increase crop yields. Moreover, conventions related to eco-friendly farming practices that do not require artificial chemical feeders include zero-budget natural farming (ZBNF). This helps to improve crop quality, cut costs, and sustain crop production under different environmental conditions. Rural farmers can contribute to food security, economic stability, and sustainable development of their villages and countries by applying innovative and enhanced farming methods	1. Precision Farming in Africa, using drip irrigation and GPS technology to increase crop yields 2. ZBNF in India, teaching sustainable farming practices without chemical fertilizers	Improves agricultural productivity and ensures sustainability, leading to better economic outcomes for farmers (Begazo *et al.*, 2023; Holleman, 2023)

areas is unattractive due to poor working conditions, which makes it challenging to attract skilled teachers to such areas. This means that students tend to be taught by inexperienced or low-qualified teachers, resulting in poor quality education and poor performance results. There is a lack of qualified teachers who could offer education to reduce the high dropout rates of learners and to improve rural education, with the situation being worse compared with urban schools. Solving this problem entails focusing more on recruitment strategies, incentivising qualified teachers to join rural schools, and improving teachers' professional development programs (Mitchell, 2010).

12.5.3 Cultural barriers

Cultural barriers significantly challenge education and skill development in rural communities. Traditional culture and established practices lock out some students, most especially girls and other vulnerable groups in society. More often than not, traditional gender roles dictate what women and girls can or cannot do, especially in rural settings. This excludes most women and girls from any form of education, and thus the cycle of inequality in the labor market continues (Colclough *et al.*, 2000). Such practices may also prioritize short-term economic returns over long-term educational returns, hence poor delivery and high school dropout rates. Also, cultural barriers and other biases stemming from prejudice may act as a bar for some sections of society to engage in skill development programs or to pursue education. For these barriers to be broken down, community members must be involved and awareness campaigns encouraged, creating good, acceptable, and relatively fair policies that advocate equal education for both genders within cultural acceptability (Malik, 2018).

12.6 Potential Solutions

Addressing rural education challenges requires potential solutions that promote sustainability and inclusivity. Community participation ensures local engagement and ownership of education programs, while capacity building strengthens the skills of teachers and community leaders. Leveraging technology provides innovative approaches to overcome geographic barriers and enhance educational access in remote areas (Gorain, 2022; Lakshmanan *et al.*, 2022).

Table 12.2. The role of innovative educational models in transforming rural education (Wang *et al.*, 2003; Simonson *et al.*, 2019; Ojobanikan *et al.*, 2024; Upadhyaya, 2024).

Educational model	Description	Key benefits
Mobile classrooms	Mobile classrooms are unique ideas developed to provide students in remote rural areas with an opportunity to study despite the challenges of transport and other infrastructure hurdles in such regions. Usually, mobile learning spaces are old vans or buses tailored with learning materials, computers, and other relevant learning tools. They give an improvised educational structure for children to learn basic skills. Using mobile classrooms is especially beneficial in the fight against educational segregation, as it provides a quality learning experience regardless of age and location. Such initiatives mainly target needy groups of students, such as girls and other disadvantaged groups, which acts as a catalyst in reducing the education gap in rural regions	Provides education access to students in isolated regions, addressing the issue of geographic barriers (Upadhyaya, 2024)
Community learning centers	Community learning centers are initiative-based institutions developed in local contexts to meet rural citizens' educational needs by offering learning options through continuing education for adults, skills development, and primary education for children and young people. These centers are usually located within the community; thus, they enable residents to engage in educational activities without necessarily having to cover long distances. They also enhance community relations and develop ways of promoting achievable learning. The community learning centers foster education by providing a place for community members to access learning aids and equipment, thereby encouraging personal development and the acquisition of skills, in addition to facilitating the development of communities	Promotes community engagement, offers flexible learning opportunities, and strengthens social cohesion (Ojobanikan *et al.*, 2024)

Continued

Table 12.2. Continued

Educational model	Description	Key benefits
Distance education programs	Distance education programs, in particular, employ technology like internet facilities, radio, or television to relay educational information to the student residents in remote or deficient socio-technological zones. This model makes it easier for learners to take courses and other instructional materials from the comfort of their homes, eliminating geographic constraints and limited physical infrastructure. Distance education is critical in rural areas as the development of regular schools may be barely possible. These programs facilitate students' self-paced learning and knowledge acquisition without daily traveling. Through this approach, access to education can be improved greatly while maintaining continuity of education and minimizing the gap between the rural and urban areas	Allows students to learn from remote locations, ensuring continued education despite physical distance from schools (Wang *et al*., 2003)

12.6.1 Community participation

Education and skill development programs need to involve community participation to enhance their success in rural areas. Education interventions are more likely to be culturally appropriate and sustainable when conducted at the community level. Most community members, such as parents, elders, and leaders, also support education initiatives and ensure that the resources are used as they should be. School development can be supported by resources donated and funding raised by the people in specific communities; role modeling can be provided through the development of schools and employment through education. Community involvement also leads to the enhancement of locals' ownership of the educational process, which in turn aids in school enrollment and dropout issues and helps reduce gender disparities. Instead of just giving a list of problems that hinder children from attending school, when the community is involved they help make decisions on issues ranging from poverty to cultural beliefs. In general, a high level of community involvement leads to active participation of community members in education, thereby enhancing education (Driskell, 2017; Nachtigal, 2019).

12.6.2 Capacity building

Capacity building mainly aims to enhance human capital in rural education and the teachers, students, and organizations involved. Regarding teachers, it entails training, new learning techniques, managerial skills, and content knowledge. Capacity-building programs empower teachers to extend their outreach and overcome limitations of the environment. In its essence, capacity building for students transcends learning encapsulated in a classroom. It embraces vocational, life, and entrepreneurship skills, whereby students get to practice what they have learnt in classrooms. Capacity building in communities is valuable as local leaders and organizations are empowered to manage educational programs properly, thus developing an independent system of rural education (Nachtigal, 2019; Ahasu *et al*., 2024).

12.6.3 Leveraging technology

Leveraging technology has immense potential in rural education by addressing the challenges of geographical isolation and limited resources. Smartphones, tablets, online courses, and edutainment apps can provide trained teachers and books or other learning materials to rural children without access to such resources. Technology also helps teachers obtain professional development and instruction resources, thus improving their ability to provide quality lessons. Moreover, using digital classrooms powered by solar systems, low-cost tablets, and radio/television makes it easier to use modern education facilities in rural areas and other places with poor infrastructure. In essence, technology can help rural regions address the challenges of education delivery, provide innovative and convenient approaches to learning, and prepare students for future employment by providing technological skills (Hillier, 2020; Ghazali and Benbrahim, 2024).

12.7 Public–Private Partnerships

Public–private partnerships (PPPs) are vital in improving education and skill development in rural areas by combining resources and expertise from various sectors. The role of governments is to provide policy support and funding, while the contribution of NGOs brings grassroots experience and tailored interventions. The involvement of educational institutions ensures quality standards and innovative practices that can be adapted to local contexts (Patrinos *et al.*, 2009).

12.7.1 Role of governments

The role of governments in PPPs for education and skill development in rural areas is pivotal. State and local governments are mandated to control policies by funding and offering infrastructural support and regulation of educational undertakings to conform to national priorities. In this regard, governments can provide subsidies, tax credits, or grants to facilitate the involvement of the private sector in financing education in rural areas (Bjärstig and Sandström, 2017). They also determine the fairness, affordability, and sustainability of PPPs in providing quality education, setting thresholds, and overseeing performance. Through central authorities, official structures, and ministries of education, national programs are developed to upgrade the training of rural teachers and, thus, enhance the implementation of educational services in remote regions. Since the government oversees different stakeholder bodies, including corporations and NGOs, a well-coordinated effort is required throughout the country to address some of the challenges facing education in rural areas (Robertson *et al.*, 2012; Ron-Balsera and Marphatia, 2012).

12.7.2 Contribution of NGOs

The contribution of NGOs is critical in enhancing education and skill development in rural areas through PPPs. NGOs operate closely with local people to ensure policies are implemented to suit their needs. They are an essential human resource pool as they have on-ground experience, understanding of cultural differences, and knowledge of the context specificity of various programs in terms of sensitization and awareness. NGOs also protect such vulnerable persons as women and disadvantaged children helping them to be able to embrace education. Governments and other private organizations must cooperate with NGOs to offer teachers training and resources for children and to introduce vocational programs to improve rural communities. They sometimes work as middlemen in adapting educational reforms to local contexts, which is important as the support of the local community is needed to ensure their success (Plummer, 2013; Bano, 2019).

12.7.3 Involvement of educational institutions

The involvement of educational institutions in PPPs contributes significantly to enhancing education and skill development in rural areas.

In many countries, universities, colleges, and research institutions collaborate with government and private organizations to develop curricula, teaching/learning resources, and trainer training. Scholastic institutions offer specialization in teaching, research, and curriculum development and, therefore, help guarantee that educational intercessions are done professionally and effectively in compliance with modern learning trends. These institutions can also assist in developing certification schemes that afford the youth from rural areas appropriate credentials that are acceptable both domestically and internationally. Their participation in PPPs creates an environment for exchanging information between the urban and rural teaching fraternities and promotes the facilitation of teaching techniques and the development of educational tools. Also, universities and institutions of comparable repute can ensure the placement of rural students in industry internships, apprenticeships, and mentorship jobs that increase their employment rate (Dunbar, 2013; Paul, 2023).

12.8 Case Studies and Success Stories

The following five case studies highlight effective strategies for improving education and skill development in rural areas, emphasizing community engagement, inclusivity, adaptability, practical skill development, and innovative technology. Local ownership ensures sustainability, while inclusivity and flexibility address unique barriers faced by marginalized groups. Additionally, skill-based training and leveraging technology are essential for creating scalable, impactful educational models in remote rural settings (Table 12.3) (Orser *et al.*, 2019).

12.9 Future Directions

Future directions in rural education and skill development focus on adapting to evolving needs and leveraging new opportunities. Recommendations for policy and practice should prioritize sustainable, inclusive approaches while addressing current gaps. Emerging trends and opportunities, such as technological advances and innovative educational models, offer promising solutions for enhancing access and effectiveness in rural education (Ludlow, 1998).

12.9.1 Recommendations for policy and practice

Future recommendations for policy and practice in education and skill development in rural areas are about sustainability based on concerted inclusion and innovation. Policies need to be implemented to incorporate community-based education systems so that rural people can be part of decision making. This can be done by decentralizing authority and giving local education committees control over resources and school performance. Donors need to align projects to the local culture, especially on issues concerning female children, through such interventions as grants, safe transport, and friendly school hours. Furthermore, the vocational training programs should be oriented to respond to local economic opportunities so that skills imparted can be practically implemented in a rural setting. Governments need to involve themselves in developing partnerships with the private sector to increase the availability of educational services with a focus on digitizing schools and classrooms. They should advocate for student-centered continuous professional learning for rural teachers to enable the teachers to use new teaching tools such as information technology. Policies should be used that capture progress in education and, as a result, make the intervention progressive and elastic to the given needs (Leach, 2000; Oviawe, 2018).

12.9.2 Emerging trends and opportunities

Technology and the jobs available define the trends and opportunities for change in rural

Table 12.3. Case studies and success stories: effective strategies for improving education and skill development in rural areas (Krishna, 2005; Azcona *et al.*, 2008; Ali, 2016; Orser *et al.*, 2019; Swargiary and Roy, 2023; Semali, 2024).

Examples of effective interventions	Country/region	Description	Lessons learned
Pratham's Read India Program	India	Focuses on improving basic literacy and numeracy through volunteer-driven teaching methods	Local ownership is crucial; using community volunteers ensures sustainability (Swargiary and Roy, 2023)
BRAC's (Base Realignment and Closure) Education Program	Bangladesh	Provides non-formal education to out-of-school children in rural areas with community support and tailored curricula	Inclusivity in education, especially for marginalized groups, is essential for impact (Ali, 2016)
Partnership to Strengthen Innovation and Practice in Secondary Education	Sub-Saharan Africa	A multistakeholder initiative improving access to quality secondary education, especially for girls	Flexibility in program design helps address cultural barriers and diverse community needs (Semali, 2024)
Barefoot College Solar Engineering Initiative	India (Rajasthan)	Trains rural women as solar engineers, focusing on practical, hands-on skills in technology	Skill-based training tied to local economic opportunities enhances community development (Krishna, 2005)
Bridge International Academies	Kenya, Uganda, Nigeria	Uses technology to provide low-cost, high-quality education in rural areas, focusing on scalable models	Leveraging technology can significantly increase access to quality education in underserved areas (Azcona *et al.*, 2008)

education. One of the trends that will be important for implementation in the future is the development of e-learning and mobile apps for education. These platforms are changing education in rural areas since students can access many amenities, from classrooms and virtual classes to vocational training. Also, solar schools and other digital centers in remote areas cut down infrastructure limitations and allow quality education even in off-grid localities. The other important trend is incorporating sustainable development goals in education. More than ever, rural education and development interventions seek to empower learners with knowledge and skills on climate change, sustainable agriculture, and renewable energy to prepare for the future. In addition, the continuity of learning and portfolio employment is also evident in rural schools as they realize the variability of work and the need to relate a skill to any economic opportunity. Finally, the emergence of ed-tech start-ups in rural segments provides new possibilities for innovation, ranging from cheap learning products and outcome-based education to AI-driven education services that level the playing field for students in rural and urban areas (Blum *et al.*, 2020; Padmaja and Mukul, 2021).

12.10 Conclusion

Education and skill development in rural areas are essential for sustainable development, and in reducing the socio-economic gap between rural and urban citizens. From this chapter, it can be concluded that material and financial support for education is still a concern due to infrastructural challenges and socio-economic constraints to quality education. However, it also highlights the importance of skill development programs, especially in different vocational training, entrepreneurship, and farming practices that benefit the rural populace. For instance, mobile classrooms, community learning centers, distance education programs, and many others hold much potential for solving the geographic and infrastructural barriers that affect rural persons. However, problems like limited funding, scarcity of experienced teachers, and cultural issues do not allow development to progress at the desired pace. Initiatives such as participation of communities, development of capacity, and use of technologies are essential to establish practical solutions and long-term educational environments in the villages. Several collaborative efforts between governments, NGOs, and educational sectors can facilitate resources and expertise sharing to improve educational processes. The following lesson can be learned from the case studies and success stories: multistakeholder partnerships and locally owned education interventions are multifactorial. Future policy and practice should be based on further developing an inclusive and sustainable education model. New opportunities such as e-learning, sustainable development education, and ed-tech solutions are promising in the future. This will ensure that the challenges discussed can be tackled comprehensively to provide rural people with all the requisite education and skills for survival in today's complex world.

References

Adekunle, T. A. (2024) Challenges associated with entrepreneurship, vocational education and self-employment in Nigeria. *Iconic Research and Engineering Journals* 7(9), 91–105.

Ahasu, E.K., Kwao, M.P., Anim, C.L. and Zang, P.L. (2024) Role of community resources in education planning and delivery in Ghana. *Acta Educationis Generalis* 14(1), 62–76. DOI: 10.2478/atd-2024-0004.

Ali, I. (2016) BRAC roles in non-formal education: A study on BRAC education for ethnic children program in Bangladesh. PhD diss, BRAC University.

Atchoarena, D., Wallace, I., Green, K. and Gomes, C.A. (2003) Strategies and institutions for promoting skills for rural development. *Education for Rural Development: Towards New Policy Responses* 239.

Azcona, G., Chute, R., Dib, F., Dookhony, L., Klein, H. *et al.* (2008) Harvesting the future: The case for tertiary education in Sub-Saharan Africa. *The Maxwell School of Syracuse University* 8.

Bano, M. (2019) Partnerships and the good-governance agenda: Improving service delivery through state–NGO collaborations. *VOLUNTAS: International Journal of Voluntary and Nonprofit Organizations* 30(6), 1270–1283. DOI: 10.1007/s11266-017-9937-y.

Begazo, T., Blimpo, M. and Dutz, M. (2023) *Digital Africa: Technological Transformation for Jobs*. World Bank Publications.

Bjärstig, T. and Sandström, C. (2017) Public-private partnerships in a Swedish rural context-A policy tool for the authorities to achieve sustainable rural development? *Journal of Rural Studies* 49, 58–68.

Blum, M.L., Cofini, F. and Sulaiman, V.R. (2020) MODULE 7: ICTs in rural advisory systems and services. *Agricultural Extension in Transition Worldwide: Policies and Strategies for Reform* 115.

Colclough, C., Rose, P. and Tembon, M. (2000) Gender inequalities in primary schooling: The roles of poverty and adverse cultural practice. *International Journal of Educational Development* 20(1), 5–27. DOI: 10.1016/S0738-0593(99)00046-2.

Collett, K. and Gale, C. (2009) Training for rural development: Agricultural and enterprise skills for women smallholders. *City and Guilds Centre for Skills Development* 24–30.

Correa, T. and Pavez, I. (2016) Digital inclusion in rural areas: A qualitative exploration of challenges faced by people from isolated communities. *Journal of Computer-Mediated Communication* 21(3), 247–263. DOI: 10.1111/jcc4.12154.

DeJaeghere, J. (2013) Education, skills and citizenship: An emergent model for entrepreneurship in Tanzania. *Comparative Education* 49(4), 503–519. DOI: 10.1080/03050068.2013.798514.

Driskell, D. (2017) *Creating Better Cities with Children and Youth: A Manual for Participation*. Routledge.

Du Plessis, P. (2014) Problems and complexities in rural schools: Challenges of education and social development. *Mediterranean Journal of Social Sciences* 5(20), 1109–1117. DOI: 10.5901/mjss.2014.v5n20p1109.

Dunbar, M. (2013) Engaging the private sector in skills development. In: *Health & Education Advice and Resource Team*. Oxford Policy Management, Oxford, England.

Echazarra, A. and Radinger, T. (2019) Learning in rural schools: Insights from PISA, TALIS and the literature. DOI: 10.1787/19939019.

Ghazali, A.E.L. and Benbrahim, L. (2024) Integration of E-learning platforms in moroccan higher education: Assessing the technological leap and addressing the digital divide among urban and rural students. *Research and Advances in Education* 3(5), 12–22.

Gorain, B.K. (2022) Leveraging physical, digital and knowledge connectivity for smart villages. In: *Smart Villages: Bridging the Global Urban-Rural Divide*. pp. 153–175. DOI: 10.1007/978-3-030-68458-7_13.

Hargreaves, A. (2014) *Handbook of Professional Development in Education: Successful Models and Practices, PreK-12*. Guilford Publications.

Hartl, M. (2009) Technical and vocational education and training (TVET) and skills development for poverty reduction–do rural women benefit. Available at: (accessed 4 October 2011).

Hillier, M. (2020) Bridging the digital divide with off-line e-learning. In: *Expanding Horizons in Open and Distance Learning*. Routledge, pp. 110–121.

Holleman, J. (2023) Cultivating the seeds of change: How intersectional identities are embedded in the spread of Zero Budget Natural Farming in Andhra Pradesh, India. Master's thesis.

Holveck, J.C., Ehrenberg, J.P., Ault, S.K., Rojas, R., Vasquez, J. *et al.* (2007) Prevention, control, and elimination of neglected diseases in the Americas: Pathways to integrated, inter-programmatic, inter-sectoral action for health and development. *BMC Public Health* 7, 1–21. DOI: 10.1186/1471-2458-7-6.

James, R., Wyn, J., Baldwin, G., Hepworth, G., McInnis, C. *et al.* (1999) *Rural and Isolated School Students and Their Higher Education Choices: A Re-Examination of Student Location, Socioeconomic Background, and Educational Advantage and Disadvantage. Commissioned Report*. Available at: http://www.detya.gov.au/nbeet/publications/# hec

Kaiser, N. and Barstow, C.K. (2022) Rural transportation infrastructure in low-and middle-income countries: A review of impacts, implications, and interventions. *Sustainability* 14(4), 2149. DOI: 10.3390/su14042149.

King, K. and Palmer, R. (2006) Skills development and poverty reduction: The state of the art. *Postbasic Education and Training Working Paper* 7.

Kraak, A., Jewison, R., Pillay, P., Chidi, M., Bhagwan, N. *et al.* (2013) Review of the current skills development system and recommendations towards the best model for delivering skills in the country. *Skills System Review Technical Task of the Human Resource Development Council of South Africa* 24.

Krishna, V.V. (2005) Education, training and skills formation for decent work in the informal sector: Case studies from Northern India. In: *Technical and Vocational Education and Training: Issues, Concerns and Prospects*. Springer-Verlag, pp. 183–214.

Lakshmanan, V.I., Murty, V.K., Chandrashekhar, V.S. and Singh, M. (2022) The role of skills development in smart villages. In: *Smart Villages: Bridging the Global Urban-Rural Divide*. pp. 185–216. DOI: 10.1007/978-3-030-68458-7_15.

Leach, F. (2000) Gender implications of development agency policies on education and training. *International Journal of Educational Development* 20(4), 333–347. DOI: 10.1016/S0738-0593(99)00082-6.

Lubowa, M., Kahara-Kawuki, A. and Ssonko, S.L. (2004) COMMUNITY RESILIENCE AND DIALOGUE (CRD) PROJECT.

Ludlow, B.L. (1998) Preparing special education personnel for rural schools: Current practices and future directions. *Journal of Research in Rural Education* 14, 57–75.

Malik, R.S. (2018) Educational challenges in 21st century and sustainable development. *Journal of Sustainable Development Education and Research* 2(1), 9–20. DOI: 10.17509/jsder.v2i1.12266.

McGrath, S. and Powell, L. (2016) Skills for sustainable development: Transforming vocational education and training beyond 2015. *International Journal of Educational Development* 50, 12–19. DOI: 10.1016/j.ijedudev.2016.05.006.

Mitchell, K. (2010) Systemic inequities in the policy and practice of educating secondary bilingual learners and their teachers: A critical race theory analysis. *Boston College). ProQuest Dissertations and Theses* 543.

Monk, D.H. (2007) Recruiting and retaining high-quality teachers in rural areas. *The Future of Children* 155–174.

Moulton, J. (2001) Improving education in rural areas: Guidance for rural development specialists. *For Charles Maguire: The World Bank*.

Nachtigal, P.M. (ed.) (2019) In: *Rural Education: In Search of a Better Way*. Routledge.

Ojobanikan, E.I., Mandah, N.C. and Foyewa, A.O. (2024) Assessment of funding approach to adult learning and education in rural communities. *Journal of Digital Learning and Education* 4(1), 61–72.

Olayiwola, S. and Omitoyin, G.O. (2022) Democray and quality of life in Nigeria. *Central Bank of Nigeria Journal of Applied Statistics* 46(3), 50–60.

Orser, B., Riding, A. and Li, Y. (2019) Technology adoption and gender-inclusive entrepreneurship education and training. *International Journal of Gender and Entrepreneurship* 11(3), 273–298. DOI: 10.1108/IJGE-02-2019-0026.

Oviawe, J.I. (2018) Revamping technical vocational education and training through public-private partnerships for skill development. *Makerere Journal of Higher Education* 10(1), 73–91. DOI: 10.4314/majohe.v10i1.5.

Padmaja, V. and Mukul, K. (2021) Upskilling and reskilling in the digital age: The way forward for higher educational institutions. In: *Transforming Higher Education Through Digitalization*. CRC Press, pp. 253–275.

Patrinos, H.A., Osorio, F.B. and Guáqueta, J. (2009) *The Role and Impact of Public-Private Partnerships in Education*. World Bank Publications.

Paul, P. (2023) Influence of Public Private Partnership Framework on Improving the Quality of Education.

Plummer, J. (2013) *Focusing Partnerships: A Sourcebook for Municipal Capacity Building in Public-Private Partnerships*. Routledge. DOI: 10.4324/9781849771368.

Ra, S., Ahmed, M. and Teng, P.S. (2019) Creating high-tech 'agropreneurs' through education and skills development. *International Journal of Training Research* 17(sup1), 41–53. DOI: 10.1080/14480220.2019.1629736.

Rao, M.S.R. (2018) A study on rural women empowerment in India: Through the eyes of entrepreneurship and skill development. *International Journal of Research in Social Sciences* 8(3), 500–515.

Robertson, S., Mundy, K. and Verger, A. (eds) (2012) *Public Private Partnerships in Education: New Actors and Modes of Governance in a Globalizing World*. Edward Elgar Publishing.

Ron-Balsera, M. and Marphatia, A.A. (2012) Do public private partnerships fulfil the right to education? An examination of the role of non-state actors in advancing equity, equality and justice. In: *Public Private Partnerships in Education*. Edward Elgar Publishing. DOI: 10.4337/9780857930699.00020.

Salemink, K., Strijker, D. and Bosworth, G. (2017) Rural development in the digital age: A systematic literature review on unequal ICT availability, adoption, and use in rural areas. *Journal of Rural Studies* 54, 360–371. DOI: 10.1016/j.jrurstud.2015.09.001.

Sanginga, P.C., Chitsike, C.A., Njuki, J., Kaaria, S. and Kanzikwera, R. (2007) Enhanced learning from multi-stakeholder partnerships: Lessons from the Eabling Rural Innovation in Africa programme. In: *Natural Resources Forum*, Vol. 31, No. 4. Blackwell Publishing Ltd, Oxford, UK, pp. 273–285. DOI: 10.1111/j.1477-8947.2007.00166.x.

Sattar, T., Yasin, G. and Afzal, S. (2011) Socio-political and economic barriers of development in education sector of Southern Punjab (Pakistan). *British Journal of Economics, Finance and Management Sciences* 3(1), 51–71.

Semali, L. (2024) Leveraging multistakeholder partnerships in technical, vocational education, and training. *Vocation, Technology & Education* 1(1). DOI: 10.54844/vte.2024.0558.

Shabaya, J. and Konadu-Agyemang, K. (2004) Unequal access, unequal participation: Some spatial and socio-economic dimensions of the gender gap in education in Africa with special reference to Ghana, Zimbabwe and Kenya. *Compare: A Journal of Comparative and International Education* 34(4), 395–424. DOI: 10.1080/0305792042000294805.

Shahidul, S.M. and Zehadul Karim, A.H.M. (2015) Factors contributing to school dropout among the girls: A review of literature. *European Journal of Research and Reflection in Educational Sciences* 3(2), 26–36.

Shonhe, L. (2019) A consolidation of challenges faced by school libraries in developing countries. *Library Philosophy and Practice* 2467, 1–12.

Simonson, M., Zvacek, S.M. and Smaldino, S. (2019) *Teaching and Learning at a Distance: Foundations of Distance Education*, 7th edn.

Swargiary, K. and Roy, K. (2023) *Empowering Minds: Adult Education in India*. Lambert Academic Publishing.

Teare, R. (2013) Personal viability–the journey to self-reliance and financial independence. In: *Lifelong Action Learning for Community Development*. Brill, pp. 99–132.

Trivelli, C. and Morel, J. (2021) Rural youth inclusion, empowerment, and participation. *The Journal of Development Studies* 57(4), 635–649. DOI: 10.1080/00220388.2020.1808194.

Upadhyaya, H. (2024) *Digital Education and Economic Transformation: Bridging the Gap*. MEADOW PUBLICATION.

Wang, R., Li, K., Martonosi, M. and Krishnamurthy, A. (2003) Distance learning technologies for basic education in disadvantaged areas. In: *Proceedings of the 8th Global Chinese Conference on Computers in Education*.

13 Rural Fintech and G20 Financial Outreach Programs: Advancing SDG through Digital Financial Inclusion

Prihana Vasishta[1*], Pooja Aggarwal[2] and Manisha Dhir[3]
[1]Centre of Management and Humanities (CMH), Punjab Engineering College, Chandigarh, India; [2]School of Business Management and Commerce, Jagat Guru Nanak Dev Punjab State Open University, Patiala, India; [3]Coventry Business School, Coventry University, Coventry, UK

13.1 Introduction

It is becoming more evident that digital finance is one of the tools that can be used to achieve the United Nations' sustainable development goals (SDGs), particularly in rural areas where people encounter challenges in accessing conventional ways of providing finance. Other tools include leveraging rural fintech with G20 financial outreach programs to provide innovative ways of connecting unbanked low-income populations with the requisite financial services. Rural areas experience a significant lack of access to financial services, perpetuating a vicious cycle of being unbanked and thereby stalling economic and social progress (Hasan *et al.*, 2023). More often than not, the low uptake of financial services in these regions results from similar factors, including the inadequacy of infrastructure, awareness, and availability, which all impede income accruement. Solutions are designed and focused in these areas by creating inclusive financial services to embrace the rural citizenry. The use of technology such as mobile banking, microloans, and other financial technologies has the potential to change the rural economy for the better. Non-banking financial institutions operating the fintech concepts can also include low-cost financial services encouraging improvements in the economy and reducing poverty while supporting the idea of inclusion in finance. One cannot underestimate the importance of fintech for all the unbanked and underserved populations of the world (Vasishta and Singla, 2024). The potential of such fintech models as mobile money and microloans is that they can help close the gap and facilitate access to financial services for all. These initiatives work best in less developed geographic locations where there are no physical banking services. With the use of information and communications technology, financial technologies can serve the unbanked communities. This, in turn, encourages social inclusion since people can afford to educate themselves, improve their health, and own businesses, which in turn helps eradicate poverty and economic growth.

Equally important are the financial outreach programs of the G20 countries that help with such reforms. These programs aim to

*Corresponding author: info.prihana@gmail.com

DOI: 10.1079/9781836992646.0013

achieve sustainable social and economic benefits by stimulating digital financial inclusion. Funds to improve the network and the new products of the G20 countries are aimed primarily at the promotion of financial inclusion. Such programs equip underserved populations with the necessary skills and information and foster the conditions within which fintech advances can be applied. The G20 outreach programs work toward establishing a more comprehensive and sustainable financial system by engaging governments, financial institutions, and technology providers. This effort principally aims at promoting SDGs through digital financial inclusion. Financial resource provision to underrepresented populations would go a long way in achieving SDGs as such empowerment would enable the populations in question to fully participate within the economy, improving their livelihoods and become more economically stable. Long-term benefits are immense with strong digital frameworks and inclusive financial systems – not just hastening the SDGs but, more importantly, bringing about long-term societal and economic benefits, making it truly equitable and sustainable for everyone. The integration of rural fintech with G20 financial outreach programs is a must for pushing forward digital financial inclusion and, by consequence, achieving the SDGs. More targeted initiatives and innovative solutions would likely change even more underserved populations' lives, thereby affecting economic growth, reducing poverty, and building an all-inclusive world economy.

13.1.1 Background

India has focused considerable efforts on financial inclusion, and despite nearly 80% of the people opening their own savings bank accounts, challenges still exist: financial illiteracy and inactivated accounts. Fintech companies have played the most critical role in pushing the development of financial inclusion. There were more than 1300 start-ups in 2014–2018, which propelled India to second position in the global fintech adoption index. There is a need for collaboration among fintechs, banks, and regulators to deepen financial inclusion further because achieving the SDGs of the United Nations is indispensable. For instance, the National Strategy for Financial Inclusion aims to improve access to financial services, especially for the underbanked. Some of the better models for financial inclusion in fintech include e-KYC, smart villages, and instant money transfers. Beyond this, fintech agility and innovation can accelerate economic development in rural contexts. Financial inclusion not only advances the economic mainstream but also reduces poverty and inequalities and thus is urgently required to achieve the sustainability vision. Under this stance, rural fintech and G20 financial outreach programs play a key role through digital financial inclusion in India. The intersection of the financial services sector and technology is changing the face of rural economies as it opens up villages to banking and financial services that otherwise would never have reached those areas. Fintech solutions bridge the gap by providing access through digital payments, credits, insurance, and investments to underserved populations across rural areas. The transformation is supported by G20 financial outreach programs, which emphasize inclusion in growth through global cooperation, sharing of knowledge, and investment in digital infrastructure. These initiatives are aligned with the SDGs, supporting economic growth, reducing inequalities, and increasing financial literacy in rural areas (Table 13.1). Rural fintech enables people and communities to have access to digital platforms that can not only expand economic opportunities but also empower the people for sustainable development and contribution toward global goals.

13.1.2 Importance of financial inclusion in national development

Around 1.4 billion people lack accounts in banks worldwide; there is still a huge gap in terms of financial access, but it is largely in developing economies. The World Bank and Global System for Mobile Communications Association (GSMA) have said that access to funding for informal micro-enterprises is the only way the gap in development can be closed. Micro, small, and medium enterprises (MSMEs) account for the highest proportion

Table 13.1. Addressing financial challenges in rural areas through digital financial inclusion and financial outreach programs.

Current financial challenges in rural areas	Primary solutions through financial outreach programs	Secondary solutions through financial outreach programs	Aspects of digital financial inclusion supported	Stakeholders involved	Potential impact on SDGs
Limited access to banking services	Expansion of banking infrastructure, mobile banking units, and agent banking networks	Improving financial literacy through community programs	Access to financial services, geographic coverage	Government, banks, fintech companies	SDG1: No poverty SDG2: Zero hunger SDG8: Decent work and economic growth
Lack of financial literacy and awareness	Financial education campaigns, workshops, and digital literacy programs	Partnership with local educational institutions	Financial literacy, digital literacy	Government, NGOs, educational institutions	SDG4: Quality education SDG10: Reduced inequality
High transaction costs for financial services	Introduction of low-cost digital transaction platforms and services (e.g. mobile wallets, unified payments interface)	Subsidies or incentives for adopting digital payment methods	Affordability, transaction cost reduction	Government, fintech companies, telecom providers	SDG1: No poverty SDG9: Industry, innovation, and infrastructure
Limited access to credit and financing options	Microfinance schemes, digital lending platforms, and credit scoring based on alternative data	Collaboration with fintech companies for innovative credit solutions	Access to credit, credit scoring and risk assessment	Banks, microfinance institutions, fintech companies	SDG1: No poverty SDG8: Decent work and economic growth
Inadequate insurance coverage	Promotion of digital insurance products tailored for rural populations	Integration of insurance services with agricultural activities	Access to insurance, risk management	Insurance companies, government, fintech companies	SDG3: Good health and well-being SDG13: Climate action
Poor connectivity and digital infrastructure	Investment in improving rural internet connectivity and digital infrastructure	Government–private sector partnerships for infrastructure development	Connectivity, infrastructure development	Government, telecom providers, private sector	SDG9: Industry, innovation, and infrastructure SDG10: Reduced inequality

Continued

Table 13.1. Continued

Current financial challenges in rural areas	Primary solutions through financial outreach programs	Secondary solutions through financial outreach programs	Aspects of digital financial inclusion supported	Stakeholders involved	Potential impact on SDGs
Challenges in savings and investment	Introduction of digital savings products and investment platforms	Incentives for rural savings and investment programs	Savings mobilization, investment opportunities	Banks, fintech companies, government	SDG1: No poverty SDG2: Zero hunger SDG8: Decent work and economic growth
Difficulty in accessing government benefits and subsidies	Implementation of direct benefit transfer systems via digital platforms	Ensuring robust grievance redressal mechanisms	Direct benefit transfer, transparency, efficiency	Government, banks, fintech companies	SDG1: No poverty SDG16: Peace, justice, and strong institutions

of businesses while sustaining more than 70% of employment globally. Fintech can provide low-cost digital services involving mobile money transfer and microloans, thus spurring economic growth and diminishing poverty. Such investment into network infrastructure and innovative financial products can bridge the gap and increase financial inclusion for emerging markets. Financial inclusion in India includes bringing available and affordable financial services to disadvantaged and low-income groups, particularly in rural and hilly areas, as a stride toward sustainable development and poverty reduction. It has been found to play a vital role in the building up of women's financial literacy and empowerment; targeting women only enhances the effectiveness of these schemes. The initiatives made by governments to get poorer families into financial mainstream services, such as insurance, have had a great impact. Improving access to financial literacy and services help women achieve financial independence and increase savings with higher returns on investment, thereby allowing financial power and freedom.

The National Strategy for Financial Inclusion (NSFI) is one such initiative that could bridge the gap in financial inclusion and further the cause of growth and development in a financial system that is becoming increasingly more inclusive. The Reserve Bank of India's FI Index for 2022 showed improvement from 56.4 in March 2022 to 60.1 in March 2023, indicating growth in access, usage, and quality of financial services. It captured better usage and quality enhancements. The Financial Inclusion Dashboard, Antardrishti, was released in June 2023 to monitor progress toward financial inclusion and identify exclusion areas. During the G20 Indian presidency, the focus was on climate change, digitalization, and the Global South. Moreover, an Indian presidency underlined global economic cooperation and rebalanced multilateralism with development finance, debt vulnerabilities, and financial inclusion. The Reserve Bank of India is focused on ensuring the availability of banking services to all sections of society and enhancing credit delivery, especially for agriculture and micro/small enterprises.

13.1.3 Scope and objectives of the study

The study this chapter is based on highlighted the challenges of digital financial inclusion, which need to be addressed to positively expedite equitable economic development and realization of the SDGs in the rural Indian contexts. This study also underlined aspects of improved digital literacy, increased investments in infrastructure, and strengthened collaborations between government and fintech sectors. Focusing on these areas, India can work towards achieving inclusive growth, whereby digital financial services benefit all segments of society, particularly those located in rural areas and who have been hitherto underserved. The research examined the current landscape of digital financial inclusion in rural India and assessed the impact of government outreach programs along with fintech innovations on achieving SDGs. The financial inclusion road in India is an important avenue to equal opportunities and national development. Despite this, the digital landscape of financial inclusion remains underdeveloped, especially in rural regions, hampering sustainable development. The study also examined the success of other large-scale initiatives such as Pradhan Mantri Jan Dhan Yojana (PMJDY) and Digital India, which have made rapid progress in expanding access to finance but are limited by lags in digital literacy in rural areas and inadequate infrastructure. A systematic literature review and a bibliometric analysis were conducted to reveal the key trends, challenges, and opportunities in accessing rural digital financial inclusion. For the analysis, keywords such as fintech, financial services, rural innovation, and digital financial inclusion were used. Central to this chapter is analysis on how the G20 outreach financial programs spearhead digital financial inclusion and prompt supportive sustainable developments in rural areas. Actionable recommendations to policy makers and stakeholders are made on how the existing programs can be re-engineered to develop effective strategies to bridge rural India's digital divide, moving towards sustainable development.

13.2 Literature Review

13.2.1 Definition and scope of digital financial inclusion

Digital financial inclusion refers to the use of information and communication technologies in the provision of financial services, especially to the vulnerable (or financially excluded) groups of population (CGAP (Consultative Group to Assist the Poor), 2015). The concept of digital financial inclusion is a convergence of digital inclusion, social inclusion, and financial inclusion (Aziz and Naima, 2021). It is considered to be an important mechanism for reducing poverty and attaining financial growth that is inclusive and sustainable (Pande *et al.*, 2012).

The expansion of mobile technologies laid the necessary infrastructure for cheaper and efficient execution of financial inclusion within developing countries (Gabor and Brooks, 2020). The literature reflects a strong association between the extent of mobile phone penetration and the level of financial inclusion within Indian states (Andrianaivo and Kpodar, 2012; Ghosh, 2016). The literature has identified various factors that increase digital financial inclusion: (i) environmental factors, such as government support, financial infrastructure, regulatory framework, mobile and internet penetration; (ii) technological factors such as service quality, performance, risk, and security concerns; (iii) organizational factors such as resources, networks, agent quality, and credibility; and (iv) individual factors such as awareness, empowerment, education, financial and digital literacy, and whether rural or urban (Bonhoure, 2024). On the other hand, several studies have identified the barriers to digital financial inclusion, including high costs of products and services, financial illiteracy, lack of geographic access (Schuetz and Venkatesh, 2020), poor internet connectivity, digital illiteracy (Grohmann and Menkhoff, 2017; Aziz and Naima, 2021), and an inadequate regulatory framework (Yawe and Prabhu, 2015), highlighting the underutilization of financial services. It can be reasonably argued that the technological advancements and establishment of suitable regulatory frameworks and economic policies by the governments of developing countries are instrumental in contributing to sustainable digital financial inclusion (Gallego-Losada *et al.*, 2023).

13.2.2 Overview of financial inclusion

The World Bank Group defines financial inclusion as an affordable access to key financial products and services, provided to individuals and businesses in a sustainable manner (World Bank Group, 2014). Financial inclusion refers to a process that ensures availability, accessibility, and usage of financial products and services by all sections of society (Sarma and Pais, 2008; Allen *et al.*, 2016). It is regarded as a cornerstone of financial development that catalyzes inclusive economic growth (Ghosh and Sahu, 2020) and enables achievement of seven of the 17 SDGs, including no poverty, zero hunger, good health and well-being, gender equality, decent work and economic growth, industry, innovation and infrastructure, and reduced inequalities (World Bank Group, 2014; CGAP (Consultative Group to Assist the Poor), 2016; Ma'ruf and Aryani, 2019; Khan, 2024). More specifically, access to financial services results in improving children's learning outcomes (Koomson and Afoakwah, 2023), creating jobs, and increasing the profitability of financial institutions (Ozturk and Ullah, 2022). At the country level, financial inclusion is measured by constructing an index reflecting use, access, and depth (Sha'ban *et al.*, 2020), whereas at the individual level, it is measured using the dimensions of account ownership, savings, and credit (Demirguc-Kunt *et al.*, 2013), along with the use of technology in financial transactions (Xu, 2020). Some studies, for instance Amoah *et al.* (2020), Ha and Nguyen (2024) and Sayed and Shusha (2019), employed customized surveys to examine financial access in different regions and/or for different sections of the population, rendering unique perspectives on financial inclusion.

Access to a transaction account lays the foundation for financial inclusion by allowing access to financial services including credit, investment, insurance and risk management, besides providing a mechanism to save, send, and receive money (World Bank Group, 2014). As a result of the World Bank Group's initiative of Universal Financial Access 2020, 80% of the adults in India, comprising 77% of females and 83% of males, owned a transaction account in 2017, as against only 35% in 2011 (World Bank Group, 2021). With increasing financial access, developing countries like India and China have enhanced the usage of transaction accounts via government reforms and private sector innovation. Beyond financial access and usage, there has also been a drastic reduction in extreme poverty in India (AFI (Alliance for Financial Inclusion), 2018).

13.2.3 Digital divide

Despite substantial progress, financial exclusion remains a challenge for many developing and emerging market countries (AFI (Alliance for Financial Inclusion), 2018). In 2017, India still had 190 million people without bank accounts, primarily comprised of those with no access to mobile phones, thereby highlighting the challenge of the digital divide. Nearly 35% of the respondents aged 15 years and above made or received a digital payment in 2021, as against 28.7% in 2017 and 22.2% in 2014, clearly highlighting an improved digital financial inclusion. However, the ratio for the use of digital payments differed for rural and urban respondents in India, with 40.2% for urban and 30.4% for rural respondents (World Bank Group, 2021). Various studies have highlighted a gap in access to financial services between rural and urban areas. For instance, younger people with higher education and financial literacy use fintech more frequently, reflecting high financial inclusion in urban vis-à-vis rural areas (Ha and Nguyen, 2024). Disparities in mobile phone ownership and internet access between different age groups, gender groups, and income groups also create a digital divide. Initiatives are required to address these challenges, which include affordable internet connectivity, digital literacy programs, availability of payment services agents, and women's empowerment (Zameer and Khan, 2024).

13.2.4 Government programs and initiatives

The importance of relevant economic policies, regulations, government initiatives, and supportive infrastructure for fintech in promoting financial inclusion is widely acknowledged in the literature (Ngo, 2019). Effective collaboration between government, regulators, and the private sector (including fintech firms) is required to improve fintech adoption amongst rural areas, which can ultimately lead to financial inclusion and attainment of SDGs (Danladi *et al.*, 2023).

13.2.4.1 Pradhan Mantri Jan Dhan Yojana

Through providing universal access to banking and financial services using a single bank account, coupled with a financial literacy drive, PMJDY plays a catalytic role in inclusive growth and poverty reduction in India (Nimbrayan *et al.*, 2018). The total number of beneficiaries under this scheme totaled 531.3 million as on August 14, 2024, against 12.5 million on January 31, 2015 (PMJDY, n.d.). In order to enhance adoption and usage, PMJDY accounts were linked to several government benefits. This initiative was believed to have wider coverage, careful monitoring, brand visibility, sustainability, focus on financial literacy, simplified KYC ("know your customer"), and grievance redressal over previously launched financial inclusion schemes (Chowhan and Pande, 2014). The access and usage of PMJDY accounts was further accelerated by the Aadhaar system of the Unique Identification Authority of India (UIDAI) and the evolution of digital payment systems, specifically the unified payments interface (UPI) (Zameer and Khan, 2024). Contrary to expectations, some recent literature found no significant impact of PMJDY on economic growth in India, necessitating a better understanding of the scheme's implementation and challenges (Bahal, 2024). Despite the significant impact of PMJDY on financial empowerment, some key financial services provided under the scheme (such as overdraft facilities and transfer of funds) remain underutilized due to a lack of technical infrastructure in rural areas (Divya *et al.*, 2024). Targeted interventions by policy makers are required to enhance the efficiency and sustainability of this scheme with the evolving needs of rural population (Kumar and Shobana, 2024).

13.2.4.2 Digital India

In 2015, India joined the "Better Than Cash" Alliance with an aim to digitize payments and support the government's vision of Digital India

(Better Than Cash Alliance, 2023). Some strategies like "India Stack," encompassing a collection of technology products and frameworks, builds an ecosystem for the digital financial system by providing enabling policy environment and underlying infrastructure (AFI (Alliance for Financial Inclusion), 2018).

13.2.5 Role of fintech innovations

Fintech represents the application of emerging technologies in the field of finance (Arner *et al.*, 2016). Fintech innovations have a great potential in accelerating financial inclusion, especially in developing economies (Telukdarie and Mungar, 2023). The United Nations and the G20 countries highlighted the significance of fintech in reducing financial exclusion and income inequality (Demir *et al.*, 2022). Fintech developments have significantly increased the access of finance to masses by decreasing the cost of financial intermediation and turnaround time (Ferrarini, 2017), however, it comes at the cost of changing the market structure of providers of financial services (Aleemi *et al.*, 2023). Fintech providers provide relatively cheaper financial services via digital applications and platforms. Besides internet banking and mobile banking services, "mobile money" – one of the fintech innovations – is considered to have had the greatest impact on the financially excluded population in developing countries (AFI (Alliance for Financial Inclusion), 2018; Gosavi, 2018; Senyo *et al.*, 2022). Some other studies have argued that by accelerating financial inclusion, fintech reduces income inequality more prominently in higher income countries (Demir *et al.*, 2022). Empirically, fintech and financial inclusion were found to have a significantly positive effect on sustainable development in the African region (Tidjani and Madouri, 2024).

Additionally, fintech start-ups play a crucial role in providing affordable and easily accessible financial products and services to unbanked populations due to their ability to create innovative business models, identify consumer demands, and fill market gaps (Umarov, 2021). More recently, fintech also encompassed the emergence of "techfins" in redefining the financial services and promoting financial inclusion (AFI (Alliance for Financial Inclusion), 2018). The 2018 Alliance for Financial Inclusion (AFI) special report on fintech for financial inclusion suggested four pillars to harness fintech for financial inclusion: building digital identity, developing digital payment infrastructure, linking provision of government services to account opening, and designing digital financial markets and systems.

13.3 Methodology

13.3.1 Research design

This study adopted a bibliometric analysis approach to exploring the intersection of rural fintech, G20 financial outreach programs, and digital financial inclusion in India. Systematic review and analysis of the existing body of literature was engaged upon to unearth the key trends, influential papers, and emerging topics that contribute towards the progress of SDGs through digital financial inclusion in the rural space. It based its research on the use of statistical and mathematical techniques in analyzing trends in academic literatures. In this bibliometric study, the following were some of the primary objectives:

1. Identify and categorize the key themes, topics, and trends within the literature on rural fintech and digital financial inclusion in India.
2. Track how the focus of research on rural fintech and financial inclusion has evolved from 2015 to 2024.

13.3.2 Data collection

The Scopus database was used to collect the data for this study. Scopus is an international scientific database that is considered one of the best available today, providing quality coverage in the entire range of peer-reviewed literature within a broad scope of bibliographic resources. Thus the search strategy focused on publications within the year range of 2015–2024 so as to achieve breadth in research studies on rural fintech and financial inclusion in India.

The keywords were chosen to represent the central themes of the research study, including themes that pertain to fintech, financial inclusion, and rural development in India. The authors designed the search strategy to be both broad and specific enough to catch all the relevant literature that was suited to the objectives of the study. The use of Boolean operators helped to ensure that search results included all aspects of the topic under research.

The study included articles from the period 2015–2024 because these reveal recent developments and trends, especially in the context of fintech and digital financial inclusion within the parameters of SDGs and India's financial outreach programs. The study only considered papers written in English so that the studies were comparable and a global audience of scholars can access the literature. The search was limited to peer-reviewed research papers such as articles, reviews, and conference papers. To accomplish this, the sources included reports, book chapters, and editorials to avoid non-academic sources that would lower the scholarly rigor of the analysis.

13.3.3 Data analysis

The research design included a detailed analytical framework that systematically guided the study through several key steps. Initially, data collection was carried out by extracting the dataset from the Scopus database using a carefully predefined search strategy. This dataset encompasses a range of bibliographic details, including titles, authors, abstracts, keywords, publication years, journals, and citation counts.

Following data collection, a thorough data cleaning and preprocessing stage was conducted to ensure the accuracy and reliability of the analysis. This involved removing duplicates, irrelevant entries, and incomplete records. Additionally, preprocessing steps were taken to standardize author names, keywords, and institutions, thereby ensuring consistency across the dataset and facilitating a more accurate analysis. Once the data was prepared, a descriptive analysis was performed to generate basic statistics that provided an overview of publication trends. This analysis included examining the number of publications per year, the distribution of papers across various journals, and the geographic distribution of research activities within the field.

The study then moved into more advanced bibliometric analysis, starting with a frequent keywords analysis. This step identified the most commonly occurring keywords to determine the main themes and focus areas within the literature. To complement this, a word cloud was generated to visually represent the prominence of these keywords, offering a quick and intuitive understanding of the central concepts in the research field. Further analysis included tree map visualization, which was used to analyze the hierarchical structure and relationships among the most frequent keywords. Additionally, the study examined word frequency over time to track the temporal evolution of key topics, identifying emerging trends and shifts in research focus. Finally, a network cluster analysis was conducted to explore the co-occurrence of keywords and collaborations among authors and institutions, thereby uncovering the structure and dynamics of research networks in the field.

13.4 Analysis and Results

13.4.1 Frequent keywords

The analysis of frequent keywords within the selected literature highlighted the primary themes and focus areas in the research on rural fintech and digital financial inclusion in India (Fig. 13.1). The keywords, which were extracted from the titles, abstracts, and keywords sections of the papers, reveal the concepts most commonly explored by researchers in this field. As depicted in the image, "financial inclusion" emerged as the most frequently occurring keyword, with a total of 77 occurrences. This indicates that the

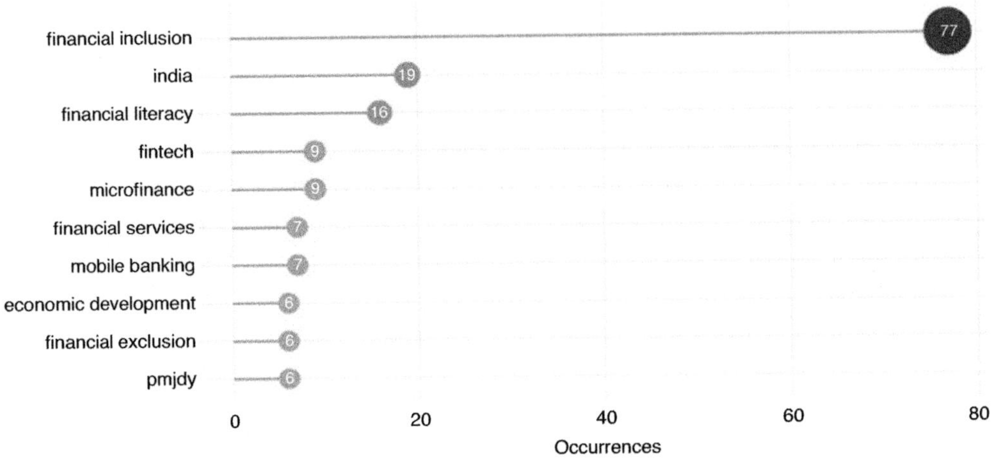

Fig. 13.1. Frequent keywords used by authors.

majority of the studies within this dataset are primarily focused on exploring various aspects of financial inclusion, particularly in rural settings. The prominence of this keyword underscores the importance placed on ensuring that financial services are accessible to underserved populations, which is a critical aspect of achieving broader economic development and financial stability. The second most frequent keyword was "India," with 19 occurrences, reflecting the geographic focus of the studies included in the analysis. This highlights that the majority of the research is specifically contextualized within the Indian setting, which is particularly relevant given India's ongoing efforts to enhance financial inclusion through various government programs and fintech innovations. "Financial literacy" was the third most common keyword, with 16 occurrences. This suggests that a significant portion of the research emphasizes the role of financial education in enabling individuals, especially in rural areas, to effectively use financial services. Enhancing financial literacy is seen as a crucial step towards empowering individuals to make informed financial decisions and fully benefit from the available financial products and services. Keywords such as "fintech" and "microfinance," each with nine occurrences, indicate a strong interest in the technological and micro-level financial solutions that are being implemented to drive financial inclusion in rural areas. The presence of "financial services" and "mobile banking," each with seven occurrences, further highlights the research focus on the delivery mechanisms of financial products, particularly through mobile and digital platforms. The keywords "economic development," "financial exclusion," and "PMJDY" each occurred six times, pointing to the broader economic goals linked to financial inclusion, as well as the specific challenges of financial exclusion and the impact of government initiatives like PMJDY in promoting financial access.

13.4.2 Word cloud

The word cloud generated from the bibliometric analysis offers a visual representation of the most prominent terms in the literature on rural fintech and digital financial inclusion in India. This visualization provides an intuitive understanding of the key concepts that dominate the research landscape.

Figure 13.2 depicts that the term "India" appears prominently at the center, signifying its central role in the studies under review. The large size of this word indicates that the research is heavily focused on the Indian context, reflecting the geographic scope of the studies included in the analysis. Other terms such as "financial

Fig. 13.2. Word cloud based on authors' keywords.

services," "financial inclusions," and "financial system" are also prominently displayed, underscoring their significance in the literature. The frequent occurrence of these terms highlights the emphasis on exploring how financial systems and services are being developed and deployed to enhance financial inclusion, particularly in rural areas. The focus on "financial inclusions" suggests a strong research interest in understanding how to bring underserved populations into the formal financial system, which is critical for achieving broader economic goals.

"Microfinance" and "banking" are also notable in the word cloud, indicating the importance of micro-level financial interventions and banking services in the pursuit of financial inclusion. These terms suggest that a significant portion of the literature is dedicated to examining the role of microfinance institutions and banks in providing financial access to rural populations. Additionally, terms like "economic development," "rural finance," and "poverty alleviation" are present, though in slightly smaller font sizes, highlighting the connection between financial inclusion and broader socio-economic objectives. The inclusion of terms such as "financial literacy" and "empowerment" reflects the literature's focus on the need to educate and empower individuals to effectively use financial services, which is a critical component of sustainable financial inclusion. The word cloud also features terms related to specific strategies and tools, such as "credit provision," "decision making," and "regression analysis," suggesting that the literature covers a range of methodologies and approaches to studying financial inclusion.

13.4.3 Tree map

In the tree map, the keyword "India" occupies the largest area, representing 12% of the total occurrences (Fig. 13.3). This emphasizes the predominant focus on the Indian context in the research under review. The large size of this block further highlights that India serves as the primary geographic focus in studies examining the intersection of fintech, financial inclusion, and rural development. Following "India," the keywords "financial services" and "financial inclusions" are also prominently featured, occupying 8 and 6% of the total occurrences, respectively. These terms underline the centrality of exploring how financial services are structured and delivered to promote financial inclusion, particularly in

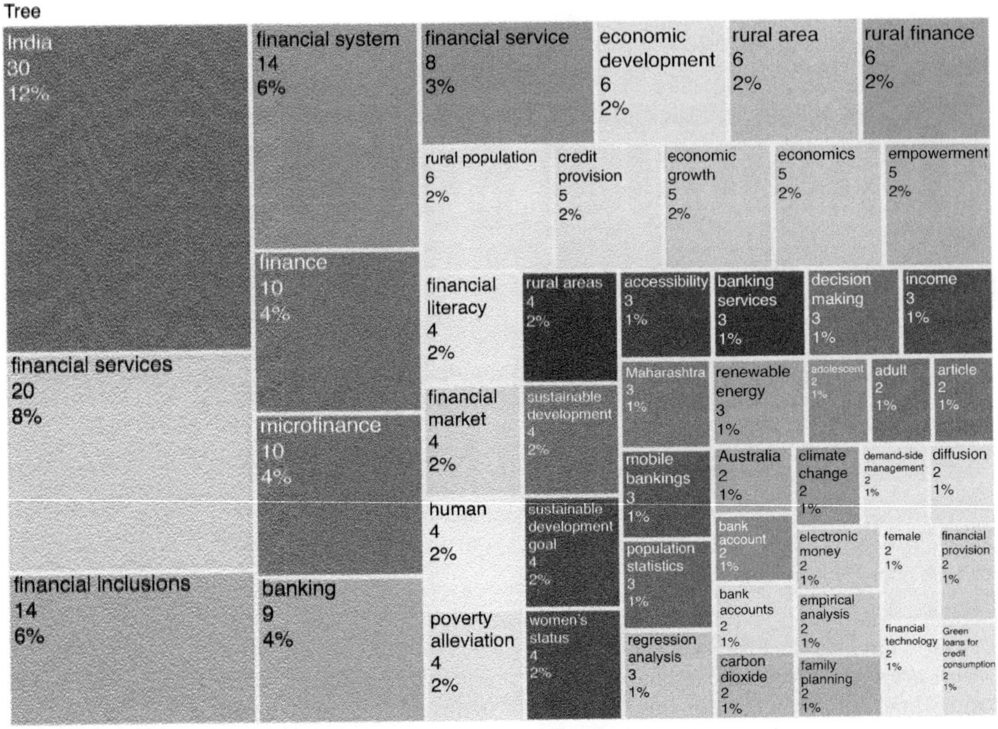

Fig. 13.3. Tree map based on authors' keywords.

underserved rural areas. The significant size of these blocks indicates a strong research emphasis on understanding and improving the accessibility and quality of financial services to foster greater inclusion.

Other notable keywords include "financial system" (6%), "microfinance" (4%), "finance" (4%), and "banking" (4%). These blocks illustrate the breadth of financial topics explored within the literature. The presence of "microfinance" and "banking" suggests that a substantial portion of the research is dedicated to exploring specific financial instruments and institutions that play a critical role in advancing financial inclusion in rural regions. Keywords like "economic development," "rural finance," and "rural area," each occupying 2% of the occurrences, indicate a focus on the broader socio-economic impact of financial inclusion. These blocks suggest that the literature also considers how financial inclusion initiatives contribute to the overall economic development of rural areas, aligning with broader development goals.

Smaller blocks such as "financial literacy," "accessibility," "empowerment," and "decision making," each occupying 1–2% of the occurrences, highlight the attention given to the human and educational aspects of financial inclusion. These keywords reflect the importance of financial literacy and empowerment as critical components in enabling individuals to effectively use financial services, thereby enhancing the overall impact of financial inclusion efforts. The tree map, with its visual hierarchy, effectively showcases the dominant themes and subthemes within the literature. It not only highlights the primary focus areas but also reveals the diversity of topics that researchers have explored in the context of rural fintech and financial inclusion in India.

13.4.4 Word frequency over time

The analysis of word frequency over time provides valuable insights into the evolving focus of research on rural fintech and digital financial inclusion in India from 2015 to 2024. Table 13.2 illustrates the occurrence of key terms across the years, highlighting the shifting priorities and emerging trends within the academic literature.

- *Financial inclusion.* The keyword "financial inclusion" shows a significant and steady increase in frequency over the years, starting from eight occurrences in 2015 and reaching 77 occurrences by 2024. This trend indicates a growing research interest in financial inclusion, particularly in the context of rural areas. The increasing focus on this term suggests that financial inclusion has become a central theme in discussions about economic development and social equity, especially in emerging economies like India.
- *India.* Similarly, the keyword "India" also shows a notable upward trend, rising from two occurrences in 2015 to 19 in 2024. This reflects the increasing concentration of research on the Indian context, likely driven by India's ambitious financial inclusion programs and the country's large rural population, which presents unique challenges and opportunities for fintech innovations.
- *Financial literacy.* "Financial literacy" begins to appear more prominently from 2019 onwards, with a noticeable increase from one occurrence in 2019 to 16 occurrences in 2024. This rising trend indicates an increasing recognition of the importance of financial education in empowering individuals to effectively use financial services, which is essential for achieving meaningful financial inclusion.
- *Fintech and microfinance.* The keywords "fintech" and "microfinance" also show an upward trajectory, although their occurrences are relatively lower compared with "financial inclusion" and "India." "Fintech" increases from no occurrences in the early years to nine in 2024, reflecting the growing role of technological innovations in financial services. "Microfinance," on the other hand, shows a more consistent presence, peaking at nine occurrences in 2024, indicating its sustained relevance as a tool for financial inclusion in rural areas.
- *Financial services and mobile banking.* "Financial services" and "mobile banking" demonstrate a gradual increase in their occurrences, reaching seven occurrences each by 2024. These trends suggest that while traditional financial services remain a significant area of research, there is also a growing interest in mobile banking as a crucial channel for delivering financial services to underserved populations.
- *Economic development and financial exclusion.* The keywords "economic development" and "financial exclusion" show moderate but consistent occurrences throughout the years, reflecting ongoing research interest in the broader socio-economic implications of financial inclusion. The steady presence of these terms suggests that researchers are continually exploring how financial inclusion initiatives impact overall economic growth and address issues of financial exclusion.
- *PMJDY and financial inclusion index.* The keyword "PMJDY" began to appear from 2021 onwards, reflecting the impact of this major Indian government initiative on the research landscape. Similarly, "financial inclusion index" shows a gradual increase in occurrences, indicating the growing importance of measuring and assessing the effectiveness of financial inclusion programs.
- *Digital payments and economic growth.* Finally, the keywords "digital payments" and "economic growth" appear more prominently in the later years, particularly from 2022 onwards. This suggests an emerging research focus on the role of digital payment systems in promoting economic growth, aligning with the broader trend of digital transformation in financial services.

Overall, the word frequency analysis over time reveals the dynamic nature of research on rural fintech and digital financial inclusion in India. The increasing occurrences of key terms reflect the evolving priorities in this field, with growing attention to financial inclusion, fintech

Table 13.2. Key word frequencies over time.

Year	Financial inclusion	India	Financial literacy	Fintech	Micro finance	Financial services	Mobile banking	Economic development	Financial exclusion	PMJDY	Banking	Financial inclusion index	Digital payments	Economic growth
2015	8	2	0	0	2	0	0	0	6	0	1	0	0	0
2016	16	5	0	0	2	0	4	1	6	0	1	2	0	0
2017	21	6	0	0	3	0	4	1	6	0	2	3	0	0
2018	26	7	0	1	3	0	4	1	6	0	2	3	0	1
2019	30	9	1	1	5	0	4	1	6	0	2	4	1	1
2020	36	10	4	3	5	1	6	2	6	0	3	4	1	1
2021	47	10	9	4	6	1	6	3	6	0	3	4	1	2
2022	57	13	10	4	8	3	6	4	6	3	4	4	1	3
2023	65	15	13	6	8	5	7	5	6	4	4	5	4	4
2024	77	19	16	9	9	7	7	6	6	6	5	5	4	4

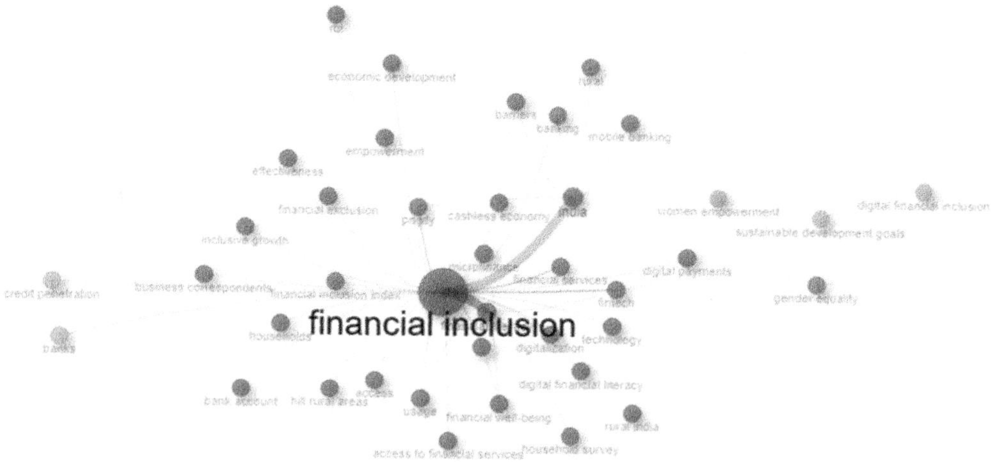

Fig. 13.4. Network cluster analysis based on authors' keywords.

innovations, financial literacy, and the specific challenges and opportunities within the Indian context.

13.4.5 Network cluster analysis

The network cluster analysis presented in Fig. 13.4 provides a visual representation of the relationships and connections between various keywords in the literature on rural fintech and digital financial inclusion in India. This analysis helps to identify the core themes and subthemes within the research field and reveals how different topics are interconnected. At the center of the network, the keyword "financial inclusion" is prominently displayed as the largest and most connected node. This indicates that financial inclusion is the central theme around which the majority of the research is organized. The numerous connections radiating from "financial inclusion" demonstrate its strong links to various other topics within the literature, highlighting its pivotal role in discussions about financial services, rural development, and economic empowerment. The surrounding clusters of keywords provide insight into the different aspects of financial inclusion that have been explored in the literature such as:

- *India and rural development.* The cluster including "India," "rural," and "mobile banking" shows a strong connection to "financial inclusion," indicating a significant focus on the Indian context and the role of mobile banking in enhancing financial access in rural areas. The connection to "barriers" suggests that researchers have also examined the challenges to financial inclusion in these regions.
- *Digital financial inclusion and gender.* Another cluster includes keywords like "digital financial inclusion," "sustainable development goals," "gender equality," and "women empowerment." This cluster reflects the growing research interest in the role of digital technologies in advancing financial inclusion and their impact on achieving broader social objectives, such as gender equality and empowerment.
- *Microfinance and financial services.* The connection between "financial inclusion," "microfinance," and "financial services" indicates that microfinance institutions and traditional financial services are key areas of focus within the literature. These connections suggest that much of the research examines how these financial tools can be leveraged to include underserved populations in the financial system.
- *Economic development and empowerment.* The cluster involving "economic development," "empowerment," and "inclusive growth" suggests that the literature often explores

the broader economic impacts of financial inclusion, emphasizing its role in fostering economic growth and empowering individuals, particularly in rural areas.

- *Technology and innovation.* Keywords like "fintech," "digital payments," and "technology" are connected to "financial inclusion," reflecting the significant interest in how technological innovations are transforming financial services and facilitating greater financial inclusion.
- *Regulatory and policy aspects.* Keywords such as "RBI" (Reserve Bank of India) and "PMJDY" indicate the importance of regulatory frameworks and government initiatives in the literature. These connections highlight how policy measures are critical in shaping the landscape of financial inclusion in India.

The network cluster analysis illustrates the complexity and interrelatedness of the research topics within the field of rural fintech and financial inclusion. The centrality of "financial inclusion" underscores its importance as the foundational theme, while the various clusters reveal the multifaceted nature of the research, encompassing technological, social, economic, and policy dimensions.

13.5 Implications

The findings from the bibliometric analysis of literature on rural fintech and digital financial inclusion in India have several important implications for policy makers, financial institutions, researchers, and development practitioners.

13.5.1 Implications for policy makers

The central focus on financial inclusion, particularly within the Indian context, underscores the critical role of government policies in shaping the financial landscape. Programs like the PMJDY and the Reserve Bank of India's initiatives have been instrumental in promoting financial access. The increasing emphasis on digital financial inclusion suggests that future policies should continue supporting technology integration in financial services, particularly in rural areas. Policy makers should consider the scalability and adaptability of successful models across different regions, and the regulatory frameworks should be continually updated to accommodate emerging fintech innovations.

13.5.2 Implications for financial institutions

The significant presence of keywords related to fintech, microfinance, and traditional banking in the literature indicates that financial institutions play a pivotal role in advancing financial inclusion. For banks and microfinance institutions, there is an opportunity to leverage digital technologies to expand their reach in underserved rural areas. The shift towards digital payments and mobile banking also suggests a need for these institutions to invest in digital infrastructure and enhance the digital literacy of their customers. Additionally, financial institutions should collaborate with fintech companies to develop innovative products that cater to the specific needs of rural populations, thereby fostering greater financial inclusion.

13.5.3 Implications for researchers

The analysis reveals a growing interest in financial literacy, gender equality, and economic empowerment, highlighting the multifaceted nature of financial inclusion research. For researchers, there is a clear need to explore these interconnected themes in greater depth. Future studies could focus on the long-term impact of financial literacy programs, the role of digital technologies in empowering women, and the broader socio-economic effects of financial inclusion. Additionally, the emergence of digital financial inclusion as a key theme suggests that further research is needed to understand the barriers to digital adoption in rural areas and to develop strategies to overcome these challenges.

13.5.4 Implications for practitioners and NGOs

For development practitioners and NGOs working in rural areas, the findings emphasize the importance of integrating financial inclusion into broader development strategies. Financial inclusion is about providing access to financial services and ensuring that these services contribute to sustainable economic development and social empowerment. Programs that focus on financial literacy, particularly digital financial literacy, are essential to ensure that rural populations can effectively utilize financial services. Furthermore, development efforts should aim to address the gender disparities highlighted in the literature, ensuring that financial inclusion initiatives also promote gender equality and women's empowerment.

13.5.5 Implications at the global level

While this study focuses on India, the themes identified have broader implications for other developing countries seeking to enhance financial inclusion. The strategies and policies that have been successful in India could serve as models for other nations facing similar challenges. However, it is important to adapt these models to local contexts and consider the unique socio-economic and cultural factors in different regions. International development organizations and policy makers in other countries can draw lessons from India's experience to design effective financial inclusion strategies that are tailored to their specific needs.

13.6 Future Research Direction

The range of areas this work brings attention to for further analyses is significant. For example, from Table 13.3 it can be clearly seen that much research is still required in formulating other alternative financial inclusion models that would address the new challenges redefining finance and microfinance. Comparative studies conducted on different regions of India or among other developing countries may also bring important insights on regional differences and best practices. Along with this, further research is needed on the integration of advanced technologies in financial services like blockchain and AI and their impact on their inclusion in finances. Furthermore, it encompasses enquiries related to broader socio-economic outcomes, which include gender equality, sustainable development, and poverty reduction. These directions, informed by observed trends and gaps in current literature, outline the roadmap for advancing the field of rural fintech and digital financial inclusion.

13.7 Conclusion

This chapter underscores the critical importance of digital financial inclusion in driving equitable economic development and achieving the SDGs in rural India. Indeed, while initiatives such as PMJDY and Digital India have brought significant positive developments, many challenges remain to be addressed, especially in bridge building between urban and rural areas. Despite much promise, issues such as a low level of digital literacy, deficits in infrastructure, and disconnection persist despite fintech innovations to increase access to finance. From the full potential of digital financial inclusion to be reached, common challenges need to be addressed through targeted policy interventions, enhanced collaboration between government and fintech, as well as more investments in rural digital infrastructure. The priority focus in these areas can be the first step towards achieving inclusive growth and bringing digital financial services to all segments of society, especially to rural populaces who are generally left out of such initiatives.

Table 13.3. Future research directions based on the thematic analysis of the keywords.

Theme	Description	Keywords/concepts	Observations	Future research directions
Central theme: financial inclusion	The core focus of the research, emphasizing access to financial services for underserved populations	Financial inclusion, financial services, financial system, financial exclusion	"Financial inclusion" is the most frequent keyword, central to discussions on economic empowerment and inclusive growth. The research consistently emphasizes the importance of integrating rural populations into the financial system	Explore innovative models for financial inclusion that address emerging challenges such as digital divides and technological barriers in rural areas. Assess the long-term impact of financial inclusion initiatives on poverty reduction and economic resilience
Geographic focus: India	Specific emphasis on the Indian context, exploring local initiatives and challenges in financial inclusion	India, rural, PMJDY (Pradhan Mantri Jan Dhan Yojana), RBI (Reserve Bank of India), rural development	The keyword "India" shows significant occurrences, reflecting the geographic concentration of the studies. Research often discusses government programs like PMJDY and the role of regulatory bodies like the RBI in promoting financial inclusion	Comparative studies on financial inclusion across different states in India to identify best practices and regional disparities. Examination of the scalability of successful Indian financial inclusion models in other developing countries
Technological innovation and digital finance	Exploration of how technology is used to promote financial inclusion, especially in rural areas	Fintech, digital payments, mobile banking, digital financial inclusion, technology	Research increasingly focuses on the role of fintech and digital platforms in delivering financial services to rural areas. Keywords like "mobile banking" and "digital payments" are linked to discussions on expanding financial access through technological innovations	Investigate the impact of emerging technologies (e.g. blockchain, AI) on enhancing financial inclusion in rural areas. Study the challenges and opportunities of digital literacy in the adoption of fintech solutions among rural populations

Continued

Table 13.3. Continued

Theme	Description	Keywords/concepts	Observations	Future research directions
Socio-economic impact and development	The impact of financial inclusion on broader socio-economic outcomes, including empowerment and growth	Economic development, economic growth, empowerment, gender equality, inclusive growth, sustainable development goals	Studies often explore how financial inclusion contributes to economic development, with a particular focus on empowering marginalized groups. Keywords like "economic development" and "empowerment" suggest that financial inclusion is seen as a tool for achieving broader social and economic objectives, including gender equality and sustainable development	Explore the relationship between financial inclusion and sustainable development, particularly in the context of climate change and environmental sustainability. Further research into the role of financial inclusion in reducing income inequality and promoting social justice
Microfinance and traditional banking	The role of microfinance institutions and traditional banking in advancing financial inclusion	Microfinance, banks, credit penetration, banking, financial services, business correspondents	The presence of "microfinance" and "banks" as keywords indicates ongoing research into how traditional financial institutions contribute to financial inclusion. Microfinance is frequently discussed as a means of providing access to finance in rural areas, complementing the role of traditional banks	Assess the effectiveness of microfinance models in the digital age and explore the integration of microfinance with digital platforms. Examine the evolving role of traditional banks in the context of increasing fintech penetration in rural areas

Continued

Table 13.3. Continued

Theme	Description	Keywords/concepts	Observations	Future research directions
Educational and literacy initiatives	The importance of financial literacy and education in enabling effective use of financial services	Financial literacy, financial education, empowerment, digital financial literacy	The increasing frequency of "financial literacy" suggests a growing recognition of the need to educate individuals, particularly in rural areas, to enhance their ability to use financial services effectively. This theme is critical to ensuring that financial inclusion efforts translate into real improvements in financial well-being	Develop and test innovative financial literacy programs tailored for rural populations with low digital literacy. Investigate the long-term impact of financial literacy initiatives on financial behaviors and economic outcomes
Regulatory and policy frameworks	The influence of government policies and regulations on the promotion of financial inclusion	RBI, PMJDY, financial inclusion index, policy, regulatory frameworks	Government policies, particularly those initiated by the RBI and through programs like PMJDY, are frequently highlighted as key drivers of financial inclusion. The focus on "financial inclusion index" indicates an interest in measuring the effectiveness of these initiatives	Examine the impact of recent regulatory changes on financial inclusion, particularly in the wake of the Covid-19 pandemic. Research on the effectiveness of financial inclusion indices as tools for policy making and their potential improvements

References

AFI (Alliance for Financial Inclusion) (2018) Fintech for financial inclusion: A framework for digital financial transformation. Special report. Available at: https://www.ada-microfinance.org/sites/default/files/2020-12/fintech-for-financial-inclusion-a-framework-for-digital-financialtransformation.pdf (accessed 24 May 2023).

Aleemi, A.R., Javaid, F. and Hafeez, S.S. (2023) Finclusion: The nexus of Fintech and financial inclusion against banks' market power. *Heliyon* 9(12), e22551. DOI: 10.1016/j.heliyon.2023.e22551.

Allen, F., Demirguc-Kunt, A., Klapper, L. and Peria, M.S.M. (2016) The foundations of financial inclusion: Understanding ownership and use of formal accounts. *Journal of Financial Intermediation* 27, 1–30. DOI: 10.1016/j.jfi.2015.12.003.

Amoah, A., Korle, K. and Asiama, R.K. (2020) Mobile money as a financial inclusion instrument: What are the determinants? *International Journal of Social Economics* 47(10), 1283–1297. DOI: 10.1108/IJSE-05-2020-0271.

Andrianaivo, M. and Kpodar, K. (2012) Mobile phones, financial inclusion, and growth. *Review of Economics and Institutions* 3(2), 30.

Arner, D.W., Barberis, J. and Buckley, R.P. (2016) The evolution of FinTech: A new post-crisis paradigm? *Georgetown Journal of International Law* 47(4), 1271.

Aziz, A. and Naima, U. (2021) Rethinking digital financial inclusion: Evidence from Bangladesh. *Technology in Society* 64, 101509.

Bahal, S. (2024) Economic growth and financial inclusion: Assessing the impact of the Pradhan Mantri Jan Dhan Yojana Initiative in india. Master's thesis, Georgetown University.

Better Than Cash Alliance (2023, March 29) Our work in india. Available at: https://www.betterthancash.org/news/exploring-the-work-of-the-alliance-in-india

Bonhoure, E. (2024). Indigenous knowledge and digital financial inclusion: A comparison between Europe and Sub-Saharan Africa. *Information Technology for Development*, 31(4), 1-29.

CGAP (Consultative Group to Assist the Poor) (2016) Achieving the sustainable development goals the role of financial inclusion. Available at: https://www.cgap.org/sites/default/files/Working-PaperAchieving-Sustainable-Development-Goals-Apr-2016_0.pdf, 2016

CGAP (Consultative Group to Assist the Poor) (2015) Digital financial inclusion: Implications for customers, regulators, supervisors, and standard-setting bodies. CGAP. Available at: http://www.cgap.org

Chowhan, S. S., & Pande, J. C. (2014). Pradhan Mantri Jan Dhan Yojana: A giant leap towards financial inclusion. *International Journal of Research in Management & Business Studies*, *1*(4), 19-22.

Danladi, S., Prasad, M.S.V., Modibbo, U.M., Ahmadi, S.A. and Ghasemi, P. (2023) Achieving sustainable development goals through financial inclusion: Collaborative approaches to fin-tech adoption in developing countries. *Sustainability* 15(17), 13039.

Demir, A., Pesqué-Cela, V., Altunbas, Y. and Murinde, V. (2022) Fintech, financial inclusion and income inequality: A quantile regression approach. *The European Journal of Finance* 28(1), 86–107.

Demirguc-Kunt, A., Klapper, L. and Singer, D. (2013) *Financial Inclusion and Legal Discrimination Against Women: Evidence from Developing Countries (No.6416)*. The World Bank.

Divya, U., Jogish, D., Mankar, M. M., & Meti, M. N. G. (2024). Impact of PMJDY scheme on the Financial empowerment of rural Poor–A study across Bangalore rural district in Karnataka, India. *Journal of Informatics Education and Research*, *4*(2).

Ferrarini, G. (2017) European economy banks. *Regulation, and the Real Sector*. Available at: https://european-economy.eu/wp-content/uploads/2018/01/EE_2.2017-2.pdf

Gabor, D., & Brooks, S. (2020). The digital revolution in financial inclusion: International development in the fintech era. In *Material cultures of financialisation* (pp. 69-82). Routledge.

Gallego-Losada, M.J., Montero-Navarro, A., García-Abajo, E. and Gallego-Losada, R. (2023) Digital financial inclusion. visualizing the academic literature. *Research in International Business and Finance* 64, 101862. DOI: 10.1016/j.ribaf.2022.101862.

Ghosh, S. (2016) Does mobile telephony spur growth? Evidence from Indian States. *Telecommunications Policy* 40(10–11), 1020–1031.

Ghosh, S. and Sahu, T.N. (2020) How financially inclusive the Asian's are-an empirical analysis, rajagiri manag. *Rajagiri Management Journal* 14(1), 49–69. Available at: https://www.emerald.com/insight/0972-9968.htm

Gosavi, A. (2018) Can mobile money help firms mitigate the problem of access to finance in Eastern Sub-Saharan Africa? *Journal of African Business* 19(3), 343–360.

Grohmann, A. and Menkhoff, L. (2017) Financial literacy promotes financial inclusion in both poor and rich countries. *DIW Economic Bulletin* 7(41), 399–407.

Ha, D. and Nguyen, K. (2024) Unlocking financial access in a developing country amidst COVID-19: The impacts of financial literacy and fintech. *Journal of the Asia Pacific Economy* 1–37.

Hasan, M., Noor, T., Gao, J., Usman, M. and Abedin, M.Z. (2023) Rural consumers' financial literacy and access to fintech services. *Journal of the Knowledge Economy* 14(2), 780–804.

Khan, M. (2024) Analyzing the effects of financial inclusion on poverty and income inequality in SAARC developing nations. *Journal of the Asia Pacific Economy* 29(1), 1–24. DOI: 10.1080/13547860.2024.2361988.

Koomson, I. and Afoakwah, C. (2023) Can financial inclusion improve children's learning outcomes and late school enrolment in a developing country. *Applied Economics* 55(3), 237–254. DOI: 10.1080/00036846.2022.2086683.

Kumar, J.S. and Shobana, D. (2024) A study on evaluation and implementation of Indian social security programs and initiatives in India. *International Journal of Research and Innovation in Social Science* 8(3), 794–807.

Ma'ruf, A. and Aryani, F. (2019) Financial inclusion and achievements of sustainable development goals (SDGs) in ASEAN. *Journal of Business and Economic Review* 4, 147–155. DOI: 10.35609/%20 jber.2019.4.4(1).

Ngo, A.L. (2019) Index of financial inclusion and the determinants: An investigation in Asia. *Asian Economic and Financial Review* 9(12), 1368–1382. DOI: 10.18488/journal.aefr.2019.912.1368.1382.

Nimbrayan, P.K., Tanwar, N. and Tripathi, R.K. (2018) Pradhan mantri jan dhan yojana (PMJDY): The biggest financial inclusion initiative in the world. *Economic Affairs* 63(2), 583–590.

Ozturk, I. and Ullah, S. (2022) Does digital financial inclusion matter for economic growth and environmental sustainability in OBRI economies? An empirical analysis. *Resources, Conservation and Recycling* 185, 106489. DOI: 10.1016/j.resconrec.2022.106489.

Pande, R., Cole, S., Sivasankaran, A., Bastian, G. and Durlacher, K. (2012) Does poor people's access to formal banking services raise their incomes?–A systematic review. In: *DFID Systematic Review EPPI-Centre, Social Science Research Unit, Institute of Education*. University of London, London.

PMJDY (Pradhan Mantri Jan Dhan Yojana) (n.d.) Pradhan Mantri Jan Dhan Yojana Progress Report. Available at: https://pmjdy.gov.in/account (accessed 22 August 2024).

Sarma, M. and Pais, J. (2008) Financial inclusion and development: A cross country analysis. In: *Annual Conference of the Human Development and Capability Association*, New Delhi.

Sayed, M.N. and Shusha, A. (2019) Determinants of financial inclusion in Egypt. *Asian Economic and Financial Review* 9(12), 1383–1404. DOI: 10.18488/journal.aefr.

Schuetz, S. and Venkatesh, V. (2020) Blockchain, adoption, and financial inclusion in India: Research opportunities. *International Journal of Information Management* 52, 101936. DOI: 10.1016/j.ijinfomgt.2019.04.009.

Senyo, P.K., Karanasios, S., Gozman, D. and Baba, M. (2022) FinTech ecosystem practices shaping financial inclusion: The case of mobile money in Ghana. *European Journal of Information Systems* 31(1), 112–127. DOI: 10.1080/0960085X.2021.1978342.

Sha'ban, M., Girardone, C. and Sarkisyan, A. (2020) Cross-country variation in financial inclusion: A global perspective. *The European Journal of Finance* 26(4–5), 319–340. DOI: 10.1080/1351847X.2019.1686709.

Telukdarie, A. and Mungar, A. (2023) The impact of digital financial technology on accelerating financial inclusion in developing economies. *Procedia Computer Science* 217, 670–678. DOI: 10.1016/j.procs.2022.12.263.

Tidjani, C. and Madouri, A. (2024) Fintech, financial inclusion, and sustainable development in the African region. *Frontiers in Applied Mathematics and Statistics* 10, 1276218. Available at: https://www.frontiersin.org/journals/applied-mathematics-and-statistics/articles/10.3389/fams.2024.1276218/full

Umarov, HS. (2021) The role of fintech startups in contemporary society. *Finance Credit* 27, 1911–1930. DOI: 10.24891/fc.27.8.1911.

Vasishta, P. and Singla, A. (2024) Emerging trends in FinTech and financial inclusion: A review and bibliometric analysis. *African Journal of Science, Technology, Innovation and Development* 16(46), 1–14.

World Bank Group (2014) *Financial Inclusion*. Available at: https://www.worldbank.org/en/topic/financialinclusion/overview (accessed 21 August 2024).

World Bank Group (2021) *Universal Financial Access 2020. Lessons for the future*. Available at: https://ufa.worldbank.org/en/country-progress/india#9 (accessed 21 August 2024).
Xu, X. (2020) Trust and financial inclusion: A cross-country study. *Finance Research Letters* 35, 101310. DOI: 10.1016/j.frl.2019.101310.
Yawe, B. and Prabhu, J. (2015) Innovation and financial inclusion: A review of the literature. *Journal of Payments Strategy & Systems* 9(3), 215–228.
Zameer, S. and Khan, S.U. (2024) With special reference Pradhan Mantri Jan Dhan Yojana (Pmjdy). *Migration Letters* 21(S3), 719–734.

14 ElasticRun: Redefining Rural Business

Swati Bhatia[1]*, Sonam Pandey[1] and Katherine Boxall[2]
[1]Asian Business School, Noida, India; [2]Istituto Marangoni, London, UK

14.1 Introduction

The digital transformation in retail has changed the retail landscape worldwide, but in rural pockets of developing nations it still does not mean much. ElasticRun is a disruptive logistics platform providing India with a crowdsourced model connecting urban suppliers directly with rural consumers. The company has been in operation since 2016 and uses local transportation networks to create a decentralized and efficient delivery system that reaches even remote pockets across India. The platform is a critical component in the process of bridging the gap between urban markets and rural consumers, integrating digital tools to optimize logistics, track deliveries in real time, and predict demand, enhancing a market that has traditionally been underserved by e-commerce giants (Vanity Stardom, 2023).

Retail is normally not able to reach the rural markets owing to underdeveloped infrastructure, high operations cost, and problems in logistics (Patel and Chavda, 2013). ElasticRun addresses this since it takes advantage of all the capabilities that small business owners and local transporters have within the rural markets to ensure last-mile connectivity. This model not only adds to the overall business environment in those regions but also offers economic opportunities to rural entrepreneurs, thereby increasing access to consumers goods for a rural population (Sengottuvelu *et al.*, 2022). Rural retail in India is estimated to be worth more than US$100 billion per year and is largely fragmented and informal. It therefore offers more relevance to the ElasticRun approach. The company had expanded its reach to more than 100,000 villages across India by 2021 by partnering with over 300 consumer goods companies to reach customers through its rural kirana store network and local delivery partners. In the logistics segment of rural areas, the innovative model resolves various problems. It increases product availability to rural customers and, at the same time, provides additional revenue to regional businesses by reducing the infrastructural overheads and increased coverage (Shrivastava *et al.*, 2018). ElasticRun claimed that it was working with more than 500,000 small store owners and delivery partners as of 2023, which hugely raised employment and entrepreneurship in the rural areas (Sengottuvelu *et al.*, 2022).

Above all else, ElasticRun is bridging the divide in the e-commerce channel between urban and rural spaces and providing a future prospect to stimulate further overall economic development in underdeveloped areas (Vanity Stardom, 2023). The services provided by this company help in formalizing the rural economy by letting rural retailers tap into a larger supply

*Corresponding author: swatb1404@gmail.com

 Agripreneurship, Digital Inclusion and Sustainability: Rural Innovation and Growth in the Global South (eds Dutta and Trivedi)
DOI: 10.1079/9781836992646.0014

chain, thus having a nod in the direction of the overall promotion of small-scale businesses. That is important in a nation like India, where the agrarian economy forms about 46% of the national income.

The objective of this chapter is to explore the transformational impact of ElasticRun on rural businesses and the broader economic implications to which they are subject. It will examine the operating strategy of the company, assess its effectiveness in streamlining rural logistics, and explore the role it plays in the general economic development of rural India (Sridharan *et al.*, 2014). Having already garnered $330 million investment in 2022 and clearly showing considerable investor confidence, this chapter examines whether the company's model can be replicated elsewhere in the developing world to upgrade rural distribution networks toward offering scalable solutions to one of the perennial challenges of last-mile delivery in rural areas.

14.2 The ElasticRun Model

ElasticRun has revolutionized rural logistics in India through the use of a huge, decentralized, local transportation resources base. To this day, the company has partnered with over 500,000 small shop owners and delivery partners across more than 100,000 villages in India. This distributed delivery network has really reduced overhead costs and improved last-mile connectivity to places that other logistics companies conventionally had difficulty reaching (Sengottuvelu *et al.*, 2022).

The economic impact is high for this model (Srinivasan *et al.*, 2021). According to ElasticRun, its delivery partners can have an additional income of INR 10,000–15,000 per month with the benefit coming principally to rural household incomes. This additional income is quite significant considering the average monthly household income in rural India.

The company has managed to keep capital expenditure relatively low as it uses existing infrastructure. This is known to be an asset-light model, which has been of great use to the company in expanding quickly and reaching a valuation of $1.5 billion as of 2022 – a true example of the efficiency and scalability of their approach. For example, ElasticRun utilized more than 500 local transporters to service a rural district in Maharashtra, India, making deliveries 30% faster and cutting logistics costs by 25%. This model helped scale operations quickly and efficiently, and it was servicing more than 100,000 rural retailers within a year (Sengottuvelu *et al.*, 2022).

14.2.1 Digital integration

The backbone of the operations in ElasticRun is their digital platform: this integrates under one roof all the threads of logistics management. The digital platform processes above 2 million orders in a day, testifing to the gigantic capacity that this company can cater to. It is the digital backbone that has allowed ElasticRun to reach a delivery accuracy rate of 98%, which is much higher than the average of around 85–90% that the rural delivery industry espouses (Sengottuvelu *et al.*, 2022).

For example, at one end, this company's predictive analytics and machine learning algorithms have made demand predictions 30% more accurate than traditional models; thus, it has managed to reduce the holding costs of inventory from rural retailers. This has also allowed the reduction of delivery times by 25% as compared with traditional logistics models that work in the rural network. More importantly, ElasticRun's digital platform brings on board more than 300 consumer goods companies, including giants like Coca-Cola, Procter & Gamble, and ITC. These relationships have expanded product offerings to rural consumers with the average rural store in the ElasticRun network now carrying 40% more stock-keeping units (SKUs) than stores not participating. Investments in the efficiency of the platform have been substantial, with ElasticRun raising $330 million in funding in 2022. Such financial backing has allowed the company to continuously optimize their technology; recent upgrades reduced order processing time by 50% and increased the accuracy of delivery time predictions to 95% (Bhatnagar and Yadav, 2022). Delivery accuracy improved by 40%, and stockouts in rural stores have reduced by 20%. The result has significantly increased sales and consumer satisfaction. When combined with advanced digital integration, this crowdsourced logistics model has been able to put up a system that not only overcomes the woes of rural logistics but also turns out to be a sizeable contributor to the development of the rural economy. The success of the company demonstrates that technology-driven, locally anchored solutions can be leveraged as a tool to smash long-standing barriers limiting rural access to markets and economic growth (Srinivasan, 2021).

14.3 Impact on Rural Market Access

14.3.1 Increased access to consumer products

ElasticRun has significantly transformed the availability of fast-moving consumer goods in rural India (Thakur, 2022). It covers more than 100,000 villages and serves nearly 130 million rural consumers so far. This massive reach has brought in a tremendous increase in product variety and availability (Goswami and Jatana, 2021).

- *Product range.* Average SKUs on shelves doubled to 40%, thereby increasing the range of SKUs in participating stores in terms of units. Numbers increased to 280–350 SKUs from about 200–250 SKUs in a usual rural store.
- *Brand access.* The portal has enabled access for over 300 consumer goods companies, from big names like Coca-Cola, Procter & Gamble, and ITC, to rural markets, bringing previously urban-centric brands within their reach. It means that against 30% of SKUs offered in urban locations, 70% are now available for access by rural consumers.
- *Less stockouts.* Demand-forecasting algorithms from ElasticRun has reduced stockouts by 35%. This ensured that the much-needed products were available and helped to increase customer satisfaction because of an improvement of up to 40% in rural consumers' perception of product availability.
- *Fresh inventory.* The average age of inventory in rural stores has been reduced by 20% due to a more streamlined supply chain, which translates into fresh products for the consumer.

14.3.1.1 Case study: Maharashtra and Gujarat, India

ElasticRun tackled the significant hurdles the rural retailers in Maharashtra and Gujarat faced in keeping up with a wide range of consumer products because of irregular supply chains. With the huge network of suppliers from cities, ElasticRun offers the seamless supply of a wide range of products to these rural retailers. The sales of rural retailers increased by 30% and the availability of products increased by 50% within the first 6 months of entering the ElasticRun network, which influenced consumer satisfaction and market development in those geographies (Virmani *et al.*, 2024).

14.3.2 Economic empowerment

ElasticRun's model has presented rural entrepreneurs with considerable economic opportunities (Likhi, 2021; Thakur, 2022).

- *Additional income.* There is surplus income available for the delivery partners within the platform of ElasticRun at INR 10,000–15,000 per month. In contrast, the average monthly income of rural households in India was only INR 10,218 as of 2021.
- *Creation of jobs.* By 2023, ElasticRun had tied up with over 500,000 small-store owners and delivery partners. This network has seen the creation of an estimated 150,000 new jobs in rural areas, and the rural unemployment in those areas was reduced to 2%.
- *Women's empowerment.* The network comprises nearly 30% female entrepreneurs as the rural partners. On average, these women have been contributing to a 60% increase in household income due to becoming a part of ElasticRun. This is

reflective of the bigger trends of economic participation of women across rural areas according to Tillmar *et al.* (2022).

- *Financial inclusion.* ElasticRun has enabled close to 200,000 rural retailers to avail formal credit by partnering with fintech companies. The average value of loans is INR 50,000 with a repayment rate of 95%, indicating the economic soundness of the business opportunities (Abramova *et al.*, 2021).
- *Skill development.* ElasticRun provides training in digital literacy and business management to partners. Under these programs, more than 300,000 rural entrepreneurs have been trained, who reported that their business efficiency improved by 25%, which is on par with most other studies conducted on digital literacy as it impacts rural businesses.
- *Formalization of rural economy.* So far, ElasticRun has formalized more than 400,000 small businesses by bringing rural retailers on to a digital platform. This in turn creates 15% tax compliance among them, thereby improving the local economic development.
- *Entrepreneurship.* ElasticRun enables rural store owners to participate in the network through which they earn commissions on each delivery; as a consequence, they can diversify their earnings.

ElasticRun has a deep and complex impact on both economic empowerment and access to rural markets. This company is revolutionizing the structure of rural retail and is also contributing to a larger and more holistic rural economy by making products more available as well as developing diverse economic opportunities. This model shows how locally oriented, technology-supported solutions can be applied to solve age-old issues regarding rural development and market access.

14.3.2.1 Case study: rural Mississippi, USA

Rural logistics in the USA face a different set of challenges, including reduced population density and increased operational costs. A similar crowdsourced business model was proposed to be operated by a company called Roadie. Similar to ElasticRun, Roadie operates local carriers to deliver products into rural areas with an objective of decreasing operational costs by utilizing existing transportation assets. Delivering a 20% reduced delivery cost to the businesses involved and an average increase in income of $200 per month for the participating carriers of rural Mississippi, the model increased economic activity in the region and made essential products more readily available (Srinivasan, 2021).

14.3.3 Impact on profitability

14.3.3.1 Cost efficiency

The model of ElasticRun is very cost-effective essentially because it is exploiting existing resources without investing a lot of resources in new infrastructure. This is the reason, therefore, that this company can offer the best value for money to partners and customers alike. For instance, cost-effectiveness on the part of ElasticRun allowed it to make a product 10% cheaper than a traditional supply chain sell to rural retailers in Karnataka, India, thus leaving room for greater profit margin and acceptable pricing to end consumers.

The model that ElasticRun has developed significantly increases cost efficiency in the rural supply chain (Somashekhar, 2014; Sengottuvelu *et al.*, 2022).

- *Operational costs.* The company has brought down the last-mile delivery cost by as much as 30% compared with conventional models of logistics through the utilization of local *in situ* resources. This works out at mean savings of INR 5–7/kg delivered (Singh, 2007).
- *Inventory management.* The algorithms used in demand forecasting by the firm have enhanced inventory turnover rates by 40%, thus bringing about a decline in holding costs for rural retailers in the order of 15–20%.
- *Returns and wastage.* The distribution network of ElasticRun has helped reduce product returns and wastage by 25%. This alone saves fast-moving consumer goods

(FMCG) companies INR 100 crore every year.

- *Technology integration.* There has been a 50% decrease in the processing costs of orders on the digital platform, from an average of INR 20 per order to INR 10 per order.
- *Rural retailer margins.* With such efficiencies, the rural retailers who partner with ElasticRun gain an increase in their profit margins, which have gone up by 15–20%, with the traditional margin of 8–10% now standing at 12–14%.
- *Consumer pricing.* The savings from the reduced transport and inventory costs have given room for better pricing. In stores affiliated with ElasticRun products are sold at a lower average price of 5–8% as compared with those of their competitors who are not associated with ElasticRun but sell at the stores in the same village.

14.3.3.2 Market penetration

ElasticRun has greatly contributed to consumer product business companies gaining entry into many dispersed rural markets that would otherwise have been challenging to reach because of complex logistical problems. Stepping on the heels of Deliveroo, in the development of a delivery network for rural Scotland, ElasticRun has aided firms selling consumer products to significantly expand their market coverage across rural India (Srinivasan *et al.*, 2021).

The measures listed here have significantly boosted the consumer goods companies' market reach by ElasticRun.

- *Rural coverage.* As of 2023, the company's network covered more than 100,000 villages, or nearly 15% of India's population. This has helped FMCG companies extend their rural footprint with an average increase of about 30%.
- *New customer acquisition.* Companies associated with ElasticRun report acquiring around 20–25% new rural customers within the first year of collaboration.
- *Net reach of product.* SKUs increased by 40% in rural areas; whereas an average ElasticRun-affiliated store used to have about 200–250 SKUs before joining the scheme, they now stock 280–350.
- *Sales growth.* The FMCG companies show an average sale volume increase in rural areas of 35–40% after tie-up with ElasticRun. In some categories, including personal care and home care products, it has reached as high as 50–60%.
- *Market share.* The average market share of rural markets for leading FMCG companies has improved by 5–7% after integrating with the ElasticRun network.
- *Brand penetration.* Newer brands and products have got 15–20% market penetration in rural areas within just 6 months of launch through ElasticRun, whereas it takes 2–3 years through traditional channels.
- *Data insights.* ElasticRun's data analytics give deep insights into the behavior of rural consumers, which helps in better development and marketing. New product launch success rates in rural markets have increased by 25%.
- *Seasonal flexibility.* Because of elastic networks, companies are now capable of increasing their distribution channels by about 40–50% during peak seasons like harvest time or festivals, thereby capturing seasonal demand better.
- *Rural-specific products.* With ElasticRun's networks and data, more than 100 rural-specific SKUs have been launched so far by companies, catering to specific needs and preferences of the rural market.
- *E-commerce partnerships.* Partnership with the biggest e-commerce platforms has enabled rural consumers access to online marketplaces, which is growing the rural e-commerce market by almost INR 5000 crore every year.

The profitability of all players in the rural supply chain has increased manifold through a combination of cost efficiency and improved market reach. Rural retailers have averaged a 30–35% increase in their yearly revenues, and FMCG companies can realize a 20–25% increase in profitability coming from their rural business once they partner with ElasticRun. It is this win–win situation that encapsulates the transformative power of ElasticRun's innovative approach to unlock economic potential within the rural markets (Desai and Manoharan, 2024).

14.4 Challenges

14.4.1 Wider competition

ElasticRun's success in rural logistics has invited serious competition from the big e-commerce players as well as start-ups. Amazon's Project Udaan and Flipkart's rural initiative have made substantive inroads to touch more than 100,000 and 80,000 villages, respectively (Kumar, 2019). Budgets of INR 1000–1500 crore for each of the big e-commerce companies for rural expansion in the next 3–5 years are quite a challenge for ElasticRun. However, ElasticRun's solid local integration and focus on small entrepreneurs gives it a competitive edge (Ibekwe *et al.*, 2021.

Local start-ups have also entered the rural distribution space, and have captured a 5–7% market share in some regions. This increased competition has forced ElasticRun to innovate and differentiate its offerings. Traditional FMCG companies have not sat back either. Large players like Hindustan Unilever and ITC have also consolidated their rural distribution channels, adding 15–20% direct touches to their businesses in the last 2 years (Maria and Santhi, 2020). This push from established players adds yet another layer of competitiveness to the competitive landscape. ElasticRun, however has taken a multipronged approach to maintain its competitive advantage. It has deepened its interest in the existing markets through 70% market penetration within its top 1000 villages, which is significantly more than its competitors. Diversification into new verticals such as fintech in rural and agri-tech sector has ensured a more fully-fledged ecosystem for rural businesses. Strategic partnerships have been an important component of the competitive strategy for ElasticRun (Mei *et al.*, 2022). The company has partnered with 15 regional rural banks and 25 agricultural cooperatives, which have boosted the position of the company at the local level and has earned them trust in the villages. These collaborations have given ElasticRun the capacity to maintain its unique position in the market (Srinivasan *et al.*, 2021).

14.4.2 Data security

As the digital platform of ElasticRun works with huge volumes of sensitive information, it now has the important challenge of providing robust security (Meyyappan *et al.*, 2019). The digital platform processes over 10 million daily transactions and maintains about 5 terabytes of data. This includes the financial information of more than 500,000 rural partners as well as the buying patterns of 130 million rural consumers, making it an attractive target for cybercriminals (Niranjanamurthy and Chahar, 2013). The rural tech sector showed a 30% increase in cyber attacks in 2024, emphasizing the importance of constant vigil and improvement in security measures. So far, there are no reported breaches at ElasticRun, and the company has created a security strategy to address these concerns, ensuring that its data transmission is encrypted end-to-end. Classified information calls for a 256-bit advanced encryption standard (AES) to secure classified information. Multifactor authentication has been a must-do for all user accounts and the company has reduced unauthorized access attempts by 75%. Third party cyber security firms are performing regular security audits, and in the last audit ElasticRun was able to attain a staggering 99.9% compliance rate. The company has also anonymized personal identifiable information for analysis purposes and access is limited to only 0.5% of their employees (Mandler *et al.*, 2013).

ElasticRun's innovative approach includes the adoption of blockchain technology for secure, transparent record-keeping of transactions. The program, as of today, is utilized for 10% of high-value transactions, promising great potential in terms of data integrity and traceability (Sengottuvelu *et al.*, 2022).

Part of the security strategy of ElasticRun is employee training. The mandatory cyber security training for everyone recorded a 95% pass rate on the last audit, thus ensuring that the human element of security is not forgotten (Cardoso, 2019). These measures have helped ensure a decline of 40% in security incidents within the last 2 years. However, cyber threats change constantly, and ElasticRun is obliged to evolve continuously, making an investment of 8–10% of its annual budget in security improvement or cyber security (Afzal *et al.*, 2024). It spends heavily on cyber security and raised its expenditure on data protection by up to 25% in 2024; this resulted in a significant fall in data breaches of 50% compared with the same period the previous year, as cited by reports from the company (Cardoso, 2019).

The dual task of maintaining competitiveness in the rural ecosystem while ensuring strong data security will be crucial to further growth and trust by ElasticRun's stakeholders (Stühmer and Stojanovic, 2011). As ElasticRun expands into new areas and more services, it will only remain successful if it can address these issues whilst revolutionizing the rural logistics and e-commerce landscape of India (Mandler *et al.*, 2013).

14.5 Future Development Trend of E-Commerce Within Rural India

14.5.1 Technical innovations

The future of e-commerce in rural India will depend considerably on the technological breakthroughs presently underway. AI and machine learning (ML) form one end of this change, holding unprecedented opportunities to streamline the flow of operations and enhance the quality of experience for the customer (Ray, 2011).

AI/ML-based demand forecasting is estimated to make a big difference to inventory management in rural areas; ElasticRun's current 92% accuracy is expected to rise to 97–98% within 5 years. This level of accuracy will cut stockout and overstock events dramatically. Rural retailers will save inventory holding costs as much as 25–30%. Hyperpersonalized product recommendations will be taken to the next level, increasing cross-selling opportunities by 40–50% in the rural e-commerce platforms (Kumari *et al.*, 2024).

ML algorithms are likely to be able to efficiently optimize the planning of routes for last-mile delivery in rural areas. Optimization in terms of route planning, with advanced models of ML that encompass weather conditions, quality of roads, and seasonal changes, is likely to increase the efficiency by another 15–20% in terms of delivery time and to reduce costs by 10–15% on fuel consumption. This will not only help in increasing the efficiency levels but also help in decreasing the carbon footprint of all rural logistics processes (Raja *et al.*, 2021).

Blockchain technology also has enormous scope for application in rural e-commerce. It is estimated that up to 50% of all e-commerce transactions in rural areas could eventually end up being recorded on blockchain platforms, with proper transparency and traceability through the entire supply chain (Sengottuvelu *et al.*, 2022). This could reduce instances of counterfeit goods by 80–90% in the rural market, which is a cause of worry in rural areas where the verification of products is difficult. The internet of things (IoT) is also likely to be of great importance to the scenario of rural logistics. IoT-based intelligent packaging would mean that product conditions would always be monitored during transit. This is particularly important for perishable items. This technology can reduce spoilage by about 30–40% of perishables in their transit phase, which is particularly important when post-harvest losses are about 40% for fruits and vegetables in India (Ray, 2011).

Underlying 5G technology will enable real-time monitoring and communication even in the most far-flung and disconnected places. This may lead to a reduction of 50–60% in delivery inaccuracy as well as reduced failed deliveries by 30% due to better coordination between partners and customers (Kumar *et al.*, 2022).

The use of virtual try-ons for clothes and accessories and augmented reality (AR)-based product visualization can increase the conversion rates of specific product categories by a huge 20–25% in rural places since online and offline experiences are bridged together (Merchán *et al.*, 2021). Voice-based e-commerce, leveraging natural language processing (NLP) in regional languages, will account for 25–30% of rural e-commerce engagement by 2026. This technology intervention will probably be most visible in districts where literacy rates are lower, hence allowing an additional 50–60 million rural consumers to enter the digital commerce ecosystem (Ray, 2011).

14.5.2 Policy support

The future course of rural e-commerce in India will be determined by governmental policies.

Once the Digital India initiative was launched in 2015, the basic groundwork of the digital infrastructure was laid down in rural areas. However, more focused policies are now needed that can properly actualize the potential of rural e-commerce (Dana and Sharma, 2023).

The BharatNet Project is a backbone for rural digital infrastructure, involving 250,000 gram panchayats (village democracies). By 2023, it is estimated that around 180,000 gram panchayats were connected (Zehra *et al.*, 2024). The project is scheduled to be completed by the year 2025 and this would add 200–250 million rural consumers online, which could enormously expand the rural e-commerce market (Sridhar, 2019).

Jan Dhan Yojana is the agenda adopted by the government for financial inclusion, making it possible for bank account penetration in rural areas to have reached 80% in 2023 (Sharma *et al.*, 2023). With more push from the policy over digital financial literacy, the digital payment system could hit the ceiling of 35% adoption in rural areas and reach up to 65% adoption by 2026, which would eventually facilitate e-commerce transactions (Gupta and Shrivastava, 2022).

The National Education Policy 2020, with its focus on digital literacy, aims to make the rural citizenry more tech-savvy. As estimated, in 2030, there could be 70–75% of the rural youth being digitally literate, as against the existing 45–50%, constituting a larger population engaged in rural e-commerce (Paul, 2024).

The government has implemented policies encouraging rural entrepreneurship, especially the launch of the Startup India program, and it is estimated that 500,000 new rural tech-enabled businesses will emerge by 2028. Suppliers and customers of the e-commerce ecosystem in the rural areas will be affected, fostering innovation and competition (Dana and Sharma, 2023). This might spur the production of rural-specific products. It is projected that by 2027, 20–25% of products sold through rural e-commerce will be locally manufactured (Bishnoi, 2019).

Data protection laws in the form of a Personal Data Protection Bill have regulated the reliance of rural consumers on digital platforms (Palaniappan *et al.*, 2016). Proper guidelines regarding data localization and user privacy would increase the rural consumer's reliance on e-commerce by 30–40%. This can further boost adoption levels.

ElasticRun could witness 30–35% of rural last-mile deliveries using electric vehicles if, by 2030, sustainable logistics policies with incentives for using electric vehicles in such deliveries are adopted. That would not only reduce environmental damage but would also reduce operational costs in the long term (Jiang *et al.*, 2023).

Thus, the Indian government's plans for developing 100,000 digital villages by 2025 could well create hubs of e-commerce activity in rural areas. Digital villages can easily become models of rural digital transformation, raising e-commerce penetration in the surroundings to 60%. As rural e-commerce matures, this would increase the need to tailor skilling initiatives and offer corresponding support through policies. Government plans to create a ready talent pool for e-commerce undertakings in the rural sector will upgrade 2–3 million young people by 2028 and serve the demand for this still nascent industry (Dana and Sharma, 2023).

The curve of growth of e-commerce in rural India is going to be explosive, with far-reaching advancements in technology and government policies coming in as supporting hands (Ray, 2011). This momentum may help increase the size of the rural e-commerce market by \$30 billion to \$100–120 billion by 2030, which is a paradigm shift in the consumption patterns and economic activity that currently prevails in rural regions (Sharma *et al.*, 2023). Companies at the cutting-edge end of this change, such as ElasticRun, would continue to innovate and adapt to these dynamics, thereby holding a competitive edge in this dynamic landscape (Akther, 2023).

14.6 Conclusion

ElasticRun's pioneering strategy in rural logistics and e-commerce has immensely supplemented the entry of markets in the rural regions of India, promoting economic empowerment and reviving livelihoods. It connects more than 500,000 small store owners with delivery partners and helps metro-area suppliers to connect with consumers in rural areas. With the introduction of this methodology, the costs of last-mile delivery have been greatly diminished, and the ease of product supply has improved, which in turn has increased rural revenues. ElasticRun faces several issues, including an ever-increasing number of challenges created by competition alongside issues related to data security. The future success of the company in India's changing rural setting depends on its capacity to innovate, make strategic use of emerging technologies, and cooperate strategically.

References

Abramova, I., Nedilska, L., Kurovska, N. and Martynyuk, H. (2021) Financial inclusion in the context of sustainable development of rural areas. *Management Theory and Studies for Rural Business and Infrastructure Development* 43(3), 328–336.

Afzal, M., Meraj, M., Kaur, M. and Ansari, M.S. (2024) How does cybersecurity awareness help in achieving digital financial inclusion in rural india under escalating cyber fraud scenario. *Journal of Cyber Security Technology* 9(6), 1–39.

Akther, F.. (2023) E-commerce in India: Trends, hurdles, and growth opportunities. *Formosa Journal of Science and Technology* 2(10), 2871–2880.

Bhatnagar, S. and Yadav, R. (2022) Can multiple channels coexist?: The case of the indian FMCG sector. In: *Multidisciplinary Perspectives Towards Building a Digitally Competent Society*. IGI Global, pp. 70–86.

Bishnoi, Vishal. (2019) Make in india initiative: A key for sustainable growth. *South Asian Journal of Marketing & Management Research* 9(3), 21–27.

Cardoso, L. (2019) Pedagogical innovation and literacy teaching. In: *Int'l Conference Proceedings*, p. 46.

Dana, L.-P. and Sharma, N. (2023) SMEs and start-ups: The growth engine of India. In: *Indian SMEs and Start-Ups: Growth through Innovation and Leadership*. pp. 1–18.

Desai, A. and Manoharan, A.P. (2024) Digital transformation and public administration: The impacts of India's digital public infrastructure. *International Journal of Public Administration* 47(9), 575–578.

Goswami, M. and Jatana, R. (2021) An analytical study on the functioning of eNAM (with special reference to Rajasthan). *International Journal of Research Culture Society* 5(1), 5.

Gupta, P. and Shrivastava, V. (2022) Role of Pradhan Mantri Jan Dhan Yojana in enhancing the paradigm of financial inclusion in rural India. *Integrated Journal for Research in Arts and Humanities* 2(2), 32–37.

Ibekwe, A.O., OgbonniaAnusui, A. and Ibekwe, A.I. (2021) Financial inclusion and entrepreneurship development in Nigeria. *Journal of Emerging Trends in Management Sciences and Entrepreneurship* 3(1), 90–104.

Kumar, Vijayant. (2019) Revolution of e-commerce in rural market. *International Journal of Trend in Scientific Research and Development* 3(5), 2062–2065.

Kumar, S.K.A., Ihita, G., Chaudhari, S. and Arumugam, P. (2022) A survey on rural internet connectivity in India. In: *2022 14th International Conference on Communication Systems & Networks (COMSNETS)*, IEEE, pp. 911–916.

Jiang, S., Yadav, O. and Nguyen, H. (2023) Electric vehicles in rural areas using synchronized routing and resupply strategies – Project 12: Deliverables and reporting requirements for UTC grants awarded in 2023. Available at: https://fersc.utk.edu/wp-content/uploads/sites/8/2024/08/Project-12.pdf

Kumari, T.L., Bambuwala, S. and Rajalakshmi, M. (2024) Ai-driven solutions for supply chain management. *Journal of Informatics Education and Research* 4(2), 861–868.

Likhi, Abhilaksh. (2021) Farmers and agricultural marketing opportunities through electronic trading platforms: Issues and challenges. *Journal of Lal Bahadur Shastri National Academy of Administration (LBSNAA)*.

Mandler, B., Antonelli, F., Kleinfeld, R., Pedrinaci, C., Carrera, D. *et al*. (2013) COMPOSE--a journey from the internet of things to the internet of services. In: *2013 27th International Conference on Advanced Information Networking and Applications Workshops*, IEEE, pp. 1217–1222.

Maria, P.R.D. and Santhi, V.E. (2020) A study on rural marketing management of Indian FMCG product. *Shanlax International Journal of Management* 7(4), 69–79.

Mei, T.H., Sjarif, M.A., Mulia, N., and JayantiT.B. (2022) The challenges of starting and operating an e-commerce business in rural areas. In: *3rd Tarumanagara International Conference on the Applications of Social Sciences and Humanities*, Atlantis Press, pp. 1944–1950.

Merchán, M.J., Merchán, P. and Pérez, E. (2021) Good practices in the use of augmented reality for the dissemination of architectural heritage of rural areas. *Applied Sciences* 11(5), 2055.
Meyyappan, R.S., Nivash, T.N. and Ariharan, J. (2019) Rural e-commerce and last mile delivery: Challenges and solutions in the Indian context. *International Journal of Management, IT and Engineering* 9(9), 164–170.
Niranjanamurthy, M. and Chahar, D. (2013) The study of e-commerce security issues and solutions. *International Journal of Advanced Research in Computer and Communication Engineering* 2(7), 2885–2895.
Palaniappan, G., Ramanigopal, C.S. and Subhashini, S. (2016) Role of Panchayati Raj institutions towards consumer protection and awareness among rural people in Tamil Nadu. *Journal of Rural and Industrial Development* 4(2), 1.
Patel, B. and Chavda, K. (2013) Rural entrepreneurship in India: Challenge and problems. *International Journal of Advance Research in Computer Science and Management Studies* 1(2), 28–37.
Paul, S., Sr (2024) Technology integration in education: Evaluating the role of the NEP 2020 in promoting digital literacy.
Raja, R., Mukherjee, I. and Sarkar, B.K. (2021) A machine learning-based prediction model for preterm birth in rural India. *Journal of Healthcare Engineering* (1), 6665573.
Ray, Sarbapriya. (2011) Emerging trend of e-commerce in India: Some crucial issues, prospects and challenges. *Computer Engineering and Intelligent Systems* 2(5), 17–35.
Sengottuvelu, C., Habib, W. and Dhivya, P. (2022) A study on elastic logistics in supply chain management. *PMIS Review* 1(1), 1–13.
Sharma, S., Jha, K.K. and Thangjam, B. (2023) Digital transformation in India: Perspectives, challenges, and future. *Research Journal of Agricultural Sciences* 15(1), 31–37.
Shrivastava, R., Sahu, R.K. and Siddiqui, I.N. (2018) Indian rural market: Opportunities and challenges. In: *Proceedings of National Conference of Innovative Solutions for Rural Development of Chhattisgarh*, pp. 23–24.
Singh, S. (2007) Leveraging contract farming for improving supply chain efficiency in India: Some innovative and successful models. In: *II International Symposium on Improving the Performance of Supply Chains in the Transitional Economies*, pp. 317–324 (Vol. 794).
Somashekhar, L. (2014) Agriculture supply chain management: a scenario in india. *International Journal of Food, Agriculture and Veterinary Sciences* 4(2), 83–89.
Sridharan, S., Maltz, E., Viswanathan, M. and Gupta, S. (2014) Transformative subsistence entrepreneurship: A study in India. *Journal of Macro Marketing* 34(4), 486–504.
Sridhar, V. (2019) *Emerging ICT Policies and Regulations: Roadmap to Digital Economies*. Springer Nature.
Srinivasan, R. (2021) *Platform Business Models*. Springer Singapore.
Srinivasan, R., Lakshmipathy, S. and Joseph, P. (2021) GRAB.inTM: GRAB.in. In: Srinivasan, R. (ed.) *Platform Business Models: Frameworks, Concepts and Design*. Springer, pp. 15–33.
Stühmer, R. and Stojanovic, N. (2011) Large-scale, situation-driven and quality-aware event marketplace: The concept, challenges and opportunities. In: *Proceedings of the 5th ACM International Conference on Distributed Event-Based System*, pp. 403–404.
Thakur, V. (2022) A review on growth of e-commerce in India. *Innovative Research Thoughts* 8(1), 56–64.
Tillmar, M., Sköld, B., Ahl, H., Berglund, K. and Pettersson, K. (2022) Women's rural businesses: For economic viability or gender equality? –a database study from the Swedish context. *International Journal of Gender and Entrepreneurship* 14(3), 323–351.
Vanity Stardom. (2023) ElasticRun founders: Reshaping last-mile logistics in rural India. Available at: https://vanitystardom.com/elasticrun-founders-reshaping-last-mile-logistics-in-rural-india/ (accessed 2 September 2024).
Virmani, D., Vij, S., Juyal, J., Sharma, A. and Mittal, V. (2024) Navigating the impact of COVID-19 on Indian economies through a lens of sustainability. In: *International Conference On Innovative Computing And Communication*, Springer Nature Singapore, pp. 149–159.
Zehra, F., Khan, F.S., Mazhar, S.S. and Khan, A.S. (2024) Revisiting digital India 2.0: Exploring new horizon and insights. In: *New Innovations in AI, Aviation, and Air Traffic Technology*. IGI Global, pp. 292–317.

Index

Printed and bound by CPI Group (UK) Ltd, Croydon, CR0 4YY
16/09/2026
14963198-0001